egradation

The GeoJournal Library

Volume 58

Managing Editors: Herman van der Wusten, University of Amsterdam,
The Netherlands
Olga Gritsai, Russian Academy of Sciences, Moscow,
Russia

Former Series Editor:
Wolf Tietze, Helmstedt, Germany

Editorial Board: Paul Claval, France
R.G. Crane, U.S.A.
Yehuda Gradus, Israel
Risto Laulajainen, Sweden
Gerd Lüttig, Germany
Walther Manshard, Germany
Osamu Nishikawa, Japan
Peter Tyson, South Africa

Land Degradation

Papers selected from Contributions to the Sixth Meeting of the International Geographical Union's Commission on Land Degradation and Desertification, Perth, Western Australia, 20–28 September 1999

edited by

ARTHUR J. CONACHER

University of Western Australia, Perth, Australia

KLUWER ACADEMIC PUBLISHERS
DORDRECHT / BOSTON / LONDON

A C.I.P. Catalogue record for this book is available from the Library of Congress

ISBN 0-7923-6770-7 ✓

Published by Kluwer Academic Publishers,
P.O. Box 17, 3300 AA Dordrecht, The Netherlands.

Sold and distributed in North, Central and South America
by Kluwer Academic Publishers,
101 Philip Drive, Norwell, MA 02061, U.S.A.

In all other countries, sold and distributed
by Kluwer Academic Publishers,
P.O. Box 322, 3300 AH Dordrecht, The Netherlands.

Printed on acid-free paper

Cover illustration:
A land degradation model for high population density areas of
the Eastern Cape.

Printed in the Netherlands.

7-92, -363 386

CONTENTS

PREFACE AND ACKNOWLEDGMENTS

Land degradation is an issue of increasing concern to most countries. It threatens not only the viability of agriculture, but also water quality, human health, biodiversity and the fundamental ecological processes on which all life depend.

The problem has been recognised for many years by the International Geographical Union (IGU) through the work of various Commissions, including earlier ones on drought and natural hazards. In 1992, a group headed by Maria Sala formed the IGU Study Group on Erosion in Regions of Mediterranean Climates. That group produced the multi-authored book on *Land Degradation in Mediterranean Environments of the World* which was published in 1998 (Conacher and Sala 1998). The Study Group explicitly sought to understand not only the mechanisms responsible for the problems but their social and economic implications, the means of overcoming the problems, and the policy instruments whereby remedial measures may be implemented.

These objectives are incorporated in the IGU's Commission on Land Degradation and Desertification (COMLAND), which developed from the Mediterranean Study Group in 1996 but with its interests expanded geographically to encompass land degradation and desertification wherever they are found. The Commission's first chair was Maria Sala. COMLAND's sixth meeting, which focused on the role of agriculture in land degradation and desertification, was held at the University of Western Australia in September, 1999. Seventy papers (including posters) were presented, 35 of which were submitted for publication. All were sent to at least two referees, and the final representation in this book reflects that process.

The breadth of COMLAND's objectives is well reflected in the contributions presented here, which will interest all professionals as well as academics and students working on land degradation. Papers covering a geographical spread of regions - by no means restricted to warm, semi-arid environments - and work on the socio-economic, remedial and policy aspects of the problem, are represented in the five parts. Part One groups papers on land degradation processes. The effects of land use practices and land use change are considered in Part Two. The third part discusses interactions between land degradation and society. Rehabilitation of degraded land is dealt with in Part Four, and three papers consider land degradation and policy in Part Five. The meeting was very ably opened and closed by Michael Stocking's Keynote Address and Closure, respectively; unfortunately it is impossible to do justice here to the strikingly effective colour illustrations which accompanied his presentations.

Several contributors, notably Michael Stocking in his Keynote Address, tackle the problem of definition. Although some writers here and elsewhere consider that land degradation and desertification are interchangeable terms, most regard land degradation as a broader concept which incorporates desertification but extends well beyond it. Degradation of vegetation and water quality, for example, are not usually considered as being forms of desertification. A previous definition of land degradation has been slightly modified in a later contribution in this book to read:

alterations to all aspects of the biophysical environment by human actions to the detriment of vegetation, soils, landforms, water, ecosystems and human well-being (Conacher and Conacher, this volume), and most, if not all, of the contributions would fit this quite comfortably.

No one could undertake the task of editing a book without the assistance of referees. The refereeing was done thoroughly and competently and I record my deep appreciation to them by listing them below: many uncomplainingly refereed more than one paper. I also thank the contributors who (nearly all) responded promptly and carefully to referees' suggestions and editorial promptings. Petra van Steenbergen, Kluwer's publishing editor (Geosciences) and her assistant, Manja Fredriksz, are thanked for their helpful cooperation and assistance. Finally, the University of Western Australia facilitated editing the volume by providing an invaluable period of study leave.

List of Referees

Abu Ali, Ramon Batalla, Harry Butler, Dan Carter, Adrian Chappell, Jeanette Conacher, Donald Davidson, Andrew Dougill, Ian Douglas, Nick Drake, Trish Fanning, Richard George, Guðrún Gísladóttir, Geoff Humphreys, Moshe Inbar, Marnie Leybourne, Lisa Lobry de Bruyn, Neil MacLeod, Grant McTainsh, Susan Marriott, Mike Meadows, David Mitchell, Jim Noble, Gregor Ollesch, David Pannell, John Pickard, Alan Pilgrim, Kate Rowntree, Maria Sala, Susanne Schnabel, Ed Skidmore, Marino Sorriso-Valvo, Michael Stocking, Les Ternan, Alan Terry, John Thornes, Frank Vanclay, Steve Warren, Helen Watson

Reference

Conacher, A.J. and Sala, M. (eds.) (2000) *Land Degradation in Mediterranean Environments of the World*, Wiley, Chichester.

ARTHUR J. CONACHER
Associate Professor of Geography
University of Western Australia
Nedlands, WA 6907, Australia
ajconach@geog.uwa.edu.au

August 2000

CHAPTER 1: KEYNOTE PAPER

AGRODIVERSITY: A POSITIVE MEANS OF ADDRESSING LAND
DEGRADATION AND SUSTAINABLE RURAL LIVELIHOODS

MICHAEL STOCKING

1. Abstract

'Agrodiversity' is the many ways in which farmers use the natural diversity of the environment for production, including their choice of crops, and management of land, water and biota as a whole. The 1998 Conference of Parties to the *Convention on Biological Diversity* asked countries to demonstrate the importance of biological diversity in supporting rural communities. Promoting agrodiversity and understanding how it functions in tropical smallholder farming systems is a key means of showing that importance.

Linking agrodiversity to land degradation control and sustainable rural livelihoods can provide a *win-win* scenario for rural development. Policies that encourage biological diversity in areas of land use will not only meet countries' responsibilities under the *Convention* but also address land degradation problems and support rural livelihoods. Biodiversity is important to people in terms of functionality, intergenerational equity, resilience and stability of the local environment, flexible responses to changing needs and the life support role of ecosystems. Sustainable rural livelihoods are best seen in a capital assets framework that shows how rural households may use various components to control land degradation. To enable agrodiversity to be implemented practically, it has been codified into core elements. (1) *biophysical diversity*, the diversity of the natural environment that controls the resource base for food production; (2) *management diversity*, which embraces the practices (many of them indigenous) of farmers, such as live hedges, soil amendments and ridge tillage techniques; (3) *agro-biodiversity*, which is the diversity of crop, plant and animal combinations; and (4) *organisational diversity*, the way that farms are owned and operated, and the way that capital assets are allocated. Each element is then systematically related to show how agrodiversity controls land degradation and how it promotes food security. The maize-*Mucuna* system in South America demonstrates the value of agrodiversity to both soil conservation and to smallholder farmers' livelihoods.

2. Introduction

The United Nations *Convention on Biological Diversity*, signed at the *Earth Summit* in Rio de Janeiro in 1992, and ratified by most developed countries (but not the United States) in 1994, set out a global agenda for the conservation and sustainable use of biodiversity and its components. It affirmed the sovereign rights of states over their biological resources and, crucially, it stated a principle that the benefits of development of biological resources should be shared fairly and equitably on mutually agreed terms. This paper is about these very benefits of biological diversity. It concentrates on the lives of disadvantaged people – those living in degraded, poor and marginal environments, for whom global benefits may seem an irrelevant luxury. It highlights the situations of those living in underdeveloped, often tropical, parts of the

1

A.J. Conacher (ed.), Land Degradation, 1–16.
© 2001 *Kluwer Academic Publishers. Printed in the Netherlands.*

world, for whom development aid is usually targeted.

Biological diversity has traditionally been a central concern of ecologists. Yet for many non-ecologists, it remains conceptually elusive (Rodda 1993). Treated with suspicion by some, 'biodiversity' means different things to different people. Some observers look upon it as a bandwagon for their pet agendas, while others bemoan the loss of precision and rigour in debates on conservation priorities (Blaikie and Jeanrenaud 1997). Referring to the variety, or the number, frequency and variability of living organisms and the ecosystems in which they occur, biodiversity itself encompasses diverse aspects. It includes: species diversity, which is the number and frequency of species of plants, animals and micro-organisms; intra-species and genetic diversity, the variety between plants of the same species and of genes within each species; and ecosystem diversity, the different types of animal-plant assemblages and their associated ecological processes. Assertions as to the threats to humans and global life-support mechanisms by loss in biodiversity (for example, Wilson and Peters 1988), and supporting arguments on the benefits and potential values of biodiversity (for example, McNeely 1988) always came from ecologists. It was only in the early 1990s, with the rise of ecological economics and political ecology (for example, Swanson 1995, an edited compilation of contributions equally from socio-economists and natural scientists), that biodiversity entered a very different arena – one that would attract pragmatic, immediate and purposeful reasons why protection of biodiversity is essential. So it was that the 1998 Third Conference of Parties to the *Convention on Biological Diversity*, meeting in Bratislava (COP3/CBD), asked countries to document and demonstrate the importance of biological diversity in addressing problems of resource depletion and supporting rural communities. This paper arises from this fundamental shift in thinking: from biodiversity being an ethical but esoteric concern of natural science to biodiversity being a foundation for human livelihoods and sustainable development.

Taking up the challenge of COP3, this paper sets out to provide an early but evident demonstration of the central importance of biological diversity to addressing the old problem of the depletion of environmental resources and the more newly articulated problem of poor people's survival. It brings together three terms – agrodiversity, land degradation and rural livelihoods - which are all receiving substantial, but usually separate, attention in current debates about global futures and how human society can attain sustainable development.

3. A Framework for Sustainable Rural Livelihoods and Environmental Protection

In parallel to the change in thinking about the importance of biodiversity, approaches to human welfare in relation to the biophysical environment have also undergone radical shifts (World Bank 1992). No longer is development simply the transfer of technology or delivery of economic support from an altruistic donor to a grateful recipient. It is far more complicated. Yet there are still searches for some 'magic' formula that will ensure that the agendas of both the recipients of aid (e.g. economic development and poverty-relief) and the givers (e.g. biodiversity conservation) are met. One such is the appealing notion that the biophysical environment can be looked after while simultaneously providing for the immediate and felt needs of human populations – the World Bank's *win-win* scenario. It has driven international development policy on some of the seemingly most intractable human problems, such as population growth and environmental decline in Africa (Cleaver and Schreiber 1994). In part, the notion is a reaction to the widespread environmental pessimism of the 1960s and 1970s. Partly also, it has been fuelled by burgeoning evidence that environmental recovery has, in places, happened spontaneously (Tiffen *et al.* 1994) and that policies and practices have, maybe inadvertently, had positive effects on old problems such as deforestation (Fairhead and Leach

1996) or soil erosion (Critchley 1999). Some traditional soil conservation methods such as trashlines in drier parts of Kenya, maintained against professional advice which would rather have terraces dug, have only lately been shown to be both economic and technically-efficient (Kiome and Stocking 1995). If such positive things have happened without (or even despite) planning, imagine – the argument goes – what could happen if the ingredients for a *win-win* could be identified. Then, their effects could be analysed and the techniques could be appropriately supported. There is now much attention in the literature to the search for alternatives, asking questions such as: "Can agricultural growth be compatible with natural resource conservation and at the same time create the conditions for improving livelihoods and reducing poverty?" (Ellis-Jones 1999:179). Much of this attention has been devoted to indigenous technologies, especially in the more difficult, arid parts of the tropics (Kerr and Sanghi 1992).

Figure 1 presents a *win-win* scenario for sustainable rural livelihoods, with components involved in land degradation and biological diversity. Environmental protection is essentially part of the global agenda, enshrined in two global *Conventions* and reflecting primarily the needs of society. The rest of this paper will look first at these two elements of the global agenda and how attention to one may benefit the other (the lower shaded box of Fig. 1). It will then use evidence from recent research to show how these two global agenda items may address sustainable rural livelihoods (SRL) for poor people.

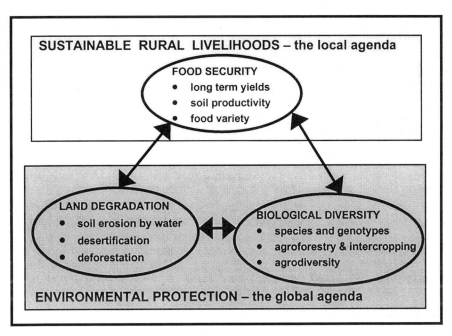

Figure 1. A *win-win* scenario. Situating agrodiversity and land degradation in the sustainable rural livelihood debate and global environmental agendas. (*adapted from* Tengberg and Stocking, 2000)

4. Land Degradation – Part of the Global Agenda

Under the UN *Convention to Combat Desertification (CCD)*, desertification is defined as land degradation in arid, semi-arid and dry sub-humid areas. The term 'desertification', however, is increasingly being avoided. Pagiola (1999) cites two reasons; the vagueness and imprecision of its definition, and the controversy over whether a process which has its major emotional impact in images of advancing sand dunes should be described as a real phenomenon. Article 1 of the *CCD* describes 'land' as the terrestrial bio-productive system which comprises soil, vegetation, other biota, and the ecological and hydrological processes which operate within the system (World Bank 1996). 'Land degradation' therefore means the reduction or loss of biological and economic productivity of major land uses (rainfed cropland, irrigated cropland, rangeland, forest) resulting from processes arising from human activity, such as soil erosion, change in physical, chemical or biological soil properties, and long-term loss of natural vegetation. In giving priority to Africa, the *CCD* recognises that the most intractable degradation problems are in the dryland agricultural and livestock practices of subsistence land users.

Land degradation causes problems at three levels (Pagiola 1999). At the field level, it results in reduced productivity. This will be discussed later in the context of food security and the local SRL agenda. At the national level, land degradation has large off-site effects, including flooding, sedimentation, effects on water quality and reductions in water supply. In developed countries such as the USA, these off-site impacts are calculated as many times larger than on-site changes in soil productivity (Crosson and Stout 1983). In less developed areas such as Java, the opposite seems to pertain (Magrath and Arens 1989). Caution does, however, need to be exercised in these national estimates because economic valuation approach and the perspective adopted substantially affect such calculations (Enters 1998).

At the third level (global), land degradation has a number of complex direct and indirect interaction effects, including:

- *climate change through emissions of greenhouse gases* – agriculture contributes a third of total carbon dioxide emissions, the vast majority of which derive from tropical deforestation.
- *reduction in terrestrial carbon sinks* – the world's topsoils store about 60% of the total stock of carbon that enters the biological cycle. Land degradation is the principal threat to the size of the soil carbon stock;
- *diminished effectiveness of carbon cycling* – reduced production of vegetation results in lower subsequent carbon inputs, which in turn promotes further land degradation;
- *reduction in biodiversity* – below ground, soil organisms are vastly reduced in degraded soils, while above ground, degraded landscapes are unable to support the total amounts and varieties of biological species that occupy fertile areas, and
- *pollution and sedimentation of international waters* – an off-site effect with global, or at least international, implications which has attracted projects under the global funding mechanism, the GEF.

Many of these global aspects of land degradation have major implications for its control through the use of biological means. This link will be explored in the following two sections; the first in another global agenda item, and the second through food security presented as part of the local agenda of SRL.

5. Biological Diversity - the Global Agenda

Biodiversity is characterised by variety and variability. Although usually associated with wild nature and protected areas, the *Convention on Biological Diversity* has encouraged a much broader and pragmatic interpretation of the term to mean the variety of all life on earth, whether wild or domesticated (Koziell 2000). This variety is both spatial and temporal. Living things interact within and between themselves, and also with the inorganic environment of atmosphere, soil and water. These interactions give rise to constantly changing networks of relationships, as living things adapt in response to natural or human-induced pressures. Human society taps into these interactive relationships, primarily through its utilisation and modification of food chains, and the attendant processes such as photosynthesis. Biodiversity is a fundamental life-support mechanism, sustaining the interconnectedness of life and adaptation to changing conditions. Table 1 presents a partial view of how different components of biodiversity are significant to life-support, showing how the human significance may impact at all levels from the local to the global.

The presentation in Table 1 of the importance of biodiversity to human society is a somewhat standard view. Many observers, even those working directly on biodiversity issues, are sometimes dismissive of the conservation of biodiversity as "important only up to a point for the sustainability of agricultural production" (Tisdell 1999:9). Different authors variously advance five main reasons for why biodiversity is so important to human society:

Table 1. Types of biodiversity and their human significance (adapted from Koziell, 2000)

Type of diversity	Human significance	Individual Components
Ecosystem diversity – biodiversity at landscape level, with distinct groupings of living things	Interactive processes within and between species, providing ecological resilience; interactions with physical environments, enabling hydrological and nutrient cycling, and major accumulations of energy.	Species combinations at multiple levels; from ecosystems occupying different parts of landscapes, to species within a single ecosystem.
Species diversity – biodiversity in its classical sense, being the distribution and abundance of species within a community or an area. Includes species richness.	Provision of a range of biological resources, required for survival. Includes food, fodder, and fuel. Reduction in risk, if one species should fail.	Variety in species of birds, mammals, invertebrates, plants, fungi, micro-organisms. Sub-species and varieties, including cultivars, landraces and domesticates.
Genetic diversity – biodiversity in frequency and variety of genes and/or genomes within populations and species. Includes differences between genes on the same chromosome.	Creates hereditary variation within species, offering opportunity to choose genetic characteristics which relate to yield, taste, palatability, nutrition, drought resistance, etc.	Alleles, phenotypes, DNA sequences

- **greater functionality:** diversity within and between species gives them differing characteristics, enabling more roles and functions to be undertaken;
- **intergenerational equity:** future human societies may become dependent on living resources preserved now, without apparent current productive benefit;
- **better resilience and stability:** agro-ecological systems are more resilient and sustainable if they possess greater genetic diversity, though empirical evidence is far from conclusive (Perrings *et al.* 1995);

-   **flexible response to changing needs:** biodiversity enables living organisms to adapt to changing conditions and external pressures;
-   **ecosystem and life-support service role:** energy conversion and material exchanges, which are necessary for maintaining a productive, healthy and stable environment, are supported by biodiversity.

Biodiversity, then, is the source of many goods and services for society at a variety of functional levels. Of particular practical interest at the global and international levels are:

- **Conservation** of genetic material of future importance to society: medicines and chemicals not yet discovered – i.e. future use value.
- **Preservation** of genetic material currently useful to society but not yet widely disseminated. Germplasm trading and exchange could become vital to developing countries.
- **Risk reduction** in material provision during times of environmental stress and scarcity.
- **Production** benefits through selection of more productive species and varieties. Continued access to genetic resources is crucial.
- **Diversification** by increasing global production of a greater range of crops.

Returning to the *win-win* scenario in Figure 1, biological diversity is well equipped to provide benefits both to society at large and to local people. Most of the benefits considered above can equally be considered as positive aspects for rural people. Furthermore, they are aspects, which are maintained by these beneficiaries, when they manage their biophysical resources of plants, soil, water, climate and land. This is what this paper will be calling *agrodiversity*. But before considering these complex linkages, the final component of the *win-win* framework is sustainable rural livelihoods (SRL), which is covered in the following section.

6. Sustainable Rural Livelihoods (SRL) – the Local Agenda

In the last five years a new focus for development aid assistance has arisen – the SRL approach. It is being embraced by development agencies, such as the UK's Department for International Development (Carney 1998). The approach is distinct from previous donor efforts such as 'Integrated Rural Development Planning' (IRDP) of the 1970s and sectoral project planning of the 1980s. Above all, it is people-centred, rather than focussing on geographical areas or needs and resources. It targets poverty, complexity, local knowledge and skills, while attempting to relate policy with practice through partnerships and participation. It takes the opportunity to mainstream environmental concerns (such as land degradation or loss in biodiversity) as part of livelihood development, rather than just adding it as another donor-led requirement. A good illustration of the different emphasis is the typical location of SRL projects in contrast to its predecessors. IRDPs were stand-alone projects with separate management and defined geographical areas. Sectoral projects were usually located in a host organisation, often a government department or ministry, building competence in service functions such as soil survey or planning. SRL projects are *within* a partner organisation (not a 'host') such as an NGO, development authority, local government or private sector company.

A framework for SRL has recently been developed in order to concentrate the focus on people, their needs and their opportunities (Scoones 1998). Intended to be a simple but practical tool, the framework concentrates on five different assets upon which individuals may draw in order to build their livelihoods (Fig. 2). Table 2 develops examples of how the availability or unavailability of these capital assets may affect land degradation positively or negatively. Thus, for example, one important asset component affecting land degradation is 'social group'. Disruptions to society through conflicts and wars are well known to cause environmental stress through refugee movements and people's unwillingness to invest in land improvements when the future is so uncertain (Black and Sessay 1997). Alternatively, long-settled and cohesive communities and social groups tend to develop their own tried and tested technologies, including a wealth of conservation measures such as ridges, planting pits and mulching methods (Tengberg *et al.* 1998).

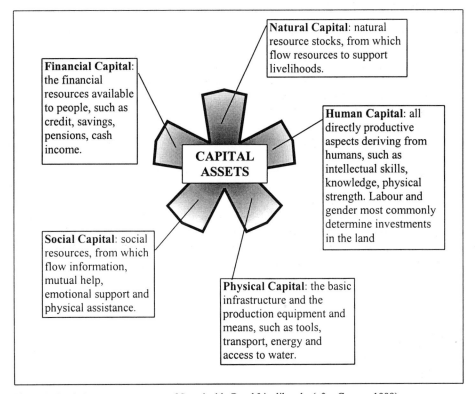

Financial Capital: the financial resources available to people, such as credit, savings, pensions, cash income.

Natural Capital: natural resource stocks, from which flow resources to support livelihoods.

CAPITAL ASSETS

Human Capital: all directly productive aspects deriving from humans, such as intellectual skills, knowledge, physical strength. Labour and gender most commonly determine investments in the land

Social Capital: social resources, from which flow information, mutual help, emotional support and physical assistance.

Physical Capital: the basic infrastructure and the production equipment and means, such as tools, transport, energy and access to water.

Figure 2. Capital assets components of Sustainable Rural Livelihoods. (after Carney, 1998)

Only in the 1990s has a new style of natural resource intervention emerged that is based on holistic and participatory planning at community and village level. To illustrate one such example and how the components of the SRL approach may provide better insights into practical solutions to land degradation, consider the *gestions de terroirs* of West Africa. Loosely translated as 'management of village lands', projects using the *gestions de terroir* approach start with the participatory identification of resource problems. Guèye (1995) describes the types of tools such as resource maps, Venn diagrams and wealth classification, as well as the use of local knowledge. This leads to the development of village institutions to address resource management decisions and implementation, monitoring and evaluation of activities (Toulmin 1993). The approach emphasises a number of important aspects of sustainable rural livelihoods:

- local diagnosis of problems – using those who live with the problem to articulate a need to address the problem;
- establishment of boundaries of responsibility – those who do the work, reap the benefit;

- development of a village management plan and map – to encourage a continuing involvement and recognition that needs in one place differ from those in another, and
- regular monitoring and evaluation – an iterative cycle of learning and adjustment of activities.

Table 2. Examples of capital assets components related to land degradation.

Capital Assets Group	Component	Positive Relation to Land Degradation	Negative Relation to Land Degradation
Natural	Land	Sufficient lands to plant grass strips or shelter belts.	Insufficient land to devote to non-consumptive uses such as conservation.
	Soil	Resilient soil able to withstand continuous cultivation without yield loss (e.g. nitosol)	Sensitive soil that loses fertility easily, which has large impact on yields (e.g. luvisol)
Social	Social groups	Indigenous and adapted techniques of soil and water conservation developed.	Social disruption, conflict, wars
	Access to institutions	Effective agricultural extension and subsidy regimes.	Ineffective agricultural extension and lack of access to technologies.
Human	Labour	Sufficient labour to intensify and invest in soil improvement.	Extensification and poor agricultural practice because of insufficient labour.
	Gender	Womens' management of home gardens using composting, ridges and inter-cropping	Overburdened women unable to undertake soil improvement
Physical	Implements	Availability of ridging implements to construct earth bunds.	Conventional tillage only, encouraging excessive erosion.
	Transport to market	Market opportunities provide further investment in land.	No cash income – subsistence cultivation only and soil mining.
Financial	Cash income	Sufficient cash to invest in land and soil improvement.	Non-farm cash income sources discourage investment in soil.
	Credit	Land users enabled to plant on time and minimise erosion.	Intermittent crop failure, debt and necessity to seek off-farm income.

While certainly not a panacea, Reij *et al.* (1996) consider *gestions de terroirs* a more promising way forward than standard solutions. Village-based land management highlights the need to combine all material, social and physical resources towards achieving sustainable support to livelihoods. Because of local empowerment and central involvement by land users, *gestions de terroirs* also stand a far better chance of accurate diagnosis of land degradation problems and committed land management. They are also more likely to rely on the intrinsic biological resources available at village level, rather than external interventions and imported inputs. The next section will develop the concept of 'agrodiversity' to show how these biological resources both control land degradation and support livelihoods.

7. Agrodiversity

Biological diversity has a crucial ecological role in agroecosystems (Altieri 1999). In addition to food production, it performs ecological services in nutrient cycling, regulation of microclimate, suppression of undesirable organisms and detoxification of noxious chemicals. Most authors call this 'agricultural biodiversity' or 'agro-biodiversity' (for example, Thrupp 1998). Brookfield and Padoch (1994), however, coined the term 'agrodiversity' to describe various biodiversity-related attributes of tropical small farm systems. They gave examples of farming systems which are resilient to external pressures and able to provide assured and sustainable production through the local management of biological diversity. Since then, the concept of agrodiversity has expanded to include the variety of practices developed by farmers that ensure them of production and enable them to withstand the vagaries of often difficult environments. Agrodiversity refers to interactions between agricultural management practices, farmers' capital assets and the diversity of species (Netting and Stone 1996).

There are four main reasons why agrodiversity should be a useful focus of interest today. First, there is more plant biological diversity in areas of land use in the tropics and sub-tropics than in all protected areas. As human populations increase, protected areas will become an increasing luxury and hence even their role in maintaining the stock of biodiversity will be under threat. So, it makes sense to concentrate our efforts on keeping biodiversity in agricultural systems. Second, small-farm agricultural systems are the primary custodians of this biological diversity. A survey for the *PLEC* project (see Acknowledgments) in one site in SW China by Guo Huijin and co-workers revealed over 110 types of managed agroforestry and 220 different associations of trees, crops and livestock. In such situations, local adaptation has effectively maintained biodiversity and even enhanced it. Third, much of this knowledge and management of biodiversity is slipping away under the pressures to commercialise, mechanise and standardise. In particular, only the elderly now hold knowledge of local plants such as medicinal species; chemicals have superseded many locally developed pest management practices, and many soil fertility maintenance techniques are now reliant on inorganic inputs. Finally, however, there is an increasing recognition that agrodiversity might hold lessons for agricultural development, especially in accessing techniques that would improve the situation of the rural poor.

Agrodiversity is largely a response of resource-poor farmers to inherent spatial and temporal environmental variability (such as the case of semi-arid farmers in Kenya: Tengberg *et al.* 1998). Brookfield and Stocking (1999) have codified agrodiversity into four principal elements, all of which have major implications for livelihoods and control of land degradation:

(1)   *Biophysical diversity*. This is the diversity of the natural environment, which controls the intrinsic quality of the natural resource base for food production. Soil productivity, the biodiversity of natural plant life and of the soil biota are included. It crucially involves knowledge of earth surface processes and interactions between organic and inorganic components, such as the control of land degradation by water erosion through vegetation cover. Farmers select within this biophysical diversity and may manipulate it substantially for both livelihoods and land degradation control purposes.

(2)   *Management Diversity*. This involves all methods of managing land, water and biota for crop production and the maintenance of soil fertility and structure. Smallholder farmers have a wide range of practices using biological, chemical and physical means such as live hedges, calcium-rich soil amendments and tillage techniques. Such practices not only enable more assured production and livelihood benefits but also act as soil and water conservation techniques.

(3)   *Agro-biodiversity*. This is the management and direct use of biological species, including all crops, semi-domesticates and wild species. In agricultural systems, the diversity of crop combinations, and the manner in which these are used to sustain or increase production, reduce risk, and enhance conservation, are aspects of agro-

biodiversity important to livelihoods and environmental protection.

(4)   *Organisational Diversity*. This is the way in which farms are owned and operated, and the way that capital assets are allocated. Farmers with different resource endowments organise differently in accord with their specific circumstances. Explanatory elements include labour, household size, off-farm employment, gender roles and dependence on farm income. Flexible responses to external factors and market opportunities are made possible, thus enabling better support to livelihoods.

Brookfield and Stocking (1999) argue that these categories are fundamental to understanding the interface between natural biological diversity and human land use. This management of variation within agro-ecosystems (Almekinders *et al.* 1995) is now acknowledged as the principal way forward to achieve the *win-win* scenarios such as that described in Figure 1. Two benefits of agrodiversity will now be considered – control of land degradation and promotion of food security:

7.1. HOW AGRODIVERSITY CONTROLS LAND DEGRADATION

Through the greater use of biological processes and of techniques adapted by farmers to the local environment, agrodiversity reduces both the risk of land degradation and its potential impact on production and livelihoods. This is achieved in complex direct and indirect ways through combinations of:

- better and more continuous vegetation cover intercepting rainfall erosivity;
- promotion of water infiltration along root channels and in soil with better structural stability;
- control of runoff and sedimentation by stems and surface organic matter;
- increased organic matter and enhancement of soil productivity encouraging greater vegetation growth and better protection to the soil;
- increased soil organic matter reducing soil erodibility;
- reduction in need for costly conservation structures and other inputs, and
- better production and financial returns enabling further capital investment in land improvement.

These ways may be exemplified through the four principal elements of agrodiversity.

Biophysical diversity includes the natural resilience of the biophysical environment, which is buffered by increases in organic matter. This increased organic matter reduces land degradation. For example, Elwell (1986) has shown conclusively on high-clay Zimbabwe soils that the stability of water-stable aggregates is crucially dependent on maintaining an organic carbon level of 2% or more. Below this threshold, aggregates break down quickly, surface sealing occurs, runoff accelerates and soil erosion results. Above this level, surface aggregates keep a good surface-water storage capacity, allow adequate infiltration and ensure optimal plant growth. Keeping this natural soil resilience relies on production of sufficient biomass, having a range of different types and qualities of organic materials and their reincorporation into the soil. Only an agrodiverse strategy of agricultural production can assure these attributes in a practical and realistic way for smallholder farmers.

Management diversity is often founded on local knowledge. Indigenous and adapted techniques of soil and water conservation are the most immediate way in which this form of agrodiversity may control land degradation. For example, Bunch and López (1999) have described how local innovation based on the use of mainly biological techniques has transformed many Central American rural areas. In one small community (Pacayas) in Honduras, 16 innovations have been added to the already biodiverse farming system. This includes four new crops, two green manures, two new species of grass for use as contour barriers in vegetable fields, new chicken pens made of king-grass, marigolds to control nematodes, the beans *Dolichos lablab* and *Mucuna pruriens* to feed cattle and chickens and Napier grass to control landslips. An increasing number of studies now attempt to assess the knowledge that farmers hold and their experimentation practices in, for example, soil fertility improvement in Ethiopia by fallowing, manuring and use of crop residues (Corbeels *et al.* 2000). Management diversity in the agrodiversity framework expresses this wide variety of practices and innovations which use the attributes of many plant species. They add resilience to the whole farming system, and they increase production through conserving soil and water, improving soil fertility and controlling land degradation.

Agro-biodiversity is the single most direct way in which biological diversity can assist the control of land degradation. Having more plants over longer time periods substantially increases the vegetation cover to the soil. Having a greater variety of plants enables species to exploit the many micro-ecological niches within a farmer's field, again maximising vegetation cover. This cover has been demonstrated to be the single most influential factor in reducing erosion rates (Stocking 1994). A monocrop of maize may typically have a mean annual vegetation cover of 30%. Under the same soil conditions, an intercrop of maize-velvetbean (see next sub-section) would have about 60%. This doubling of cover reduces erosion hazard to one-sixth of the maize monocrop situation. Agro-biodiversity includes the many complex sequential and spatial intercropping practices employed by farmers. It will include the many systems of agroforestry that are so well exemplified in smallholder farms. It also embraces multi-storey cropping in some of the world's most complex and productive agricultural systems, such as the Chagga home gardens in northern Tanzania and the Kandy multi-storey gardens in the hill lands of Sri Lanka. A good review of the many beneficial processes in agroforestry systems and how they directly affect the soil and control land degradation is by Huxley (1999).

Finally, *organisational diversity* expresses the variety between farms and plots of land, and of the use of different resource endowments and of different income and labour sources. Tengberg *et al.* (1998) show how the characterisation of farmers into different resource levels largely explains broad differences in practice and land management. In this organisational diversity, the variables found most important in explaining farmers' engagement in local soil and water conservation practices were access to cash income, farm size, livestock ownership, labour availability, input use, cultivation methods and gender of head of household. Complex spatial patterns of uptake of the various techniques of conservation such as trashlines, *Fanya-juu* terraces, log-lines and stone bunds were found in Mbeere District of Kenya, depending on what this paper is calling 'organisational diversity'. Farmers' actual choice of conservation technique also related to type of soil and quality of land (biophysical diversity).

7.2. HOW AGRODIVERSITY PROMOTES FOOD SECURITY

One aspect of livelihoods presented in the framework for this paper (Fig. 1) is food security. Food security is primarily a function of maintaining assured production of food, and providing the variety of foods that society demands. This paper has already shown how agrodiversity has the potential to control land degradation. Implicit in this link is that control of land degradation, in turn, improves soil quality and increases production. Specific data on the soil property and productivity benefits of land degradation control are rare. While it is clear from most data that land degradation decreases crop yields, we can only assume that agrodiverse strategies will be at least as cost effective as the conservation measures which would have prevented land degradation.

Results from a large number of experiments sponsored by the UN Food and Agriculture Organization in Africa and South America (Fig. 3), show the exponential nature of the general

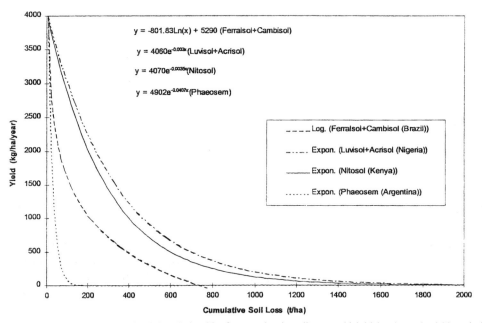

Figure 3. General erosion-productivity relationships for several major soil types, with initial maize grain yield on virgin land set to 4000 kg/ha. (Source: Tengberg and Stocking, 1997)

relationship between erosion as a measure of land degradation and crop yield as a measure of soil productivity and food security. The steepness of the initial decline in yields is largely a function of the specific soil type, which is an aspect of biophysical diversity in this paper. However, actual decline in yields in a farmer's field is far more sensitively related to specific erosion rates. In terms of aspects that are amenable to farmer's management, erosion rates are primarily a function of vegetation cover. Putting these relationships together, Figure 4 presents the example of the FAO experiments for Cambisols. This soil is the typical tropical brown earth; very attractive to smallholder farmers as it has a reasonable intrinsic fertility, is easily cultivated and has few of the problematic aspects of many tropical soils, except that it can be rather stony. The suite of curves in Figure 4 gives the best available evidence of the influence of management standard as assessed through vegetation cover (management diversity and agro-biodiversity). Clearly, the greater the degree of agrodiversity, the more likely is it that yields on Cambisols can be maintained at reasonably productive levels with low inputs.

What sort of agrodiverse practices can provide for the better cover and for meeting real livelihood benefits in smallholder farming systems in the tropics? Only one, albeit quite remarkable, example will be presented here. The maize-*Mucuna* (velvetbean) system has been achieving enthusiastic acceptance in many South and Central American countries.

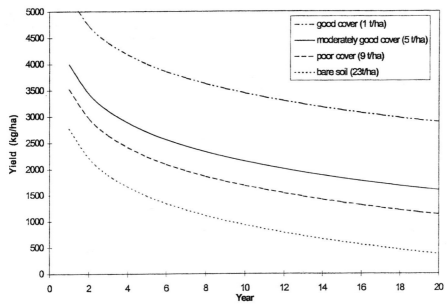

Figure 4. Maize yield scenarios with continued erosion for a Cambisol in a low input farming system with various degrees of vegetation cover. Typical annual erosion rates in tonnes per hectare for a 6% slope are shown for each cover level. (Source: Tengberg and Stocking, 1997).

On the Ferralsols (Oxisols) in the western hilly parts of Santa Catarina State in southern Brazil, many smallholders used to gain only a precarious living through arable cultivation of maize on these very erodible soils. Conservation programs had largely failed because the cost of structures was greater than any benefits to be gained in the short term. The maize-*Mucuna* system has transformed both the erosion situation and the maize-based livelihoods of local farmers. To account for this change, take the stages in a seasonal agricultural cycle. First, maize is directly drilled by hand (or with a horse-drawn drill) into the soil. Two weeks later, as the maize has started to germinate, *Mucuna pruriens,* a climbing legume bean seed, is also directly drilled. The maize grows rapidly and is harvested. The *Mucuna* is slower to grow, only reaching maximum growth rates after harvest of the maize. By the early dry season, the *Mucuna* completely covers all the maize stalks with a dense mat of creeping biomass. Over winter the *Mucuna* dies back, forming a thick black mulch over the soil surface. In the second season, the farmer simply direct drills into this mulch for the new maize crop. The benefits are manifold:

- an increased yield of maize: the *Mucuna* is largely non-competitive with the maize, and farmers in southern Brazil relate how the benefits of the whole system translate to approximately tripled yields compared with monocrop maize;
- weeds are out-competed or completely stifled by the *Mucuna* mulch; this saves considerable labour at peak times;
- tillage is unnecessary, again saving labour and cost, and
- in favourable sites, the *Mucuna* even sets sufficient seed to make replanting in future years unnecessary.

The maize-*Mucuna* system is, therefore, an excellent example of how an intercrop (agro-biodiversity), managed in a way that avoids mechanical tillage (management diversity), using the substantial labour benefits of suppressing weeds (organisational diversity), while improving soil structure and chemical fertility (biophysical diversity), can transform rural livelihoods. It is a special case that was promoted by missionaries in many South American countries rather than by the extension system. Nevertheless, it is a classic demonstration of the value of agrodiversity.

8. Conclusion

'Biological diversity' is a term of our time. Its protection in areas of land use is now one of the highest priorities on the international agenda as expressed by the government signatories to the *Convention on Biological Diversity*. Yet, biological diversity remains an enigma. It is difficult for non-ecologists to link the term with items of practical or economic relevance; modern agriculture is the antithesis of biodiversity; and rural development in the poorer parts of the tropics would seem to have far more pressing problems than protecting handfuls of obscure species and habitats. However, for the *Convention* to make headway, it is vital that biodiversity be seen in its pragmatic context, not only in ecologist's jargon as a 'life-support system' but also in a real practical context making the environment more secure and people's livelihoods better supported.

'Agrodiversity' is that practical, linking context. Through the framework presented in this chapter, it enables biodiversity to be viewed as a 'global agenda' linkage component with control of land degradation and the promotion of sustainable rural livelihoods. It encourages both environmental protection and sustainable rural livelihoods to be tackled at the same time and by the same instrument – the promotion of biological diversity in areas of land use. At the same time agrodiversity is inclusive of the techniques of management that are protective of the environment and support livelihoods. The ways in which farmers prepare their soils, plant different species and varieties together, use beneficial interactions between species, and employ novel techniques of weed management, are as important as the actual species and varieties used on-farm. This integration of the more traditional components of biological diversity with new studies into, for example, the rationality of indigenous technical knowledge and of mechanisms to cope with difficult climates or changing demographic patterns is the primary novel approach promoted under 'agrodiversity'. So, this widening of the boundaries of what may legitimately be included under 'biological diversity' is the central and fundamental means of enabling the World Bank's *win-win* scenario to be put into practical effect.

However, case studies which demonstrate these values are relatively few in number. Agrodiversity now needs to develop an empirical record as to its value, performance, sustainability and future potential. This is the objective of the *PLEC* project (see Acknowledgments below) which has demonstration sites established, where agrodiversity is being shown to land users, scientists, professionals and policy-makers as a desirable aim of policy and practice in agricultural development. Our sites are all in areas of key natural biological diversity importance in the following countries; Papua New Guinea, China, Thailand, Kenya, Uganda, Tanzania, Ghana, Guinea, Jamaica, Mexico, Brazil and Peru. We expect that over the next five years there will be a flow of examples of how agrodiversity is alive and well in many parts of the world. Our collaborators will be publishing the evidence that important pockets of 'good practice' exist to demonstrate that on-farm management of biological diversity is the most positive means of addressing the human needs of sustainable livelihoods and protection against land degradation.

9. Acknowledgments

This paper is based on the Keynote Address to the Conference of the IGU Commission on Land Degradation and Desertification, held in Perth at the University of Western Australia in September 1999. It is a contribution from the UNU/UNEP project, *People, Land Management and Environmental Change (PLEC),* which is largely funded 1998-2002 by the Global Environment Facility (i.e. the global funding mechanism set up after Rio to support developing countries' efforts in meeting their *Convention* responsibilities). *PLEC's* theme is to demonstrate the benefits of biodiversity in supporting small farmers' livelihoods, and in turn these farmers' important and often unsung roles in preserving biodiversity. The author is Associate Scientific Co-ordinator of *PLEC.* All PLEC collaborators in developing countries - now over 200 in number - are thanked for their contributions to our knowledge of biological diversity on agricultural lands. However, the opinions expressed here are the responsibility of the author alone.

10. References

Almekinders, C., Fresco, L. and Struik, P. (1995) The need to study and manage variety in agro-ecosystems, *Netherlands Journal of Agricultural Science* **43**, 127-142.

Altieri, M.A. (1999) The ecological role of biodiversity in agroecosystems, *Agriculture, Ecosystems and Environment* **74**, 19-31.

Black, R. and Sessay, M.F. (1997) Forced migration, environmental change and woodfuel issues in the Senegal River Valley, *Environmental Conservation* 24, 251-260.

Blaikie, P. and Jeanrenaud, S. (1997) Biodiversity and human welfare, in K.B. Ghimire and M.P. Pimbert (eds.), *Social Change and Conservation,* Earthscan, London, pp. 46-70.

Brookfield, H. and Padoch, C. (1994) Appreciating agrodiversity: a look at the dynamism and diversity of indigenous farming practices, *Environment* **36**, 7-45.

Brookfield, H. and Stocking, M. (1999) Agrodiversity: definition, description and design, *Global Environmental Change* **9**, 77-80.

Bunch, R. and López, G. (1999) Soil recuperation in Central America: how innovation was sustained after project intervention, in F. Hinchcliffe, J. Thompson, J. Pretty, I. Gujit and P. Shah (eds.), *Fertile Ground: the Impacts of Participatory Watershed Management,* IT Publications, London, pp.32-41.

Carney, D. (ed.) (1998) *Sustainable Rural Livelihoods: What Contribution Can We Make?* Department for International Development, London.

Cleaver, K.M. and Schreiber, G.A. (1994) *Reversing the Spiral: the Population, Agriculture and Environment Nexus in Sub-Saharan Africa,* The World Bank, Washington DC.

Corbeels, M., Shiferaw, A. and Haile, M. (2000) Farmers' knowedge of soil fertility and local management strategies in Tigray, Ethiopia, *Managing Africa's Soils* No. 10, International Institute for Environment and Development, London.

Critchley, W. (1999) Harnessing traditional knowledge for better land husbandry in Kabale District, Uganda, *Mountain Research and Development* **19**, 261-272.

Crosson, P.R. and Stout, A.T. (1983) *Productivity Effects of Cropland Erosion in the United States,* Resources for the Future, Washington DC.

Ellis-Jones, J. (1999) Poverty, land care and sustainable livelihoods in hillside and mountain regions, *Mountain Research and Development* **19**, 179-190.

Elwell, H.A. (1986) Determination of the erodibility of a subtropical clay soil: a laboratory rainfall simulator experiment, *Journal of Soil Science* **37**, 345-350.

Enters, T. (1998) Methods for the economic assessment of the on- and off-site impacts of soil erosion, *Issues in Sustainable Land Management* No. 2, International Board for Soil Research and Management, Bangkok.

Fairhead, J. and Leach, M. (1996) *Misreading the African Landscape: Society and Ecology in a Forest-Savanna Mosaic,* Cambridge University Press, Cambridge.

Guèye, M.B. (1995) The active method of participatory research and planning (MARP) as a natural resource management tool, in D. Stiles (ed.), *Social Aspects of Sustainable Dryland Management,* Wiley, Chichester, pp. 83-92.

Huxley, P.A. (1999) *Tropical Agroforestry,* Blackwell Science, Oxford.

Kerr, J. and Sanghi, N.K. (1992) *Indigenous Soil and Water Conservation in India's Semi-Arid Tropics,* Gatekeeper Series No.34, International Institute for Environment and Development, London.

Kiome, R. and Stocking, M. (1995) Rationality of farmer perception of soil erosion: the effectiveness of soil conservation in semi-arid Kenya, *Global Environmental Change* **5**, 281-295.

Koziell, I. (2000) *Diversity not Adversity: Sustaining Livelihoods with Biodiversity,* Rural Livelihoods Department, UK Department for International Development, London.

Magrath, W.B. and Arens, P. (1989) *The Costs of Soil Erosion on Java: A Natural Resource Accounting Approach,* Environment Department Working Paper No. 18, The World Bank, Washington DC.

McNeely, J.A. (1988) *Economics and Biological Diversity: developing and using economic incentives to conserve biological resources*, International Union for the Conservation of Nature and Natural Resources, Gland.

Netting, R. McC. and Stone, P. (1996) Agro-diversity on a farming frontier: Kofyar smallholders on the Benue Plains of central Nigeria, *Africa* **66**, 52-77.

Pagiola, S. (1999) The global environmental benefits of land degradation control on agricultural land, *World Bank Environment Paper* **16**, The World Bank, Washington DC.

Perrings, C.A., Mäler, K-G., Falke, C.S., Holling, C.S. and Jansson, B-O. (1995) *Biodiversity Conservation: Problems and Policies*, Kluwer Academic Publishers, Dordrecht.

Reij, C., Scoones, I. and Toulmin, C. (1996) *Sustaining the Soil: Indigenous Soil and Water Conservation in Africa*, Earthscan, London.

Rodda, G.H. (1993) How to lie with biodiversity, *Conservation Biology* **7**, 1959-1960.

Scoones, I. (1998) *Sustainable Rural Livelihoods: A Framework for Analysis*, Working Paper No.72, Institute of Development Studies, Brighton.

Stocking, M.A. (1994) Assessing vegetative cover and management effects, in R. Lal (ed.), *Soil Erosion Research Methods*, 2nd Edition, Soil and Water Conservation Society, Ankeny, Iowa, pp. 210-232.

Swanson, T.M. (ed.) (1995) *The Economics and Ecology of Biodiversity Decline: the forces driving global change*, Cambridge University Press, Cambridge.

Tengberg, A., Ellis-Jones, J., Kiome, R. and Stocking, M. (1998) Applying the concept of agrodiversity to indigenous soil and water conservation practices in eastern Kenya, *Agriculture, Ecosystems and Environment* **70**, 259-272.

Tengberg, A. and Stocking, M. (1997) *Erosion-induced Loss in Soil Productivity and its Impacts on Agricultural Production and Food Security*, Land and Water Development Division, UN Food and Agriculture Organization, Rome.

Tengberg, A. and Stocking, M. (2000) Land degradation, food security and agro-biodiversity - examining an old problem in a new way, in S. Sombatpanit et al. (eds.), *Meeting the Challenges of Land Degradation in the 21st Century*, Proceedings of the 2nd International Land Degradation Conference, Khon Kaen, Thailand, Science Publishers, Enfield NH (in press).

Thrupp, L.A. (1998) *Agricultural Biodiversity and Food Security: Predicaments, Polices and Practices*, World Resources Institute, Washington DC.

Tiffen, M., Mortimore, M. and Gichuki, F. (1994) *More People, Less Erosion: Environmental Recovery in Kenya*, Wiley, Chichester.

Tisdell, C. (1999) *Biodiversity, Conservation and Sustainable Development: Principles and Practice with Asian Examples*, Edward Elgar, Cheltenham.

Toulmin, C. (1993) Combating desertification: setting the agenda for a global convention, *Drylands Network Programme*, Issues Paper 42, International Institute for Environment and Development, London.

Wilson, E.O. and Peters, F.M. (eds.) (1988) *Biodiversity*, National Academy Press, Washington DC.

World Bank (1992) *World Development Report 1992: Development and the Environment*, Oxford University Press, New York.

World Bank (1996) *Desertification: Implementing the Convention*, Second Edition, The World Bank, Washington DC.

AUTHOR

MICHAEL STOCKING

School of Development Studies
University of East Anglia
Norwich NR4 7TJ
United Kingdom

E-mail: m.stocking@uea.ac.uk

Part One

Land Degradation Processes

Four papers are presented in this part. They are perhaps related more closely to the 'desertification' part of COMLAND's title than any of the others. The first, by Adrian Chappell and Clive Agnew, is set in the region which usually springs to mind when the term 'desertification' is introduced. They present a detailed analysis which suggests that data on spatial and temporal rainfall patterns in the West African Sahel reflect the changing spatial distribution of the recording network. If they are correct, the implications are clear: land degradation in the region since the 1930s reflects human actions not climate change. Dougill and Thomas also question a prevalent belief, namely that nebkha dunes are an indicator of desertification. Their analysis suggests that although nebkhas do reflect aeolian sediment transport, they are not necessarily directly linked to land degradation. The third paper by Greg Okin, Bruce Murray and William Schlesinger, deals with a very specific aspect of desertification in the United States. Soil chemical and remote sensing analyses are used to show that areas disturbed by human actions (abandoned pivot irrigation sites) lead to the spread of aeolian sediment downwind with distinctive geochemical signatures - perhaps a microcosm of the far more catastrophic Aral Sea situation.

Another very distinctive paper is that by Noble, Gillen, Jacobson, Low, Miller and the *Mutitjulu* Community working in central Australia. The paper is important in reminding us that while degradation affects soils and vegetation, these elements of the biophysical environment provide habitat for indigenous biota. In this instance the discussion delves further, showing the cultural significance of the extinction of a native animal species as a result of land degradation. That this paper could also have been included in Part Three also serves to demonstrate the complexity of the inter-linkages amongst the various aspects of land degradation.

CHAPTER 2

GEOSTATISTICAL ANALYSIS AND NUMERICAL SIMULATION OF WEST
AFRICAN SAHEL RAINFALL

ADRIAN CHAPPELL and CLIVE T. AGNEW

1. Abstract

The pattern of rainfall in the Sahelian zone of West Africa has been the subject of much
analysis since the major droughts of the 1970s. It is widely accepted that annual rainfall
has declined since that time and that it is a regional trend. The rain gauge network has
changed during this recent period of desiccation. Since the rainfall of the Sahelian region
is now recognised as not being homogeneous, there is some doubt about the reliability of
regional rainfall aggregated statistics. This analysis uses geostatistics to investigate the
impact of changes to the rain gauge network since 1931 and numerical simulations in an
attempt to validate the results. This approach can include most of the rainfall records,
unlike conventional methods. It seems that the configuration of the rainfall stations in the
E-W direction has accounted for the spatial variation of rainfall better since the 1970s
because of the eastward extension of rainfall stations. The rainfall stations in the N-S
direction do not sample adequately the generally large spatial variation for the period
between ca. 1945 to 1975. The only period when the distribution of rainfall stations in both
directions was adequate to sample the spatial variation in rainfall was since 1970. Thus, the
results suggest that the persistent downward trend in the aggregated annual rainfall since
the 1970s is a return to a more accurate estimate of the rainfall in this region; prior to the
1970s the aggregated annual rainfall was overestimated. The pattern of regional annual
aggregated rainfall statistics is an artifact of the annual location of stations and the spatial
variation in rainfall. Regional maps of (total summer) rainfall are based on the
geostatistical analysis which takes account of the spatial variation in the rainfall. These
maps are believed to be more reliable than those based on conventional interpolation
approaches and form the basis for a revised median annual rainfall estimate for the west
African Sahel between 1931 and 1990.

2. Introduction

Speculation about the likelihood of severe drought, desertification and famine in the west
African Sahel (Fig. 1) has been fuelled by a reported change in the climate between 1970
and 1990. The evidence comes from the decrease since 1970 in average rainfall (Fig. 2)
from varying numbers of stations (Fig. 2) in the region (Fig. 1). This decrease has been
observed for some time (Winstanley 1973; Lamb 1974; Nicholson 1979). More recently,
Copans (1983), Flohn (1987) and Druyan (1989) have suggested that the decline in rainfall

19

A.J. Conacher (ed.), Land Degradation, 19–35.

is below average and that drought is persistent. Hulme (1992) reported that rainfall is becoming more variable in the Sahel, and the downward trend more persistent. He also observed that the number of rainfall stations is declining. Ba *et al.* (1995) observed in their analysis of satellite-determined rainfall in the Sahel, that the number of stations available between 1983 and 1988 decreased from 271 to 147 (a reduction of 46%). This trend had already been observed by UNEP (1992), starting in the late 1970s and continuing through to the 1990s. Perhaps of greater importance to the aggregation of rainfall is the network of

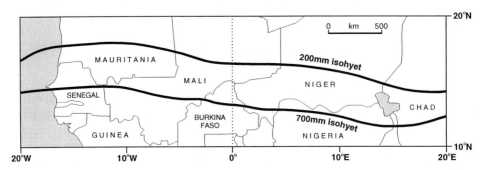

Figure 1. West African political boundaries and the Sahelian region (10˚N to 20˚N, 20˚W to 20˚E) in which rainfall records have been used.

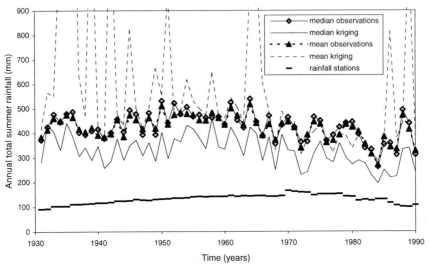

Figure 2. Average and median annual total summer (June, July, August, September) rainfall for station observations and kriging estimates and frequency of stations between 1931 and 1990 for the West African Sahel.

Sahelian rainfall stations "…. that are unevenly distributed in space, sparse in critical regions, and/or reported irregularly … it is often impossible to obtain a sufficient rainfall dataset over wide areas from a conventional rain gauge network" (Ba *et al.* 1995:412). Hence, there is concern about the ability of the rainfall network in the Sahel to assess climate change, since the number and locations of rainfall stations used to determine rainfall trends in the Sahel are uncertain.

Current approaches for analysing rainfall require continuous temporal records. Hulme (1992) selected rainfall stations for his analysis of African rainfall using the criteria advocated by the World Meterological Office (WMO). It requires stations to have at least 83% of the monthly rainfall totals in both 30-year periods (1931-1960 and 1961-1990); a continuous record of 300 months. As a result Hulme's (1992) analysis included only 572 stations (ca. 37%) of the 2099 stations in Africa. The temporal discontinuity of African rainfall records is not accounted for in conventional approaches. It is not clear how missing data should be accounted for in calculating the rainfall anomaly index (RAI). The rainfall data could be pre-processed to replace all missing data with estimated values. However, Bärring and Hulme (1991) suggest that the replacement of missing data should be conducted only if the missing observations are scarce and well scattered over time because individual rainfall records can have considerable influence on the calculation of an average or RAI.

Doubt can also be cast on spatially aggregated rainfall statistics for all, if not major parts, of the Sahel because they take no account of local variations. Most publications which include Sahelian rainfall figures have a time series of rainfall anomalies that depict changes for the entire region. Nicholson and Palao (1993:286) state that "…. the entire Sahel cannot be treated as homogeneous with respect to rainfall variability". The spatial distribution of rainfall is continuous and there are distinct patterns in its spatial variation which differ from year to year in response to local and regional climatic fluctuations. It is difficult to reconcile the calculation of a simple average value for the area that represents rainfall adequately with the complex spatial and temporal variation in its distribution.

The aim of this paper is to test the hypothesis that the reduction since 1970 in the temporal record of west African Sahel rainfall (Fig. 2) depends on the location and number of rainfall stations which contribute to the annual spatial aggregation of rainfall. Geostatistics are used to provide evidence of the structural change in the spatial variation of rainfall, to show that annual variation in the rainfall station network has altered sampling efficiency of the rainfall distribution and to map the annual variation in rainfall distribution. The geostatistical estimates of rainfall across the west African Sahel provide the ability to directly test the hypothesis using numerical simulations of rainfall station networks with variable station frequency and areal coverage.

3. Methodology

3.1. DATASET

Monthly rainfall data for the west African Sahel (Fig. 1) between 1931 and 1990 were provided from the Climate Research Unit at the University of East Anglia. The rainfall stations meet the standards set by the WMO to reduce measurement error and these data are also screened for statistical outliers (Hulme 1992) to reduce systematic errors. These data were used to calculate annual total summer rainfall (TSR) by the addition of monthly rainfall for June, July, August and September. Few stations had years in which at least one monthly rainfall total was missing, but when this occurred in the record the TSR for that year was excluded from the analysis. The number of locations which provided TSR data varied annually (Fig. 2) between 100 and 170 across the region.

3.2. GEOSTATISTICS AND SPATIAL AGGREGATION OF RAINFALL

Geostatistics is the practical application of the Theory of Regionalised Variables (Matheron 1971) which regards spatial properties as random functions. The model regards a set of spatial observations as one stochastic realisation of a probabilistic function. Although the approach is stochastic, spatial data are essentially deterministic, in other words they are a function of their position in space, but their variation is so irregular that the best way of treating them is as if they were random variables. Unlike conventional statistical approaches, geostatistics does not require a continuous temporal record, therefore rainfall records from all stations can be included. It is important to include all rainfall data because the model of spatial variation (variogram) can be unreliable when there are few samples (Webster and Oliver 1992).

The basis of the theory is that the variation in an attribute Z can be defined by a stochastic component and a constant. Following Oliver et al. (1989a) this may be written as

$$Z(\mathbf{x}) = m_r + e(\mathbf{x}) \tag{1}$$

where $\mathbf{x}$ denotes the spatial co-ordinates, m_r is the mean in a region r and the quantity $\varepsilon(\mathbf{x})$ is the spatially dependent random variable. This last component has a mean of zero,

$$E[\varepsilon(\mathbf{x})] = 0 \tag{2}$$

and a variance defined by

$$\mathrm{var}[\varepsilon(\mathbf{x}) - \varepsilon(\mathbf{x} + \mathbf{h})] = E[\{\varepsilon(\mathbf{x}) - \varepsilon(\mathbf{x} + \mathbf{h})\}^2] = 2\gamma(\mathbf{h}) \tag{3}$$

where $\mathbf{h}$ is a vector, the lag, that separates the two places $\mathbf{x}$ and $\mathbf{x}+\mathbf{h}$ in both distance and direction. Thus, the variance of $e(\mathbf{x})$ depends on the separation $\mathbf{h}$ and not on the actual position of $\mathbf{x}$. This assumes that the variable is second order stationary and that the mean, variance, and variogram exist. Matheron (1971) realised that second order stationarity was too strong for many spatial variables and reduced the assumptions to stationarity of the mean and variance of the differences. This is the Intrinsic Hypothesis. Equation (3) is then equivalent to:

$$\mathrm{var}[z(\mathbf{x}) - z(\mathbf{x} + \mathbf{h})] = E[\{z(\mathbf{x}) - z(\mathbf{x} + \mathbf{h})\}^2] = 2\gamma(\mathbf{h}) \tag{4}$$

The semi-variance, $\gamma(\mathbf{h})$, is half the expected squared difference between two values separated by $\mathbf{h}$, as above. The function that relates γ to $\mathbf{h}$ is the semi-variogram or more commonly the variogram.

The spatial variation in the Sahelian rainfall is distinctly anisotropic; the degree of variation in the east-west direction (E-W) is different from that in the north-south direction (N-S). The latter has a stronger rainfall gradient than the former. Anisotropic variation is commonly due to processes operating at differing spatial scales in different directions. Experimental variograms were computed in Genstat (Genstat 5 Committee 1992) for TSR ¡every year between 1931 and 1990 in the two principal directions of spatial variation.

The intrinsic hypothesis is taken further to assume quasi-stationarity that limits the area to a local neighbourhood v. There are situations where this does not hold. In some regions the local mean values of the variables vary predictably or deterministically from one part of the region to another. This is evidence of trend which violates the assumptions of geostatistics. Following Delfiner (1976) trend was removed by fitting low-order polynomials on the spatial coordinates using least squares regression. The variograms were then estimated from the residuals of the trend, which are random variables (Olea 1975; 1977; Oliver 1987; Chappell 1995; Chappell *et al.* 1996). Models were fitted to the re-computed variograms using weighted least squares in Genstat (Genstat 5 Committee 1992) and the best model was selected using the largest explanation of variance.

The models fitted to the experimental variograms in their isotropic form are described below and the lag **h** becomes the scalar $h = |\mathbf{h}|$ (Webster and Oliver, 1990). The quantity c is the sill variance and a is the range of the bounded models:

Spherical: $\tilde{a}(h) = c \{1.5\,(h/a) - 0.5\,(h/a)^3\}$ for $h \le a$ (5)
$\tilde{a}(h) = c$

Circular: $\tilde{a}(h) = c \{1 - (2/\eth)\cos^{-1}(h/a) + (2/\eth)(h/a)\sqrt{(1-h^2/a^2)}\}$ for $h \le a$ (6)
$\tilde{a}(h) = c$

In the case of unbounded models, the linear model is the simplest in which the semi-variance increases linearly with increasing h. The parameter w is used to control the gradient of the model:

Linear: $\tilde{a}(h) = wh$ (7)

The models are commonly used to derive parameters (nugget variance, range and sill variance) which are diagnostic of the structure of spatial variation (Chappell and Oliver 1997) and for use in estimation at unsampled locations (Chappell 1998). A nugget variance is common for sample data of a continuous variable (McBratney and Webster 1986; Chappell *et al.* 1996). Although measurement error and stochastic variation in data contribute to the nugget, the largest source of variation is commonly due to spatially dependent variation that occurs over distances much smaller than the shortest sampling interval (Webster and Oliver 1990). The nugget variance can be used to indicate the amount of 'noise' in the sampling system (Curren and Dungen 1989; Atkinson 1997; Chappell *et al.* 2000) since it explains the amount of variation that is unresolved by the sampling scheme (rainfall stations). When the difference between samples in space is at a maximum for the average separation distances, the sill variance is reached and the model is bounded.

The lag separation distance at which the variogram reaches its sill is the range; this is the limit of spatial dependence. Beyond this limit there is no relation between sampling points at the separating distance.

When data are sparse an important aim in subsequent analyses is to estimate the values of a property at unsampled locations. Kriging is the method of geostatistical estimation (Oliver et al. 1989b). It is essentially a moving weighted average, but the weights are derived from the variogram model rather than in an arbitrary way, as with most other methods of interpolation (Webster and Oliver 1990). Hence, the estimates are based on the spatial variation of the property because kriging uses the variogram model. Kriging has been shown to be one of the most reliable two-dimensional spatial estimators (Laslett et al. 1987) and it might be expected to produce more reliable estimates of regional rainfall and therefore the location of the isohyets, than simple methods of interpolation.

The anisotropic behaviour of the rainfall data was accounted for in the kriging estimation procedure by a linear transformation of the co-ordinates known as geometric or affine anisotropy (Journel and Huijbregts 1978). Using GSLIB (Deutsch and Journel 1992), the parameters of the models fitted to the variograms were used to solve anisotropic ordinary punctual kriging equations and to estimate TSR residuals from the quadratic polynomial on a 1 degree (ca. 110 km) grid across the west African Sahel (Fig. 1). The quadratic polynomials for each year were added back to the kriging estimates. Finally, isohyets were threaded through these combined estimates for every year between 1931 and 1990 (with the same isoline frequency).

The kriging estimates of TSR include the observed TSR and are assumed to provide a better estimate of an aggregated statistic for the region than the observations alone because of the large number of estimations at unsampled locations. The average and median of the TSR observation were calculated annually between 1931 and 1990. The average and median of the TSR kriging estimates were also calculated annually between the same period.

3.3. NUMERICAL SIMULATION OF RAINFALL STATION NETWORKS

The kriging estimates of TSR at unsampled locations across the Sahel every year between 1931 and 1990 provide data which are used to validate the inferences drawn from the geostatistical analysis and to directly test the hypothesis of this paper using numerical simulations. Several numerical procedures were written to simulate observations provided by a network of rainfall stations. Three experiments were performed, in which the sample locations were kept constant, to test the effect of:

1. increasing frequency of rainfall stations (80-200 stations, increasing in 20 station steps) on the aggregation of annual TSR for (a) the coastal Sahel (<5°W) and (b) for the entire Sahelian region (<20°E);
2. increasing rainfall sample area (5°W to 20°E area, increasing 5° areal steps) on the aggregation of annual TSR for (a) 80 rainfall stations and (b) 200 rainfall stations, and
3. the interaction between variable station frequency (using the actual annual rainfall station frequency) and increasing sample area (5°W to 20°E, increasing 5° area) on the aggregation of annual TSR.

These experiments were repeated 20 times using randomly selected simulated rainfall stations and the results were aggregated (average and median) to produce a robust estimation of the random selection process.

4. Results

4.1. GEOSTATISTICS AND SPATIAL AGGREGATION OF RAINFALL

As expected, the experimental variograms computed for TSR every year between 1931 and 1990 in the two principal directions of spatial variation exhibited trend (not shown) in both directions caused by the systematic variation in space of the rainfall. Trend was removed from the annual TSR data. Figure 3 shows examples for several years of the experimental variograms computed on the residuals from the trend of the annual TSR data and the fitted models. Examples of the model parameters for the variograms are shown in Table 1a and b. The majority of the variograms were fitted best by bounded models but some of the variograms were fitted best by unbounded (linear) models. The occurrence of the latter model suggested that as the area of interest increased, so more sources of variation were encountered (Webster and Oliver 1990; Chappell 1995). In other words, the TSR remained spatially dependent with increasing distance.

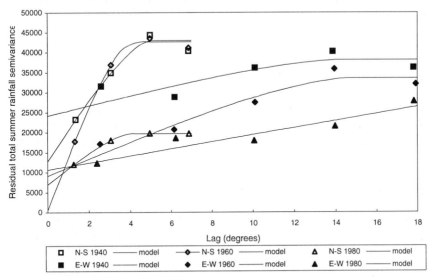

Figure 3. Variograms without trend and fitted models for total summer rainfall in the west African Sahel computed for 1940, 1960 and 1980 in the East-West and North-South directions.

Table 1. Examples of the model parameters fitted to (a) E-W variograms and (b) N-S variograms of total summer rainfall residuals of the trend removed by quadratic polynomials.

a) Year	Model Type	Model Variance	Range	Variance Sill/Gradient[1]	Nugget
1931	Spherical	58.5	7.84	15970	16767
1940	Circular	61.7	10.70	10749	16832
1950	Spherical	88.9	19.04	22392	12347
1960	Circular	93.2	12.01	21692	11452
1970	Linear1	77.7	-	1144	15670
1980	Linear1	82.3	-	1461	5676
1990	Circular	62.0	10.33	11184	3007

b) Year	Model Type	Model Variance	Range	Variance Sill	Nugget
1931	Circular	86.8	4.96	33568	4659
1940	Circular	89.4	4.96	30140	12785
1950	Circular	70.2	5.28	44067	10782
1960	Spherical	80.7	2.25	53669	1242
1970	Spherical	92.7	4.10	31372	5636
1980	Spherical	98.9	4.41	13264	6449
1990	Circular	67.0	4.50	8856	7868

[1]The linear model is unbounded and does not have a limit or range of spatial dependence, does not reach a sill but includes a gradient.

The parameters of the models fitted to the variograms in the two principal directions (N-S and E-W) for each year are plotted in Figures 4-6. The magnitude of the sill variance in the E-W direction (Fig. 4) was similar to that in the N-S direction (Fig. 4) and decreased over time. The occurrence of bounded (sill variance) and unbounded (gradient) models was very different in each direction. Notably, most models fitted to the variograms in the E-W direction were unbounded after 1974 which suggested that increasing sources of variation were present since this date. The range of the variograms in the N-S direction was considerably less variable and had a greater magnitude than the range of the variogams in the E-W direction (Fig. 5). The magnitude and variability of the nugget variance in the E-W variograms (Fig. 6) decreased over time. The magnitude of the nugget variance in the N-S variograms (Fig. 6) was similar to that of the nugget variance in the E-W variograms, but the inter-annual variation of the former was considerably larger than the latter and the general trend in both was very different.

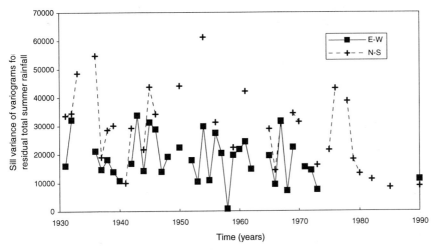

Figure 4. The sill variance of variograms without trend for total summer rainfall computed for East-West and North-South directions every year between 1931 and 1990 for the west African Sahel.

Figures 7a, b, c, d, e, f and g are examples of the TSR maps produced following the anisotropic punctual kriging procedure. The isohyets threaded through the estimates of total summer rainfall for several years showed the expected anisotropic variation, whereby the rainfall gradient was greater in the N-S direction than in the E-W direction. The N-S rainfall gradient was greater within the period 1945 to 1970 (Fig. 7c, d, e) than outside this period (Fig. 7a, b, f, g) as evident from the isohyet compression in the south of the region.

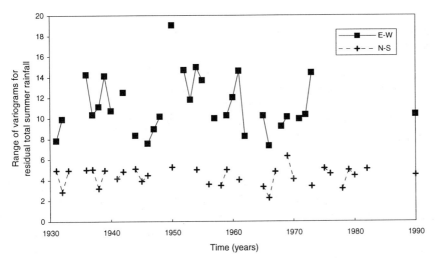

Figure 5. The range of variograms without trend for total summer rainfall computed for East-West and North-South directions every year between 1931 and 1990 for the west African Sahel (missing data are caused by unbounded variation; see text for details).

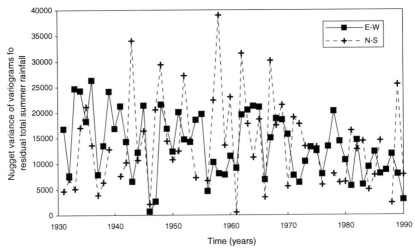

Figure 6. The nugget variance of variograms without trend for total summer rainfall computed for East-West and North-South directions every year between 1931 and 1990 for the west African Sahel.

The average annual kriging estimates of TSR had greater variation than the average observed TSR (Fig. 2). However, the kriging estimates and observed TSR averages were similar between ca. 1970 and 1984. The annual variation in the median TSR kriging estimate (Fig. 2) was similar to the median observed TSR.

4.2. NUMERICAL SIMULATIONS OF RAINFALL STATION NETWORKS

Separate simulations were conducted on the affect of the number of stations (static area) and increasing area (static number of stations) on the median TSR kriging estimates (not shown).

Within the same area, the temporal variation in the median TSR for 80 rainfall stations was very similar to that of 200 rainfall stations. In contrast, the temporal variation in the median TSR for the entire Sahel region (<20°E) resulted in a consistently higher magnitude and greater variation than that of a small area (<5°W) of the same region.

The results of simulating the effect of an interaction between the actual number of rainfall stations and an increasing area sampled by the simulated stations are shown in Figure 8. The median TSR generally decreased as the area sampled by the rainfall stations increased. However, the difference between the median TSR for each area was greater before 1960 than that after 1960. There also appeared to be a further decrease in the difference between the median TSR for each area in the period between 1960 and 1970 and after 1970.

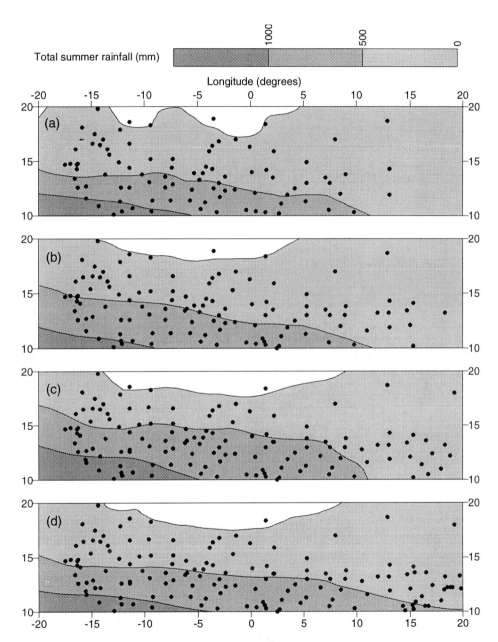

Figure 7. Ordinary (anisotropic) punctual kriging estimates of total summer rainfall (mm) and the location of rainfall stations (?) for years (a) 1931, (b) 1940, (c) 1950, (d) 1960, (e) 1970, (f) 1980, (g) 1990.

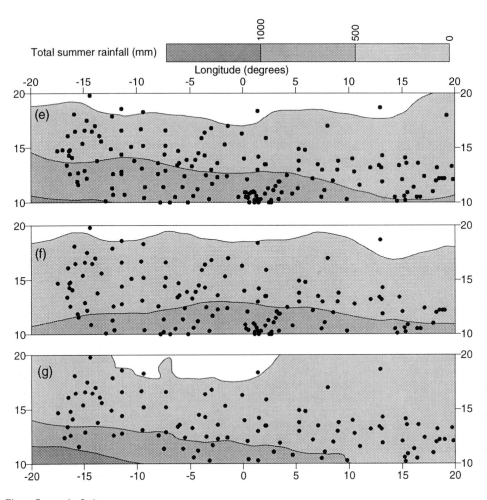

Figure 7 cont. (e, f, g)

5. Discussion

5.1. GEOSTATISTICS

The shorter range of TSR in the N-S direction (Fig. 5; ca. 4°; 440 km) than in the E-W direction (ca. 10°; 1100 km) is evidence of anisotropy in the variation. It confirms our expectation of the rapid decrease in rainfall with latitude (Davy *et al.* 1976; Farmer 1989).

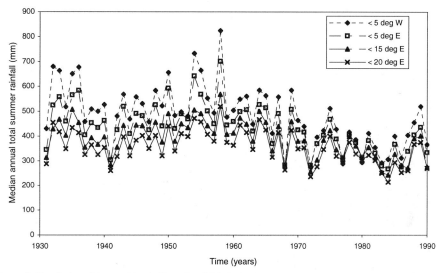

Figure 8. Simulation of the combined affect of variable (actual) rainfall station frequency and increasing areal sampling, on the calculation of the median annual total summer rainfall in the west African Sahel.

Notably, the temporal variation in the range of TSR over time, in both directions, showed a repeated pattern that appeared to have a periodicity of 4-5 years. More importantly probably, the temporal variation in the range of the variograms in the N-S direction was far smaller than that in the E-W direction. The results suggested that the steeper rainfall gradient in the N-S direction remained consistent over time but the more gentle rainfall gradient in the E-W direction was more variable. The range of TSR probably corresponded to the area covered by the summer easterly squall systems (Nieuwolt 1977; Hastenrath 1985; Reading *et al.* 1995), and the periodicity of the range of TSR suggested that the nature of these systems may have a cyclical pattern. Thus, variation in the temporal rainfall gradient in the E-W direction suggested that the distribution of rainfall was similarly variable.

The sill variance of the N-S variograms decreased over time (Fig. 4). In conjunction with a consistent range in the N-S direction, the results suggested that the variation of rainfall in this direction was becoming more uniform over time, thus questioning Hulme's (1992) assertion of increased variation in Sahelian rainfall. The sill variance of the variograms in the E-W direction (Fig. 4) fluctuates considerably over time but the trend appears to remain consistent. However, the models fitted to the variograms in the E-W direction prior to 1974 are bounded. Since this date the fitted models are unbounded (linear) which suggested that there were increased sources of variation. Thus, the results suggest that it was not variation in the *amount* of TSR, it was the size of the area that received the rainfall that increased (since 1974).

The magnitude and variability of the nugget variance in the E-W variograms (Fig. 6) decreased over time and suggested that the spatial variation of rainfall in this direction was better sampled over time. If the range of the E-W variograms showed a similar temporal variation to the nugget in the same direction, then the spatial variation of rainfall might have explained the pattern in the nugget variance of the E-W variograms; if the rainfall stations remained at the same location over time the spatial variation of rainfall would result in similar temporal variation in the nugget variance and range of the E-W variograms. That this was not the case suggested that the decrease in nugget variance of the E-W variograms was due to the extension over time of rainfall stations into the easterly end of the region (Fig. 7). This temporal and spatial variation in the location of rainfall stations was important for the calculation of an aggregated (average or median) rainfall statistic, because rainfall was small in the east as shown by the isohyets threaded through the kriging estimates of total summer rainfall for several years (Fig. 7). The isohyets were also compressed in the west of the region, which suggested that rainfall was more variable than in the east. Thus, over time an eastwards extension would locate more rainfall stations in the area where rainfall has small magnitude and low spatial variation.

There was no similarity in the temporal variation of the range in the N-S variograms (Fig. 5) and the nugget variance of the variograms in the same direction (Fig. 6), suggesting that the latter's pattern is not explained by variation in the former. The nugget and sill variance of the N-S variograms before ca. 1945 and after ca. 1970 are generally smaller than that for the period between these dates, suggesting that the spatial variation of rainfall was smaller and better sampled in the early and late periods of the rainfall record than between ca. 1945 to 1975. An examination of the rainfall maps (Fig. 7) suggests that rainfall stations were generally located in the south of the region and that over time there was little N-S variation in their location. The probable explanation for the temporal variation in nugget and sill variance of the N-S variograms is that the configuration of the rainfall stations was inadequate to sample the generally large spatial variation between ca. 1945 to 1975. This is important for the calculation of an aggregated rainfall statistic because (1) the statistic is probably controlled by the predominance of rainfall stations in the south where high magnitude rainfall existed (Fig. 7), and (2) over time the static rainfall station configuration did not adequately represent the spatial variation in rainfall along the N-S rainfall gradient.

5.2. SPATIAL AGGREGATION OF RAINFALL

A comparison between the average and the median annual kriging estimates of TSR (Fig. 2) suggests that the average statistic is strongly affected by the high rainfalls on the coast and in the west of the region. The median is a more robust statistic than the mean for measuring the temporal variation in spatially aggregated rainfall.

Before 1970 the inter-annual variation in the magnitude of the average annual kriging estimates suggests that there is considerable temporal variation in the occurrence of high magnitude rainfall in the coastal region. Between ca. 1970 and 1984 the kriging and observed TSR averages are similar (Fig. 2), suggesting that the average observed TSR provided a better estimate of the aggregated rainfall than before 1970. This is probably because the spatial variation in rainfall during this period decreased and there are no large rainfall estimates on the coast, enabling the rainfall station configuration to provide a better sample than before.

The median TSR kriging estimate appears not to be affected by large rainfall estimates

(Fig. 2) because its temporal variation is similar to the median observed TSR. However, the median observed TSR is consistently larger than the median TSR kriging estimate, probably because the former depends on fewer stations located mainly in the wetter west of the region than the latter, which comprises many estimates over the entire region. The median annual TSR kriging estimate is believed to form the best estimate of aggregated rainfall for west African Sahel between 1931 and 1990.

5.3. NUMERICAL SIMULATIONS OF RAINFALL STATION NETWORKS

The magnitude and temporal variation of the median TSR kriging estimate are affected more by an increase in area than by an increase in rainfall stations. This effect is generally greater before 1970 than it is after this date, and the downturn in median TSR since 1955 is considerably greater for the smaller sample area than it is for the entire region. The simulations suggest that a large amount of the downturn in median rainfall since 1960 can be explained by an eastwards increase in the area sampled by the rainfall stations. These results are similar to the more realistic simulation of the interaction between a variable (but actual) number of rainfall stations and an increasing area sampled by those stations (Fig. 8). The similarity shows that the main control on the spatially aggregated statistic (median TSR kriging estimate) is the eastwards extension of rainfall stations into a drier area. In all simulations the difference between the median TSR for increasing areas, decreased after ca. 1960. This represented the decrease in spatial variation of TSR as the rainfall stations extended eastwards into a drier, more homogeneous rainfall area.

6. Conclusion

The geostatistical methodology provided a robust approach to handling the erratic rainfall records of the west African Sahel. The approach included most of the rainfall records in the region over the period of study, unlike other conventional approaches which vastly reduce the rainfall records (possibly compounding the difficulty of investigating the data) because of the requirement of a continuous rainfall record.

The geostatistical analysis of the spatial structure of total summer rainfall (TSR) elucidated several key points which were not evident from classical statistical analysis of the rainfall data:

(1)   the E-W configuration of rainfall stations accounted for the spatial variation of rainfall (in other words sampling was more efficient) in recent years. The results suggested this was due to the eastwards extension of rainfall stations in recent years;

(2)   the N-S configuration of rainfall stations remained approximately static over time. However, the spatial variation of rainfall increased during the period 1945 to 1970, resulting in rainfall stations providing a poor sample of the rainfall distribution, and

(3)   the results suggested that only since 1970 has the configuration of rainfall stations been adequate to sample the spatial variation of rainfall (in the E-W and N-S directions). Ironically, this is the period when many workers have reported a persistent downward trend in the average annual rainfall.

Validation of these assertions using the simulations showed that the downturn in the spatially aggregated annual TSR since the 1960s is an artifact of the temporal variation in the location of rainfall stations and an over-reliance on simplistic aggregation statistics. The results suggested that since 1960 there has not been a persistent decline in TSR but that pre-1960 annual estimates of TSR were over-estimated. Since 1960 the annual TSR has been more accurately estimated than in the preceding years. Despite the reduction in the number of rainfall stations since the 1960s, improved estimation of TSR occurred because the rainfall station network was better able to sample the spatial variation of rainfall.

7. Acknowledgments

A. Chappell is grateful to M. Ekström for support with data manipulation, to G. Dobrzynski for drawing the location map and to M. Oliver and C. Skelly for incisive comments on an earlier version of this manuscript which did much to improve it. The authors are also grateful for the monthly rainfall data provided by the Climatic Research Unit at the University of East Anglia.

8. References

Atkinson, P.M. (1997) On estimating measurement error in remotely-sensed images with the variogram, *International Journal of Remote Sensing* **18**, 3075-3084.

Ba, M.B., Frouin, R. and Nicholson, S.E. (1995) Satellite derived interannual variability of West African Rainfall during 1983-88, *Journal of Applied Meteorology* **34**, 411-431.

Bärring, L. and Hulme, M. (1991) Filters and approximate confidence intervals for interpreting rainfall anomaly indices, *Journal of Climate* **4**, 837-847.

Chappell, A. (1995) Geostatistical mapping and ordination analyses of ^{137}Cs-derived net soil flux in south-west Niger, Unpublished PhD Thesis, University of London.

Chappell, A. (1998) Mapping ^{137}Cs-derived net soil flux using remote sensing and geostatistics, *Journal of Arid Environments* **39**, 441-455.

Chappell, A. and Oliver, M.A. (1997) Geostatistical analysis of soil redistribution in SW Niger, West Africa, in E.Y. Baafi and N.A. Schofield (eds.), *Quantitative Geology and Geostatistics* **8**, Kluwer, Dordrecht, pp. 961-972.

Chappell, A., Oliver, M.A., Warren, A., Agnew, C.T. and Charlton, M. (1996) Examining the factors controlling the spatial scale of variation in soil redistribution processes from south-west Niger, in M.G. Anderson and S.M. Brooks (eds.), *Advances in Hillslope Processes*, Wiley, Chichester, pp. 429-449.

Chappell, A., Seaquist, J.W. and Eklundh, L.R. (in press) Improving geostatistical noise estimation in NOAA AVHRR NDVI images, *International Journal of Remote Sensing*

Copans, J.(1983) The Sahelian drought, in K. Hewitt (ed.), *Interpretations of Calamity*, Allen & Unwin, London, pp 83-97.

Curran, P.J. and Dungan, J.L. (1989) Estimation of signal-to-noise: a new procedure applied to AVIRIS data, *IEEE Transactions on Geoscience and Remote Sensing* **27**, 620-628.

Davy, E.G., Mattei, F and Solomon, S.I. (1976) An evaluation of the climate and water resources for development of agriculture in the Sudan-Sahelian zone of West Africa, *WMO Special Environmental Report* **No. 9**, World Meteorological Organisation, Geneva.

Delfiner, P. (1976) Linear estimation of non stationary spatial phenomena, in M. Guarascio, M. David and C. Huijbregts (eds.), *Advanced Geostatistics in the Mining Industry*, Reidel, Dordrecht, pp. 49-68.

Deutsch, C.V. and Journel, A.G. (1992) *GSLIB Geostatistical Software Library and User's Guide*, Oxford University Press, Oxford.

Druyan, L.M. (1989) Advances in the study of sub-saharan drought, *International Journal of Climatology* **9**, 77-90.

Farmer, G. (1989) Rainfall, *IUCN Sahel Studies*, Nairobi, 1-25.

Flohn, H. (1987) Rainfall teleconnections in northern and eastern Africa, *Theoretical and Applied Climatology* **38**, 191-197.

Genstat 5 Committee (1992) *Genstat 5, Release 3, Reference Manual,* Oxford University Press, Oxford.

Hastenrath, S. (1985) *Climate and Circulation in the tropics*, D. Riedel, Boston.

Hulme, M. (1992) Rainfall changes in Africa, *International Journal of Climatology* **12**, 685-699

Journel, A.G. and Huijbregts, Ch.J. (1978) *Mining geostatistics*, Academic Press, London.

Lamb, H.H. (1974) The Earth's changing climate, *Ecologist* **4**, 10-15.

Laslett, G.M., McBratney, A.B., Pahl, P.J. and Hutchinson, M.F. (1987) Comparison of several spatial prediction methods for soil pH, *Journal of Soil Science* **38**, 325-341.

Matheron, M.A. (1971) *The Theory of Regionalised Variables and its Applications*, Cahiers du Centre de Morphologie Mathématique de Fountainebleau no. **5**.

McBratney, A.B., and Webster, R. (1986) Choosing functions for semi-variograms of soil properties and fitting them to sampling estimates, *Journal of Soil Science* **37**, 617-639.

Nicholson, S.E. (1979) Revised rainfall series for the West African subtropics, *Monthly Weather Review* **107**, 620-623.

Nicholson, S.E. and Palao, I.M. (1993) A re-evaluation of rainfall variability in the Sahel, *International Journal of Climatology* **13**, 371-389.

Nieuwolt, S. (1977) *Tropical Climatology in Low Latitudes*, Wiley, Chichester.

Olea, R.A. (1975) *Optimum Mapping Techniques Using Regionalised Variable Theory*, Series on Spatial Analysis, no. **2**, Kansas Geological Survey, Lawrence.

Olea, R.A. (1977) *Measuring Spatial Dependence with Semi-Variograms*, Series on Spatial Analysis no. **3**, Kansas Geological Survey, Lawrence.

Oliver, M.A. (1987) Geostatistics and its applications to soil science, *Soil Use and Management* **3**, 8-20.

Oliver, M., Webster, R. and Gerrard, J. (1989a) Geostatistics in physical geography. Part I: theory, *Transactions of the Institute of British Geographers* **14**, 259-269.

Oliver, M., Webster, R. and Gerrard, J. (1989b) Geostatistics in physical geography. Part II: applications, *Transactions of the Institute of British Geographers* **14**, 270-286.

Reading, A., Thompson, R.D. and Millington, A.C. (1995) *Humid tropical environments*, Blackwell, Oxford.

United Nations Environment Programme (1992) World *Atlas of Desertification*, Edward Arnold, London.

Webster, R. and Oliver, M.A. (1992) Sample adequately to estimate variograms of soil properties, Journal of Soil Science **43**, 177-192.

Webster, R. and Oliver, M.A. (1990) *Statistical Methods in Soil and Land Resource Survey*, Oxford University Press, Oxford.

Winstanley, D.W. (1973) Rainfall patterns and general atmospheric circulation, *Nature* **245**, 190-194

AUTHORS

ADRIAN CHAPPELL[1] AND CLIVE T. AGNEW[2]

[1]Telford Institute of Environmental Systems, Division of Geography, School of Environment and Life Sciences, University of Salford, Manchester, M5 4WT. UK (a.chappell@salford.ac.uk) - corresponding author

[2]School of Geography, University of Manchester, Oxford Road, Manchester M13 9PL. UK (clive.agnew@man.ac.uk)

CHAPTER 3

POTENTIALS AND PROBLEMS IN USING NEBKHA DUNES AS INDICATORS
OF SOIL DEGRADATION IN THE MOLOPO BASIN, SOUTH AFRICA AND
BOTSWANA

A. J. DOUGILL and A. D. THOMAS

1. Abstract

Nebkha dunes, formed from sediment accumulations around shrubs, have been proposed as
a reliable, rapid indicator of dryland degradation. This paper aims to investigate the
applicability of this link for the Molopo Basin, southern Africa, from sedimentological and
biochemical evidence. This study shows that sediments are largely locally derived from
interdune areas suggesting saltation and surface creep are the key aeolian transport
processes leading to nebkha formation. The local derivation of sediments implies that wind-
blown losses from neighbouring agricultural fields and regional scale movements of
suspended dust are minimal and not required for nebkha development. Nebkha sediments
are significantly enriched in available inorganic nutrients (N and P), suggesting that
sediment movements, litter inputs and/or improved conditions for nutrient mineralisation
can lead to an increase in spatial variability of soil chemical characteristics. However,
nutrient-enriched sub-canopy niches on nebkhas retain a seed resource of palatable grass
species even in intensively grazed areas. Further, the use of nebkha resources by local
communities suggests that although nebkhas do indicate aeolian transport of sediment, the
immediate association with soil degradation is questioned and thus requires further
investigation in southern Africa.

2. Introduction and Background: Soil Degradation in Southern Africa

Classifications of soil degradation are based on the basic premise that a reduction in land
agricultural productivity must occur (Abel and Blaikie 1989; UNEP 1997). The
development of rapid appraisal indicators of soil degradation has been stressed by various
development agencies (e.g. UNEP 1997; DFID 1998) as an important research challenge
facing dryland scientists. Within southern Africa, there is a recognised need to develop and
adopt simple subjective assessments of land degradation based on integrating local
knowledge and scientific understanding of soil and vegetation processes (Dean *et al.* 1995;
Thomas and Sporton 1997; Snyman 1998; van Rooyen 1998). For example, this
requirement has led to the recent publication of a rangeland health assessment guide for arid
Karoo shrublands aimed at enabling local farmers to conduct rangeland health assessments
(Milton and Dean 1996). Such a hands-on approach aims to make farmers more aware of
potential soil degradation threats and has been deemed successful for commercial ranchers

37

(Milton *et al.* 1998). Given the increasing predominance of small land-holders throughout South Africa, in association with the government policy of land redistribution, the development of such participatory action research approaches in communal farming areas is deemed vital. For this to be achieved, a clearer understanding of processes leading to soil degradation is required; and there is a need to to link these processes to visible indicators, either in terms of changes in vegetation cover or geomorphic features linked to soil degradation processes.

Semi-arid areas on the margins of the Kalahari sandveld (as classified by the continuous extent of Kalahari sand cover - Thomas and Shaw 1991), including the Molopo Basin study sites investigated here, support greater resident populations, through mixed pastoral and arable farming systems, than the pastoral systems found within the Kalahari. Particularly high populations are found in the communally farmed regions of the former black homeland of Bophutatswana, where relocation of outside populations and intensive agricultural development projects have led to greater pressures on agricultural land (Drummond 1995). This region has been classified as displaying evidence of soil degradation, through a variety of processes including both water and wind erosion (UNEP 1997). This regional-scale classification of soil degradation, based on subjective GLASOD methodologies (Oldeman *et al.* 1990), also suggests that such visible physical processes of soil redistribution are not associated with soil chemical deterioration. Particularly important in this regard, is the potential existence of positive feedbacks between soil heterogeneity and invasion by desert shrubs (Schlesinger *et al.* 1990). Localised nutrient enrichment under shrub canopies has been recognised in a range of semi-arid regions (for example, Blackmore *et al.* 1990; Schlesinger *et al.* 1990; Weltzin and Coughenour 1990; Belsky 1994; Ludwig and Tongway 1995). Predominantly, enrichment is classed as resulting from increased organic inputs to the sub-canopy habitat and the ability of bush roots to draw nutrients from a far wider radius. Where bushes also trap sediment to form nebkhas, this heterogeneity is potentially increased further by the net loss of sediment-bound nutrients from the eroding areas between bushes.

Nebkha dunes, also referred to as nabkha's (Nickling and Wolfe 1994; Khalaf *et al.* 1995) or coppice dunes (Gile 1966; Thomas and Tsoar 1990), are plant obstacle dunes, which are formed by the trapping of sand within the body of a plant (Cooke *et al.* 1993). Rather than the vegetation simply acting as an obstacle, bushes become an integral part of the dune, growing as the sediment accumulates. Some confusion remains over the exact mechanism of formation, with some authors stressing the potential for rainsplash and overland flow erosion to lower intershrub areas (Parsons *et al.* 1992). Others stress the ability of high rooting densities and termite activity under bushes to lead to plant mounds (Biot 1990). However, it is apparent from the growing number of detailed morphological and sedimentological studies in semi-arid localities bordering the Sahel and in the Middle East that nebkhas are good indicators of wind erosion and deposition (Gunatilaka and Mwango 1987; Nickling and Wolfe 1994; Tengberg 1994, 1995; Khalaf *et al.* 1995; Tengberg and Chen 1998).

This study investigates nebkha formation processes and the nutrient implications of sediment redistribution in southern African locations. In particular, the suitability of nebkhas for use as rapid appraisal indicators of processes which adversely affect agricultural potential by reducing soil fertility are investigated. To gain an improved understanding of both nebkha formation processes and the soil degradation implications, this paper aims:

1. to assess from nebkha sedimentological characteristics the likely process of formation and sediment source areas, and
2. to estimate the extent to which nebkhas represent an indicator of soil degradation in terms of associated reductions in soil fertility from either the communal grazing areas in which they are found, and/or from surrounding cleared arable fields.

3. Research Design and Methods

3.1. SITE SELECTION AND CHARACTERISTICS

To address these objectives, field studies investigated sedimentological, morphological and biochemical properties of nebkhas, interdunes and other potential source sediments in the Molopo Basin of North West Province, South Africa and Southern District, Botswana (Fig. 1). The region is semi-arid with a mean annual rainfall of *c.* 500 mm concentrated in the summer wet season (October - March). Highest wind speeds occur towards the end of the dry season (August - October) or at the time of intense convectional thunderstorms. Before agricultural clearance, the area was characterised by mixed bush savanna vegetation communities located on well drained sandy soils (Hutton series - Molope 1987). Unlike the Kalahari sandveld soils to the north, with their uniform fine sand contents of over 97 % (Dougill *et al.* 1998), the Molopo soils contain a substantial proportion of medium to coarse

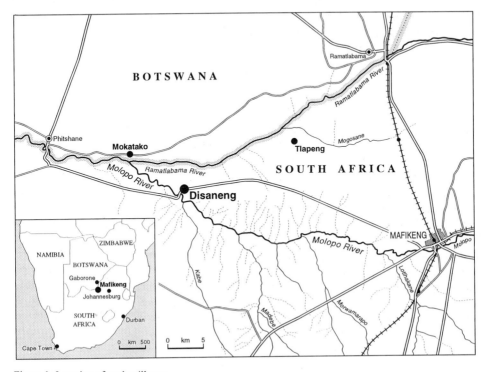

Figure 1. Location of study villages

sands (*c*. 25-30%). In undisturbed areas a weak surface crust typifies these soils. All sites were dominantly populated by Tswana tribal groups, though the different agricultural policy histories in the two countries has led to subtle differences in land use patterns, with higher proportions of arable land continuously cultivated around South African villages.

Study sites were chosen around three villages (Disaneng and Tlapeng in South Africa, and Mokatako in Botswana) all of which have a high density of nebkha dunes on rangeland areas surrounding the village. The chosen villages were also typified by a relatively equal mix of communal rangeland, used for the grazing of cattle and goats by all villagers, and of arable fields, typically farmed by a dominant 'master farmer' from the village community. The choice of Disaneng was also governed by background information available from participatory studies on drought coping strategies (Vogel and Drummond 1993). At Disaneng one study transect was placed downwind of the Agricultural Development Corporation (AGRICOR) Irrigation Scheme to assess potential inputs from large-scale intensively-farmed arable land as found around select villages in the region (Drummond 1995).

3.2. SAMPLING

Data were collected at three distinct spatial scales to allow detailed investigation of the research objectives. First, at the village scale, two 3 - 5 km long, and 30 m wide, ecological and morphological surveys were undertaken on transects orientated to the north and east of villages (corresponding to the direction of the predominant winds at the end of the dry season). Transects were surveyed for changes in nebkha density, average nebkha size and bush cover (visual estimates by species) per 100 m interval. These surveys were then combined with field geomorphological site descriptions, supported by aerial photograph analysis, to investigate links between the spatial distribution of nebkhas and geomorphological and/or land use factors, including the decreasing grazing intensity with distance from the village. Nebkha classification involved all sub-bush canopy sediment accumulations with a crest length of greater than 60 cm being classed as a nebkha, matching the minimum recorded size of nebkhas in Sahelian studies (Tengberg and Chen 1998). Inclusion of nebkhas below 1.5 m in diameter led to the inclusion of sediment accumulations around low growing *Gnidia caffra* shrubs (typical of disturbed or cleared ground), as well as larger accumulations around bushes, notably *Acacia heblecada* and *Grewia flava*. The average size of nebkhas was also calculated for every 100 m interval to enable variations in the stage of nebkha development to be recognised and linked to variations in ecological, geomorphological or land use variables.

Second, more detailed plot-scale (30 m by 30 m) studies were conducted at three or four selected sites along each of the six transects. Changes in nebkha density and size were recorded for all of these sites. Plot locations were located at set intervals along the transects to reflect differences in grazing intensity away from the villages. For all nebkhas within the demarcated plots, nebkha length, width, orientation and the dominant vegetation type were noted. Further field surveying of a minimum of ten nebkhas in each plot provided measures of nebkha height. Sediment samples were then collected from the surface of surveyed nebkhas, from interdune areas adjacent to the sampled nebkhas and from key potential source areas in the vicinity of plots (such as arable fields, pan depressions, river valleys and tracks).

Finally, at the nebkha scale, analyses of variations in grain-size characteristics and organic matter content with depth through soil profiles were conducted for 15 selected nebkhas and neighbouring interdunes. For these sites, surface sediment samples were taken together with subsurface samples from 5 cm depth. Further, for two nebkhas, detailed cut-through pits were excavated and sediment samples taken at depths of approximately 0, 5, 10, 15, 25, 30, 40 and 50 cm, to assess the likely formation processes leading to nebkha growth. In particular, the relative importance of organic inputs and accumulation of sediments from different potential aeolian transport processes, and therefore likely distance of transport, was investigated from this information.

3.3. SEDIMENT ANALYSIS

Sediment samples were air-dried prior to determination of grain size, organic matter (OM) and for selected samples, inorganic nutrient content. Grain-size distributions were determined at half-phi intervals in the range - 1.0 to + 4.0 phi (2 mm to 0.063 mm). The amount of material retained in each sieve (and the <4.0 phi receiver) after shaking was weighed and the information plotted as a cumulative frequency curve. From this, the mode (50), graphic mean (Mz) and inclusive graphic standard deviation or sorting (I) were calculated according the methods of Folk and Ward (1957). Folk and Ward parameters, although criticised because they necessitate the log-normal transformation of grain-size data (Bagnold 1973), are a powerful tool in analysis of aeolian sediments (Lancaster 1986). Their use here also facilitated the direct comparison of grain-size data with previous nebkha studies (Khalaf *et al.* 1995; Tengberg 1995) and sedimentological research from other dune environments (Livingstone 1987; Thomas 1997).

Following sieving, samples were grouped into different size classes broadly representative of the different modes of aeolian transport (creep, saltation and suspension), as assessed from a range of previous process-based aeolian studies (Bagnold 1973; Pye and Tsoar 1990; Sterk *et al.* 1996). The following size classes were used: suspension material, <0.063 mm; saltation particles, from 0.063 to 0.5 mm; and creep particles, >0.5 mm. Analysis of organic matter content (according to the loss-on-ignition method of Rowell 1994) and inorganic nutrient concentration (NO_3^--N, PO_4^{3-}-P, NH_4^+-N and exchangeable K, all analysed colorimetrically using a portable Palintest Photometer 5000) were determined from samples from each size class to assess potential nutrient and organic redistribution through each of these transport processes. This approach prevented the dispersal of sediments using Calgon and consequently the removal of all organic material prior to sieving was not possible.

4. Results and Analysis

The results are presented at different spatial scales (village, plot and nebkha) along with initial interpretation of findings. At the village scale the data were used to investigate the relationship between the nebkha characteristics and changing grazing intensities away from the villages as well as intermittant changes in geomorphology. Plot scale data provided the bulk of the detailed sedimentological and morphological information on the nebkhas, interdunes and likely sources areas in order to better understand formation processes.

Finally, detailed within-nebkha stratigraphy was used to provide historical information on nebkha formation. The wider implications of the findings are provided in the subsequent discussion section.

4.1. VILLAGE SCALE

General geomorphological, sedimentological and ecological characteristics of nebkhas around the three study villages are summarised in Table 1. Findings show the inherent variability in the morphology of nebkhas across the region caused by the different land use histories, ecological community characteristics, and the geomorphological setting with its direct impact on the nature of local source sediments. In particular, clearance of rangeland and subsequent ecological recovery on transects at Disaneng has led to the higher proportion of nebkhas formed under *G. caffra* shrubs. This factor is responsible for the smaller size of nebkhas at Disaneng, due to the significantly smaller dimensions of nebkhas formed under *G. caffra* [mean nebkha length under *Acacia spp.* = 2.63 0.67 m (n = 53); mean nebkha length under *G. caffra* = 1.08 0.13 m (n = 45)]. Further heterogeneity is introduced by the variations in geomorphological and sedimentological setting. This is particularly notable around Tlapeng, where the presence of river gravel, calcrete and pan deposits leads to a notably coarser make up of nebkhas (Table 1) and in certain areas

Table 1. Morphological, ecological and sedimentological characteristics of nebkhas in the Molopo Basin.

Village	Morphology			Sedimentology			Ecology		
	Mean nebkha length (m)	Mean nebkha height (m)	Mean organic matter content (%)	Mean grain size - M_z (ϕ)	Mean sorting - σ_1. (ϕ)	Nebkhas Acacia spp. dominant (%)	Nebkhas Grewia flava dominant (%)	Nebkhas Gnidia caffra dominant (%)
Disaneng	1.84 (n = 109)	0.11	2.27	1.91	1.15	49	9	42
Tlapeng	5.32 (n = 68)	0.24	3.78	1.80	1.24	70	26	4
Mokatako	3.33 (n = 69)	0.12	4.07	2.18	1.15	67	22	11

appears to result in sheet flow around bushes, aiding nebkha development. The heterogeneity caused by such land use and geomorphological factors means that no significant trends in nebkha density or size were recorded in relation to the distance from villages where grazing pressure is concentrated. Thus, whilst grazing pressure is an important factor removing vegetation cover, thus increasing the susceptibility of soils to wind erosion, other variables also control the nature of nebkha development and growth. It is with respect to improving understanding of the processes controlling nebkha formation, that subsequent plot-scale studies were aimed.

4.2. PLOT SCALE

Grain size data from nebkha sediments and neighbouring interdunes are summarised in Table 2. Nebkha sediments are typically coarse and poorly sorted. The grain size characteristics of the potential source area sediments are variable (Table 3). Channel sediments are much coarser than nebkhas, whilst track sediments are notably finer, and arable fields are slightly coarser than nebkhas.

The statistical similarity between the mean grain size (Mz) of nebkha and interdune sediments (Table 2) implies that the two environments are formed from similar sediment sources. This similarity suggests that nebkhas are most likely formed by local redistribution of sediment at the dune-interdune scale without a major input of deposited fines, as would be expected if large-scale dust-storms led to a significant input of suspended fine sediments. Neighbouring arable fields at both Disaneng and Mokatako are typified by slightly coarser surface sediments (Table 3) than neighbouring nebkhas and interdunes (Table 2), implying that some preferential removal of fines may have occurred due to wind erosion from these fields.

Table 2. Grain size parameters for surface sediment samples from nebkhas and neighbouring interdunes from study sites in the Molopo Basin [mean and (standard deviation) values shown].

Study Location	Sample environment	Mode -ϕ50 (ϕ)	Mean - $M_z(\phi)$	Sorting - $\sigma_i(\phi)$
All sites	Nebkha	2.05	1.91	1.18
	(n = 99)	(0.41)	(0.30)	(0.14)
	Interdune	2.15	1.94	1.33 *
	(n = 76)	(0.34)	(0.21)	(0.19 ')
Disaneng	Nebkha	2.02	1.91	1.15
	(n = 52)	(0.46)	(0.36)	(0.14)
	Interdune	2.01	1.87	1.34 *
	(n = 34)	(0.46)	(0.28)	(0.13)
Tlapeng	Nebkha	1.94	1.80	1.24
	(n = 26)	(0.33)	(0.23)	(0.14)
	Interdune	2.16	1.89	1.46 *
	(n = 19)	(0.45)	(0.26)	(0.17)
Mokatako	Nebkha	2.40	2.18	1.15
	(n = 21)	(0.17)	(0.17)	(0.16)
	Interdune	2.51 *	2.24	1.17
	(n = 23)	(0.10 ')	(0.11)	(0.14)

* - significant difference ($p < 0.05$) between mean values for nebkha and interdune sediments assessed using t-test.
' - significant difference ($p < 0.05$) between variance values for nebkha and interdune sediments assessed using F-test two sample test.

Table 3. Grain size parameters for surface sediment samples from potential source areas recognised around study sites in the Molopo Basin.

Sample Environment	Mode -ϕ50 (ϕ)	Mean - M_z (ϕ)	Sorting - σ_1 (ϕ)
Disaneng arable fields	2.15	1.81	1.30
Disaneng track	2.40	2.12	1.17
Disaneng gully	2.55	2.32	1.09
Tlapeng – Mogosane channel	1.35	1.25	1.51
Tlapeng – Ramatlabama channel	2.00	1.75	1.46
Tlapeng track	2.30	2.00	1.29
Mokatako track	2.55	2.32	1.08
Mokatako arable fields	2.45	2.17	1.12

Significant differences ($p < 0.05$) are, however, apparent between the sorting of nebkha and interdune sediments (Table 2). The better-sorted nebkha sediments are a result of the size-selective transport processes leading to their formation. The presence of a coarser interdune surface (Fig. 2) suggests removal of these saltatable sediments from interdune sites. Comparisons with research in other semi-arid areas shows the coarser nature of nebkha dune sediments compared with other nebkha and mobile dune environments (Table 4), probably resulting from the predominance of coarser sediments in the local source materials, especially the local soils. The binding of particles as aggregates and mycrophytic surface crusts will act to increase further the proportion of sediments moving by saltation and creep, with fines potentially moved as part of these larger aggregates. Consequently, it appears that suspension of fine sediments in large-scale dust storms is not necessary to explain the formation of nebkhas in southern Africa.

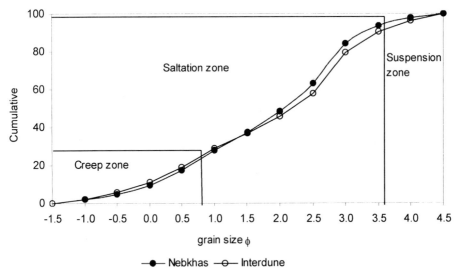

Figure 2. Mean cumulative percentage grain size curves for nebkha and interdune sediments sampled in the Molopo Basin

The importance of organic matter additions from bushes and/or termite activity on nebkha formation was investigated through assessments of the organic matter content of nebkha and neighbouring interdunes in both bulked samples and differentiated by grain size class (Table 5). The data show that bulked nebkha sediments are characterised by significantly higher (p <0.05) organic matter contents than neighbouring interdune sediments. Organic accumulation appears to be caused partly by litter inputs which are responsible for some of the higher organic matter concentrations in coarse fractions (Table 5), and partly by the accumulation of fine fractions in nebkhas (Figure 2) with their residually higher organic matter content (Table 5). This accumulation of organic material will contribute to nebkha

Table 4. Grain-size parameters of nebkhas and linear dunes crests from range of research studies.

Location	Mean (M_z)	Sorting (δ_1)	Source
Nebkhas			
Molopo Basin	1.91	1.18	Dougill and Thomas (this study)
Kuwait	2.3-4.6	1.0-1.8	Khalaf *et al.* (1995)
Burkina Faso	2.82	1.23	Tengberg (1995)
Linear dune crests			
Kalahari	2.21	1.76	Goudie (1970)
	2.37	0.57	Lewis (1936)
	2.16	0.49	Lancaster (1986)
Namib	2.15	0.54	Besler (1983)
	2.11	1.71	Goudie (1970)
	2.25	0.39	Lancaster (1983)
	2.44	0.37	Lancaster (1981)
Simpson	2.53	0.43	Folk (1971)
	2.22	0.67	Buckley (1989)
Negev	1.87	0.42	Tsoar (1978)

Table 5. Mean organic matter content (and standard deviation) of nebkha and interdune sediment samples from study sites in the Molopo Basin

Sample class	Nebkha OM content (%)	Interdune OM content (%)
Bulk samples	2.83 * (1.20; n = 21)	1.85 (0.58; n = 22)
Coarse fractions (> 1.0 ϕ)	2.81 * (1.78; n = 17)	1.19 (1.52; n = 7)
Medium fractions (4.0 – 1.0 ϕ)	1.18 (0.55; n = 16)	1.24 (0.35; n = 6)
Fine fractions (< 4.0 ϕ)	6.67 (1.19; n = 14)	6.00 (0.71; n = 6)

* - significant difference (p < 0.05) between mean values for nebkha and interdune sediments assessed using t-test.

growth; however, the low amounts involved (typically < 3 %) suggest that this is not the main formation process. The organic enrichment of nebkha sediments is well known amongst local populations who reported using nebkha soils to fertilise garden vegetable plots.

Table 6 summarises inorganic nutrient analyses on selected sediment samples (both bulked and by grain size class) from nebkhas and interdunes. These data corroborate the high variability typical of inorganic nutrient availability in nutrient-poor Kalahari sand soils (Dougill *et al.* 1999). Despite this, it is apparent from these preliminary findings that there is a significant nutrient enrichment in nebkha sediments compared with adjacent interdunes. This is especially true for nitrate (NO_3^--N) and phosphate (PO_4^{3-}-P), the main inorganic forms of the two key limiting nutrients (N and P) in southern African rangelands (Skarpe and Bergström 1986; Scholes 1990). Increased nutrient concentrations in nebkha sediments were recorded for all grain size classes (Table 6). This suggests that selective redistribution through erosion and deposition is not the sole cause of the increased inorganic soil nutrient heterogeneity with some *in situ* nutrient accumulation occurring on nebkhas. The higher organic matter content of nebkha sediments (Table 5) will also increase nutrient retention, but this also may not be sufficient to explain the nutrient enrichment observed on the medium grain size fractions which dominate nebkha development. Alternative mechanisms, such as the likely increased mineralisation rates (from organic to inorganic nutrient forms) in the moister sub-bush canopy habitats, and the input of fines as coatings on larger particles, require further investigation to assess the cause and extent of this sub-canopy inorganic nutrient enrichment.

Table 6. Mean extractable inorganic nutrient concentrations (and standard deviations) for nebkha and interdune sediment samples from study sites in the Molopo Basin

Sample class	Sample environment	NO_3^--N ($\mu g\ g^{-1}$)	PO_4^{3-}-P ($\mu g\ g^{-1}$)	NH_4^+-N ($\mu g\ g^{-1}$)	Ex. K ($\mu g\ g^{-1}$)
Bulk samples	Nebkha (n = 11)	283.9 (204.4)	115.5 (97.0)	54.2 (40.9)	5653.4 (3982.7)
	Interdune (n = 6)	45.8 (22.4)	87.5 (63.2)	102.0 (92.4)	3175.0 (1994.6)
Coarse fractions (> 1.0 ф)	Nebkha (n = 11)	190.3 (173.0)	108.0 (95.8)	257.0 (144.5)	3532.2 (1828.4)
	Interdune (n = 6)	39.6 (36.4)	41.7 (25.5)	47.8 (35.2)	2791.7 (1479.4)
Medium fractions (4.0 – 1.0 ф)	Nebkha (n = 11)	279.8 (223.8)	132.5 (100.4)	81.1 (80.7)	5217.8 (3926.3)
	Interdune (n = 6)	55.8 (21.8)	50.8 (88.4)	95.5 (93.5)	3395.8 (903.9)
Fine fractions (< 4.0 ф)	Nebkha (n = 11)	512.1 (326.9)	267.1 (81.7)	117.1 (158.9)	9450.8 (5394.6)
	Interdune (n = 6)	170.4 (133.1)	162.5 (131.3)	395.8 (522.9)	9520.8 (4254.7)

4.3. NEBKHA SCALE

Grain size characteristics of the surface and subsurface of nebkhas and interdunes are summarised in Table 7, with more detailed depth profile variations shown for two excavated nebkhas in Fig. 3. Results show the existence of a coarser surface sediment 'lag' for both nebkha and interdune sediments. For interdunes, this is indicative of the active removal of fines through erosion from interdunes. However, on nebkhas, the presence of crusting and aggregate formation associated with greater soil fauna activity may potentially contribute to the presence of coarser material following sieving.

Nebkha surface sediments are enriched in organic matter (Table 5) due to inputs from the bush canopy. This organic input will, in time, be broken down leaving the mineral sediment input to contribute predominantly to the growth of the nebkha, a factor demonstrated by the uniformly finer deposited sediments which characterise nebkhas through their depth profiles down to the former interdune surface (as shown in Fig. 3). The importance of mineral fractions in the growth and development of nebkhas was shown further by the consistently low organic matter content (< 3 %) in all subsurface nebkha sediment samples. This demonstrates that the accumulation of organic material, whilst important in binding nebkha sediments and preventing further erosion, is not the predominant process leading to nebkha formation.

Table 7. Grain size parameters for surface and subsurface (5 cm depth) sediment samples from nebkhas and neighbouring interdunes from study sites in the Molopo Basin [mean and (standard deviation) values shown].

Sample environment	Sample depth	Mode -ϕ50 (ϕ)	Mean - M_z (ϕ)	Sorting - σ_I (ϕ)
Nebkhas	Surface	2.03	1.89	1.30
(n = 13)		(0.50)	(0.30)	(0.14)
	Subsurface	2.43	2.26	1.06
		(0.35)	(0.33)	(0.17)
Interdunes	Surface	1.97	1.87	1.45
(n = 15)		(0.57)	(0.34)	(0.21)
	Subsurface	2.36	2.15	1.32
		(0.47)	(0.25)	(0.16)

5. Discussion and Conclusions

This paper aimed firstly to improve the understanding of nebkha formation in southern Africa, and secondly to assess their value as indicators of soil degradation in mixed farming systems characteristic of the region. In terms of nebkha formation, results from all scales indicate that the local scale redistribution of sediments from interdunes to nebkhas is the most important formation process. The predominance of medium and coarse grain size fractions in nebkha sediments is notable (Figs. 2 and 3), especially when compared with sediment characteristics of nebkhas and active dunes in other semi-arid environments (Table 4). Such sediment assemblages suggest that the saltation and creep of interdune sediments over distances of a few metres are the main sediment transport processes leading to nebkha formation. Intensive grazing of interdunes could increase the extent of such movements in two ways. First, by reducing the surface vegetation cover, land surface roughness is reduced and thus particle saltation is enhanced. Second, the active disturbance of the sediment surface by hooves breaks up weakly formed mycrophytic surface crusts which stabilise sand

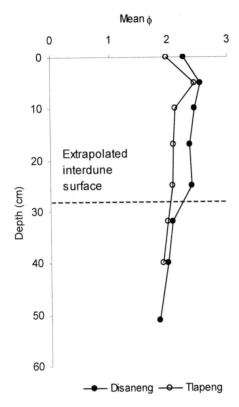

Figure 3. Variation in mean grain size (M$_z$) with depth in two sample nebkha dunes

sediment movements by saltation. The importance of such localised transport processes is further demonstrated by the existence of finer sediments bound up in aggregates which will move solely by creep or saltation, and by the only limited organic enrichment of nebkha sediments caused by surface litter accumulation (Table 5). The presence of nebkhas in this region is therefore a reliable indicator of local scale aeolian transport (erosion and deposition) of sediment.

Assessments of the potential viability of nebkhas as indicators of the occurrence, and nature, of soil degradation are less clear cut. Whilst the presence and growth of nebkhas demonstrate the existence of sediment erosion and deposition and increased local scale heterogeneity in soil nutrient availability, there remains uncertainty as to whether such changes in resource availability can be immediately associated with soil degradation, in terms of reduced potential agricultural productivity. Particularly important factors in this regard include potential ecological resilience imparted by the nutrient-enriched, and grazing protected (due to thorny impenetrable nature of bush species, especially *Acacia heblecada*), sub-bush canopy niche which nebkhas provide. The continued existence of grass biomass cover on nebkhas to the end of the dry season, provides a potential grazing reserve during drought, but more importantly, a grass seed base capable of maintaining significant grass cover in intensively grazed interdune areas following rainfall. Ecological studies, assessing the relative cover of different grass and annual species in relation to the spatial distribution of nebkhas and interdunes, are required to investigate this possible resilience mechanism further. Additionally, the value of nebkha resources to local farmers in terms of both the bush resource (potential use for fuel, fodder and fencing) and the organically-enriched sediments which nebkhas contain, needs assessment through further participatory studies focusing on such resource use issues.

The sedimentological and biochemical findings reported here provide a preliminary indication of the potential for nutrient and organic matter depletion of interdune soils as a result of the localised sediment redistribution to form nebkhas. The significant enrichment of the main inorganic fractions of nitrogen and phosphorus in nebkha sediments shows that physical redistribution of sediments could be capable of reducing soil fertility for extensive areas of communally farmed rangeland on the margins of the Kalahari sandveld. However, the possible increase in mineralisation rates under nebkha bush canopies and the positive feedback between bush canopies and nutrient availability (Schlesinger *et al*. 1990; Belsky 1994) implies that an immediate direct association with chemical soil degradation in interdune areas should be avoided. Consequently, the use of nebkhas as rapid indicators of soil degradation is questioned, with a clear requirement for further process-based studies assessing the relative importance of such transfers on the observed increased spatial heterogeneity of soil fertility.

6. Acknowledgments

The authors gratefully acknowledge the financial support provided by the Royal Society (Dudley Stamp Memorial Fund) and the University of Salford (Embryonic Research Fund). The assistance, co-operation and logistical support provided by staff at the Department of Geography and Environmental Science, University of North West, South Africa is gratefully acknowledged. Fieldwork assistance provided by Beth Maher, Gilly Sheridan, Simon Hutchinson and Nigel Trodd is much appreciated. The manuscript was improved by useful comments from Geoff Humphreys as reviewer.

7. References

Abel, N.O.J. and Blaikie, P.M. (1989) Land degradation, stocking rates and conservation policies in the communal rangelands of Botswana and Zimbabwe, *Land Degradation and Rehabilitation* 1, 101-123.

Bagnold, R.A. (1973) *The physics of blown sand and desert dunes,* 5th edition, Chapman and Hall, London.

Belsky, A.J. (1994) Influences of trees on savanna productivity - tests of shade, nutrients, and tree-grass competition *Ecology* 75, 922-932.

Besler, H. (1983) The response diagram: distinction between aeolian mobility and stability of sands and aeolian residuals by grain-size parameters, *Zeitschrift fur Geomorphologie Supplementband* 45, 287-301.

Biot, Y. (1990) The use of tree mounds as bench-marks of previous land surfaces in a semi-arid tree savanna, Botswana, in J.B. Thornes (ed.), *Vegetation and Erosion*, John Wiley, London. pp. 437-450.

Blackmore, A.C., Mentis, M.T. and Scholes, R.J. (1990) The origin and extent of nutrient-enriched patches within a nutrient-poor savanna in South Africa, *Journal of Biogeography* 17, 463-470.

Buckley, R. (1989) Grain-size characteristics of linear dunes in central Australia, *Journal of Arid Environments* 16, 23-28.

Cooke, R.U., Warren, A. and Goudie, A.S. (1993) *Desert Geomorphology*, UCL Press, London.

Dean, W.R.J., Milton, S.J. and du Plessis, M.A. (1995). Where, why, and to what extent have rangelands in the Karoo, South Africa, desertified? *Environmental Monitoring and Assessment* 37, 103-110.

Department for International Development (DFID) (1998) *Sustainable Rural Livelihoods: What contribution can we make?* DFID, London.

Dougill, A.J., Heathwaite, A.L. and Thomas, D.S.G. (1998) Soil water movement and nutrient cycling in semi-arid rangeland: vegetation change and system resilience, *Hydrological Processes* 12, 443-459.

Dougill, A.J., Thomas, D.S.G. and Heathwaite, A.L. (1999) Environmental change in the Kalahari: integrated land degradation studies for nonequilibrium dryland environments, *Annals of the Association of American Geographers* **89**, 420-442.

Drummond, J.H. (1995) Development and change: irrigation and agricultural production in Dinokana village, North West Province, South Africa, in T. Binns (ed.), *People and Environment in Africa*, John Wiley, London, pp. 239-247.

Folk, R.L. (1971) Longitudinal dunes of the north-western edge of the Simpson Desert, Northern Territory, Australia. I. Geomorphology and Grain-size relationships *Sedimentology* **16**, 5-54.

Folk, R.L. and Ward, W.C. (1957) Brazos River bar - a study in the significance of grain-size parameters, *Journal of Sedimentary Petrology* **27**, 3-27.

Gile, L.H. (1966) Coppice dunes and the Rotura soil, *Proceedings of the Soil Science Society of America* **30**, 657-676.

Goudie, A.S. (1970) Notes on some major dune types in southern Africa, *South African Geographical Journal* **52**, 93-101.

Gunatilaka, A. and Mwango, S. (1987) Continental sabkha pans and associated nebkhas in southern Kuwait, Arabian Gulf, in L.E. Frostick and I. Reid (eds.), *Desert Sediments: Ancient and Modern*, Geological Society Special Publication No. 35, Blackwell, Oxford, pp. 187-204.

Khalaf, F.I., Misak, R. and Al-Dousari, A. (1995) Sedimentological and morphological characteristics of some nabkha deposits in the northern coastal plain of Kuwait, Arabia, *Journal of Arid Environments* **29**, 267-292.

Lancaster, N. (1981) Grain-size characteristics of Namib Desert linear dunes, *Sedimentology* **28**, 115-122.

Lancaster, N. (1983) Linear dunes of the Namib sand sea, *Zeitschrift fur Geomorphologie Supplementband* **45**, 27-49.

Lancaster, N. (1986) Grain-size characteristics of linear dunes in the south-western Kalahari, *Journal of Sedimentary Petrology* **56**, 395-400.

Lewis, A.D. (1936) Sand dunes of the Kalahari within the borders of the Union, *South African Geographical Journal* **19**, 22-32.

Livingstone, I. (1987) Grain-size variation on a 'complex' linear dune in the Namib Desert, in L.E. Frostick and I. Reid (eds.), *Desert Sediments: Ancient and Modern*, Geological Society Special Publication No. 35, Blackwell, Oxford, pp. 281-292.

Ludwig, J.A. and Tongway, D.J. (1995) Spatial organisation of landscapes and its function in semi-arid woodlands, Australia, *Landscape Ecology* **10**, 51-63.

Milton, S.J. and Dean, W.R.J. (1996) *Karoo Veld: Ecology and Management*, ARC - Range and Forage Institute, Pretoria, South Africa.

Milton, S.J., Dean, W.R.J. and Ellis, R.P. (1998) Rangeland health assessment: a practical guide for ranchers in arid Karoo shrublands, *Journal of Arid Environments* **39**, 253-265.

Molope, M.B. (1987) Common Soils in the Molopo District, *Tikilogo* **1**, 65-76.

Nickling, W.G. and Wolfe, A.S. (1994) The morphology and origin of nabkhas, region of Mopti, Mali, West Africa, *Journal of Arid Environments* **28**, 13-30.

Oldeman, L.R., Hakkeling, R.T.A. and Sombroek, W.G. (1990) *World Map of the Status of Human-Induced Soil Degradation*, An explanatory note, ISRIC/UNEP, Wageningen, Netherlands.

Parsons, A.J., Abrahams, A.D. and Simanton, J.R. (1992) Microtopography and soil surface materials on semi-arid piedmont hillslopes, southern Arizona, *Journal of Arid Environments* **22**, 107-115.

Pye, K. and Tsoar, H. (1990) *Aeolian sand and sand dunes*, Unwin Hyman, London.

Rowell, D.L. (1994) *Soil Science: Methods and Applications*, Longman, London.

Scholes, R.J. (1990) The influence of soil fertility on the ecology of southern African dry savannas, *Journal of Biogeography* **17**, 415-419.

Schlesinger, W.H., Reynolds, J.F., Cunningham, G.L. Huenneke, L.F., Jarrell, W.M., Virginia, R.A. and Whitford, W.G. (1990) Biological feedbacks in global desertification, *Science* **247**, 1043-1048.

Skarpe, C. and Bergström, R. (1986) Nutrient content and digestibility of forage plants in relation to plant phenology and rainfall in the Kalahari, Botswana, *Journal of Arid Environments* **11**, 147-164.

Snyman, H.A. (1998) Dynamics and sustainable utilization of rangeland ecosystems in arid and semi-arid climates of southern Africa *Journal of Arid Environments* **39**, 645-666.

Sterk, G., Herrmann, L. and Bationo, A. (1996) Wind-blown nutrient transport and soil productivity changes in southwest Niger, *Land Degradation and Development* **7**, 325-335.

Tengberg, A. (1994) Nebkhas - their spatial distribution, morphometry, composition and age - in the Sidi Bouzid area, central Tunisia, *Zeitschrift fur Geomorphologie* **38**, 311-325.

Tengberg, A. (1995) Nebkha dunes as indicators of wind erosion and land degradation in the Sahel zone of Burkina Faso, *Journal of Arid Environments* **30**, 265-282.

Tengberg, A. and Chen, D. (1998) A comparitive analysis of nebkhas in central Tunisia and northern Burkina

Faso, *Geomorphology* **22**, 181-192.

Thomas, D.S.G. (1997) Sand seas and aeolian bedforms, in D.S.G. Thomas (ed.), *Arid Zone Geomorphology: Process, Form and Change in Drylands*, John Wiley, London, pp. 373-412.

Thomas, D.S.G. and Tsoar, H. (1990) The geomorphological role of vegetation in desert dune systems, in J.B. Thornes (ed.), *Vegetation and Erosion*, John Wiley, London, pp. 471-489.

Thomas, D.S.G. and Shaw, P.A. (1991) *The Kalahari Environment*, Cambridge University Press, Cambridge.

Thomas, D.S.G. and Sporton, D. (1997) Understanding the dynamics of social and environmental variability. The impacts of structural land use change on the environment and peoples of the Kalahari, Botswana, *Applied Geography* **17**, 11-27.

Tsoar, H. (1978) *The dynamics of longitudinal dunes*, Final Technical Report, European Research Office, United States Army, London, DA-ERO 76-G-072.

Tsoar, H. and Møller, J.T. (1986) The role of vegetation in the formation of linear dunes, in W.G. Nickling (ed.), *Aeolian Geomorphology*, Allen and Unwin, London, pp. 75-95.

United Nations Environment Programme (UNEP) (1997) *World Atlas of Desertification*, Arnold, London.

van Rooyen, A.F. (1998) Combating desertification in the southern Kalahari: connecting science and community action in South Africa, *Journal of Arid Environments* **39**, 285-297.

Vogel, C.H. and Drummond, J.H. (1993) Dimensions of drought: South African case studies, *GeoJournal* **30**, 93-98.

Weltzin, J.F. and Coughenour, M.B. (1990) Savanna tree influence on understory vegetation and soil nutrients in northwestern Kenya, *Journal of Vegetation Science* **1**, 325-334.

AUTHORS

A J DOUGILL[1] AND A D THOMAS[2]

1 - School of the Environment, University of Leeds, Leeds, LS2 9JT, UK; adougill@env.leeds.ac.uk (corresponding author)
2 - Division of Geography, Telford Institute of Environmental Systems, University of Salford, Manchester, M5 4WT, UK; a.d.thomas@salford.ac.uk

CHAPTER 4

DESERTIFICATION IN AN ARID SHRUBLAND IN THE SOUTHWESTERN UNITED STATES:

Process Modelling and Validation

GREGORY S. OKIN, BRUCE MURRAY and WILLIAM H. SCHLESINGER

1. Abstract

In the Mojave Desert of the southwestern U.S. human destruction of soil crusts and removal of vegetation have led to progressive, expanding degradation of adjacent arid shrublands. Aeolian mobilisation of dust, sand and litter triggered by anthropogenic disturbance contributes to the destruction of islands of fertility in adjacent areas by killing shrubs through burial and abrasion. This interrupts nutrient-accumulation processes and allows the loss of soil resources by abiotic transport. Thus the processes of degradation spread across the landscape driven largely by abiotic processes.

Soil chemical analyses and remote sensing observations presented here are designed to test a model hypothesis of degradation of arid shrublands. Nutrient and non-nutrient chemical species in the soil act as tracers of material transport and provide clues as to the nature of progressive anthropogenic degradation in arid shrublands. Remote sensing yields information about short- and long-term effects on the landscape as well as important constraints on the magnitude of degradation. Field, chemical and remote sensing observations argue for an extension of recent definitions and models of desertification to include the loss of islands of fertility in established shrublands. This extended model places arid shrublands in a continuum of physical and ecological processes and ecosystems that links semiarid grasslands with Sahara-like hyperarid barren lands.

2. Introduction

The Earth's expanding population is pressing into previously sparsely-inhabited regions. As a result the world's drylands, comprising nearly a third of the total land surface, are coming under increasing land use pressures. Despite their low vegetative productivity, these lands are not barren. Rather, they consist of fragile ecosystems vulnerable to anthropogenic disturbance in which degradation is often irremediable on human timescales. Any successful management plan or attempt at remediation must be firmly rooted in an understanding of natural arid zone processes and the way in which they are perturbed by human activities.

Observations in the deserts of the southwestern U.S. and elsewhere indicate that human

53

A.J. Conacher (ed.), Land Degradation, 53–70.
© 2001 *Kluwer Academic Publishers. Printed in the Netherlands.*

destruction of soil crusts and removal of vegetation lead to a progressive, expanding degradation (Campbell 1972; Bowden *et al.* 1974; Wilshire 1980; Fryrear 1981; Hyers and Marcus 1981; Khalaf and Al-Ajmi 1993; Spitzer 1993; Ray 1995; Bach 1998). Indirect disturbance of arid lands adjacent to areas of direct disturbance can extend far beyond those initially disturbed. Severe financial and societal consequences can result including property damage, increased health and safety hazards, and decreased agricultural productivity (Clements *et al.* 1963; Bowden *et al.* 1974; Fryrear 1981; Hyers and Marcus 1981; Leathers 1981; Bach 1998).

The purpose of this study is to present soil chemical analyses and remote sensing observations which are designed to test the model hypothesis presented here. Chemical species are used to trace material transport from two severely disturbed sites in the American Southwest, and to probe the loss of N and P through dust emission. Remote sensing data provide temporal and spatial information about the extent of indirect land degradation. An integrative landscape process model of arid shrubland degradation is developed in which aeolian transport of material is the primary mechanism of degradation of arid shrublands.

3. Model Hypothesis and Approach

The model hypothesis to be tested explains the progressive devegetation of areas adjacent to sites of direct disturbance and destruction of islands of fertility in these areas. The inferred sequence can be visualised as:

(1)   mobilisation by wind of dust and plant litter, depleting the soils of nutrients in areas of direct disturbance;

(2)   mechanical damage to and burial of plants by saltating sand in adjacent downwind areas;

(3)   reduction of vegetation cover in adjacent areas, leading to an expanding area in which wind removes dust and litter material, depleting the soils of nutrients;

(4)   dune formation in adjacent areas, which decreases near-surface water availability for young plants, increases temperature and albedo, and buries ecologically important cryptobiotic crust and other bacterial communities, and

(5)   reduction of effective soil moisture and depletion of soil nutrients in areas of direct and indirect disturbance.

According to the model hypothesis, aeolian mobilisation of dust, sand and litter triggered by anthropogenic disturbance contributes to the destruction of islands of fertility by killing shrubs through burial and abrasion. Sand blown from areas of direct disturbance may cover areas downwind several times the size of the initial disturbance. This interrupts nutrient accumulation processes and allows the loss of soil resources by abiotic transport processes. The resulting reduction of vegetation cover, in turn, increases runoff and wind transport, reduces latent heat flux through evapotranspiration, and results in increased surface temperatures. These feedbacks can result in continuing reduction of vegetation cover and may contribute to regional climate change. The degradation process places arid shrublands in a landscape and process continuum between semiarid grasslands and Sahara-like hyper-arid barren lands.

In this study, soil samples and remote sensing data at two sites in the southwestern United States were collected in order to test the depletion of soil nutrients by wind transport. At each site samples were collected from areas of severe direct disturbance and from adjacent areas both downwind and upwind. Sample populations from disturbed and undisturbed areas at each site were compared using t-tests to determine whether depletion had occurred. Once significant depletion of soil nutrients was identified, depletion factors were calculated by assuming that undisturbed upwind areas are representative of the pre-disturbance state for all populations. Soil surveys (Bulloch and Neher 1977; Tugel and Woodruff 1978; Meek 1990) in the field areas indicate no difference in soil classification or surface geological history over the area sampled.

4. Methods

4.1. STUDY SITE DESCRIPTIONS AND PRELIMINARY FIELD OBSERVATIONS

4.1.1. Manix Basin, Southeastern California, USA
Soil samples for laboratory analysis were collected in the Manix Basin in the Mojave Desert, about 40 km ENE of Barstow in southeastern California (centred around 34°56.5" N 116°41.5" W at an elevation of about 540 m above sea level) (Fig. 1). The basin was the site of ancient Lake Manix which existed during the peak pluvial episode of the last glaciation (14 000-18 000yr. B.P.) and drained through Afton Canyon to the east; several smaller short-lived lakes followed this episode (Meek 1989; Dohrenwend et al. 1991; Morrison 1991; Meek 1997). Much of the basin is filled with fine- to medium-grained lacustrine, fluvial and deltaic sediments capped by weak deflationary armouring consisting

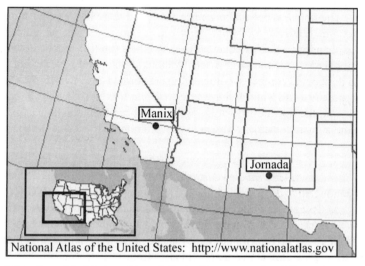

Figure 1. Locations of the Manix and Jornada Basins (from www.nationalatlas.gov)

of gravel-sized particles concentrated at the soil surface (Meek 1990; Dohrenwend *et al.* 1991). There is clear evidence of pre-modern aeolian sand mobilisation (Sharp 1966; Meek 1990; Dohrenwend *et al.* 1991; Evans 1962), indicating that wind erosion has a long history as a geological process in the area.

In the modern climate of the Manix Basin precipitation falls mostly in the winter, although there can be significant summer precipitation in some years (National Climate Data Center 1997). The average annual temperature is 19.6˚C, the average winter temperature is 9.1˚C, and the average summer temperature is 31.4˚C (Meek 1990). Average wind speed at the airport in Daggett is greater than 5 m s^{-1} at a height of 6.1 m (National Climate Data Center 1993). A total of 82.7% of the winds with velocities greater than 5.6 m s^{-1} at Daggett blow toward ENE (23.4%), E (30.2%) or ESE (29.1%).

The vegetation in undisturbed areas of the basin is dominated by an association of *Larrea tridentata* (Creosote bush) and *Ambrosia dumosa* (White Bursage or Burro Bush), with minor occurrences of *Atriplex polycarpa* (Desert Saltbush), *Atriplex hymenelytra* (Desert Holly), *Atriplex canescens* (Four-Winged Saltbush), *Ephedra* (Mormon Tea) and *Opuntia*. *A. polycarpa* can be locally abundant in disturbed areas. *Schismus* grass can be abundant in the basin in both disturbed and undisturbed areas in wet years.

There has been extensive human activity in the Manix Basin, with several phases of agricultural activity utilising groundwater recharged from the Mojave River which carries runoff from the San Bernardino Mountains to the south-southwest (Tugel and Woodruff 1978). After the mid-1970s, central-pivot irrigation agriculture became the dominant form of land use in the area, with fertiliser used to enhance the soil's productivity. Many fields have now been abandoned throughout the northern part of the basin due to increasing costs of pumping groundwater to the surface for agriculture (Ray 1995). Many abandoned fields display sand plumes downwind, where material from the field has blown on to adjacent, undisturbed desert as shown in Plate 1 (see also Ray, 1995).

This study concentrates on an abandoned central-pivot agricultural field in the Manix Basin. From a series of Landsat MSS images, we have concluded that this field underwent two phases of cultivation: from 1979 to 1981 and from 1987 to 1990, or seven years in total (Ray 1995). The field has been fallow since 1990, and there has been no significant shrub regrowth. The *Larrea tridentata-Ambrosia dumosa* shrubland immediately downwind of the abandoned field is characterised by dunes (up to 1 m tall) in the lee of shrubs, sometimes coalescing into sand sheets. Although total plant cover is not visibly lower in the area downwind compared with nearby undisturbed desert, *Ambrosia dumosa* individuals are typically dead, whilst living *Larrea tridentata* exhibit dead lower branches and abrasion scars on the upwind side of living limbs. These features (dunes, plant and limb mortality, and abrasion scars) are absent in the community upwind of the abandoned field.

4.1.2. Jornada Basin, south-central New Mexico, USA
This study was conducted at the Jornada Long Term Ecological Research (LTER) site 37 km northeast of Las Cruces, New Mexico, in the Chihuahuan Desert ecosystem (Fig. 1). It is located on the Jornada del Muerto plain, which is bounded by the San Andres Mountains on the east and the Rio Grande Valley and the Fra Cristobal-Caballo Mountain complex on the west. Elevation above sea level varies from 1180 to 1360 m. The Jornada Plain consists of unconsolidated Pleistocene detritus. This alluvial fill from the nearby

mountains is 100 m thick in places and the aggradation process is still active. Coarser materials are found near foothills along the eastern part of the study area. The topography of the study area consists of gently rolling to nearly level uplands, interspersed with swales and old lake beds (Buffington and Herbel 1965).

The climate of the area is characterised by cold winters and hot summers and displays a bimodal precipitation distribution. Winter precipitation usually occurs as low-intensity rains or occasionally as snow and contributes to the greening of shrub species in the basin in the early spring. Summer monsoonal precipitation, usually in the form of patchy, but intense, afternoon thunderstorms, is responsible for the late-summer greening of grasses. The average annual precipitation between 1915 and 1962 in the basin was 230 mm, with 52% falling between July 1 and September 30 (Paulsen and Ares 1962). Maximum temperature is highest in June, when it averages 36°C, and lowest in January, when it averages 13°C (Buffington and Herbel 1965).

The principal grass species in the study area are burrograss (*Scleropogon brevifolius*), several species of *Aristida*, and tobosa grass (*Hilaria mutica*), while major shrubs are creosote (*Larrea tridentata*), mesquite (*Prosopis glandulosa*) and tarbush (*Florensia cernua*). Soils in the basin are quite complex but generally range from clay loams to loamy fine sands, with some areas being sandy or gravelly (Soil Conservation Service 1980).

In the spring of 1991, a semi-circular plot of land was cleared and maintained clear of vegetation (Havstad, pers. comm. 1999). This 'scraped site' is situated in patchy *Scleropogon brevifolius-Hilaria mutica* grassland with scattered *Prosopis glandulosa* (mesquite) and *Yucca* spp. The soil is loamy fine sand and sandy loam of aeolian origin derived from Rio Grande floodplain sediments (Bulloch and Neher 1977). The historical wind direction in the basin is to the northeast, as indicated by the alignment nearby Holocene aeolian sand ridges.

Observations made in the summer of 1998 indicate that in the seven years since the site's establishment, a large area immediately downwind of the barren area has been adversely affected by sand blown from the scraped site. The area affected by blown sand extends at least 200 m downwind of the initial disturbance (Plate 1). Burial and abrasion of the shrubs as well as the grasses in this area have led to a significant decrease in shrub cover and the formation of dunes in the lee of remaining mesquite. Grasses, once plentiful in the adjacent downwind area, are completely absent. This pattern is strikingly similar to that seen in the Manix Basin and argues that anthropogenic disturbance and wind-erodable soil are the major contributors to the degradation of arid shrublands and the destruction of islands of fertility, even in semiarid regions of the Chihuahuan Desert.

4.2. IMAGE ACQUISITION

Data from the Jet Propulsion Laboratory's Airborne Visible Infrared Imaging Spectrometer (AVIRIS) were acquired on August 29, 1998 over the Manix Basin and on May 27, 1997 over the Jornada Basin. AVIRIS measures the total upwelling spectral radiance in the solar-reflected spectrum from 400 to 2500 nm at 10-nm intervals. These spectra are acquired at 20 m spatial resolution across a 10 km wide swath from a NASA ER-2 aircraft flying at an altitude of 20 km. AVIRIS measurements are spectrally, radiometrically and spatially calibrated and converted to units of radiance (μW/cm2/nm/sr). The imaged vegetation and soils exhibit small but characteristic molecular absorptions and scattering characteristics which AVIRIS is sensitive enough to detect.

4.3. SAMPLE COLLECTION

Soil samples were collected in order to test the depletion of soil nutrients by wind transport from areas of direct human disturbance and from adjacent, downwind areas relative to undisturbed upwind areas. Samples were collected by inserting a 5 cm diameter steel pipe into the ground to a depth of 5 cm and placing the contents of the pipe in a sealed plastic bag. Samples were collected along three 200 m transects at each site; upwind of the disturbances, on the disturbances themselves, and downwind of the disturbances. Sample locations along the transects were determined by a random number table. Transects were orientated perpendicular to the apparent average wind direction as indicated from remote sensing images of sand blow-outs (Plate 1) or, in the case of the Jornada LTER, historical wind directions as evidenced by sand streaks and ridges from remote sensing images. Locations of samples from beneath shrubs or from the bare soil between shrubs were noted.

4.3.1. Manix
Soil samples were collected on February 17, 1999 from an abandoned central-pivot agricultural field that was last cultivated in 1989 (Ray 1995). The three transects which were sampled were located 100 m upwind of the edge of the field, in the centre of the field, and 100 m downwind of the edge of the field. Thirty samples were collected along each transect and two extra samples were collected in the upwind transect. Samples in the downwind transect were obtained from locations where soil has blown off the field and accumulated in dunes and sand sheets downwind of shrubs. Locations of samples from the fine-grained dunes or the coarser interdune areas were also noted.

4.3.2. Jornada
Soil samples were collected on June 20-21, 1999 from the scraped site located at 32°34'14"N 106°45'30"W. The three transects were located 50 m upwind of the edge of the scraped site, in the centre of the site, and 100 m downwind of the edge of the site. Twenty samples were collected from the upwind and scraped site transects, and thirty samples were collected from the downwind transect. All samples from downwind of the scraped site are from dunes. Here, sand removed from the scraped site blankets the entire downwind area and has in some places mounded in mesquite coppice dunes.

4.4. LABORATORY ANALYSIS

In the laboratory, all samples were oven-dried overnight, sieved to less than 2 mm and analysed for NO_2^-, NO_3^-, NH_4^+, water-extractable PO_4^{-3}, bicarbonate-extractable PO_4^{-3}, Cl$^-$, SO_4^{-2}, K$^+$ and Li$^+$. The sum of NH_4^+-N, NO_2^--N, and NO_3^--N is considered to be an index of total available N (N_{Avail}). Bicarbonate-extractable PO_4^{-3} is considered to be an index of plant-available P. Samples from the Manix Basin were also analysed for Na$^+$, while samples from the Jornada Basin were analysed for Mg^{+2} and Ca^{+2}. Methods used in this study are described in Schlesinger et al. (1996). Concentrations of all species were converted to, and are reported in, μg (species)/g soil.

4.5. STATISTICAL ANALYSIS

Samples were divided into populations and sub-populations: upwind of the field or scraped site (sub-populations: undershrub and intershrub), on the field or scraped site, and downwind of the field or scraped site (sub-populations: undershrub and intershrub; dune and interdune). For each population and sub-population, the mean (in μg (species)/g soil), standard deviation, and coefficient of variation (C.V. = standard deviation / mean) was calculated. t-tests (α=0.01), calculated assuming that population variance is not equal, were used to deduce significant differences between means, and to determine their relative magnitude. The undershrub sub-populations were too small to be included in t-test comparisons, and for the Jornada samples intershrub population t-test results are identical to results from the whole transects and therefore are not reported or discussed. All Jornada samples in the downwind transect were from dune sands blown off the scraped site.

Pearson product-moment correlation coefficients were calculated for all species for all samples as a whole and for each transect. Correlations were tested for significance at α=0.01 against the null hypothesis that ρ=0 according to the method described by Sokal and Rohlf (1981).

5. Results

Remote sensing images from AVIRIS for both sites are shown in Plates 1A and 1B. Mean species concentrations and t-test results for all populations and sub-populations are given in Tables 1a and 1b. Significant Pearson product-moment correlation coefficients are given in Tables 2a and 2b.

5.1. SPATIAL INFORMATION: REMOTE SENSING

As shown in Plate 1, areas of progressive indirect disturbance downwind of areas of direct disturbance are visible in remote sensing images. They appear in both sites as bright plumes issuing from the downwind edge of direct disturbances which are also apparent in the images.

Since its establishment in 1979, the sand blow-out associated with the abandoned agricultural field at Manix has expanded to several times the size of the field itself. In Jornada, after only seven years since the initial disturbance at the scraped site, the sand blow-out associated with this feature is at least as large as the scraped site.

5.2. AVAILABLE NITROGEN (N_{AVAIL})

5.2.1. Manix
Samples from the abandoned field show the highest concentrations of N_{Avail}. Mean N_{Avail} concentration on the downwind transect is higher than the mean concentration on the upwind transect with 95% confidence. In the downwind transect, samples from dunes have higher mean N_{Avail} concentrations than samples from the interdune areas with 95% confidence. Samples from intershrub areas in both the upwind and downwind transects have lower mean N_{Avail} concentrations than samples from the abandoned field. The elevated

1.0 km

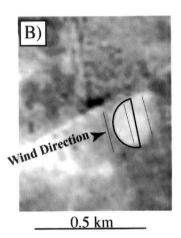

0.5 km

Plate 1. AVIRIS Images of Study Sites.

A) AVIRIS image over set of central-pivot fields in the Manix Basin, southeastern California, acquired August 28, 1997. The abandoned central-pivot agricultural field studied here is enclosed by the dark black line and the approximate locations of the sample transects are denoted by lighter lines. The redeposition of material eroded from the fields can be seen as brighter plumes downwind of the abandoned fields.

B) AVIRIS image over the scraped site, Jornada Basin, south-central New Mexico, acquired May 27, 1997. The semi-circular scraped site is enclosed by the dark black line and the approximate locations of the sample transects are denoted by lighter lines. The area downwind of the scraped site appears brighter than the surrounding desert due to sand blown off the scraped site on to otherwise undisturbed desert.

Table 1a: Means and t-test results for all populations and sub-populations for Manix samples
All numbers reported as μg/g soil. For t-test Results: **U**=Upwind, **UIs**=Upwind-intershrub, **F**=Field, **D**=Downwind, **DIs**=Downwind-intershrub, **DId**=Downwind-interdune, **Dd**=Downwind-dune.

Species	Population	Sub-population	Mean	C.V.	Species	Population	Sub-population	Mean	C.V.
N_{Avail}	Upwind		5.29	0.49	Cl	Upwind		3.92	0.38
		Intershrub	4.68	0.30			Intershrub	4.02	0.39
		Undershrub	7.63	0.60			Undershrub	3.53	0.33
	Field		14.1	0.57		Field		6.64	0.62
	Downwind		7.73	0.87		Downwind		4.41	0.54
		Intershrub	7.82	0.90			Intershrub	4.38	0.56
		Undershrub	6.71	0.24			Undershrub	4.74	0.41
		Dune	10.8	0.79			Dune	4.25	0.29
		Interdune	5.07	0.61			Interdune	4.54	0.68
t-test results	U<F>D, UIs<F>DIs, F>DId				*t-test results*	U<F>D, UIs<F>DIs, F>Dd			
Bicarb P	Upwind		3.35	0.23	SO_4	Upwind		6.27	0.37
		Intershrub	3.20	0.22			Intershrub	5.90	0.36
		Undershrub	3.99	0.19			Undershrub	7.76	0.34
	Field		4.29	0.34		Field		19.5	0.79
	Downwind		5.57	0.57		Downwind		7.73	0.43
		Intershrub	5.49	0.60			Intershrub	7.70	0.45
		Undershrub	6.59	0.32			Undershrub	8.13	0.29
		Dune	7.32	0.54			Dune	8.84	0.38
		Interdune	4.08	0.25			Interdune	6.78	0.46
t-test results	D>F>U, DIs>UIs<F, Dd>DId				*t-test results*	U<F>D, UIs<F>DIs, DId<F>Dd			
PO_4	Upwind		1.80	1.05	Na	Upwind		9.37	0.47
		Intershrub	1.47	0.90			Intershrub	9.60	0.48
		Undershrub	3.10	1.03			Undershrub	8.46	0.45
	Field		4.07	0.74		Field		28.3	1.29
	Downwind		5.12	0.73		Downwind		33.8	1.87
		Intershrub	5.09	0.76			Intershrub	36.3	1.79
		Undershrub	5.44	0.22			Undershrub	3.30	0.12
		Dune	7.52	0.54			Dune	28.1	2.70
		Interdune	3.06	0.52			Interdune	38.7	1.35
t-test results	D>U<F, DIs>UIs<F, F<Dd>DId				*t-test results*	U<F, UIs<F, F=Dd			
Mg	Upwind		n/a	n/a	K	Upwind		193	0.23
		Intershrub	n/a	n/a			Intershrub	193	0.23
		Undershrub	n/a	n/a			Undershrub	191	0.25
	Field		n/a	n/a		Field		258	0.40
	Downwind		n/a	n/a		Downwind		170	0.35
		Intershrub	n/a	n/a			Intershrub	167	0.35
		Undershrub	n/a	n/a			Undershrub	200	0.37
		Dune	n/a	n/a			Dune	155	0.48
		Interdune	n/a	n/a			Interdune	183	0.22
t-test results	n/a				*t-test results*	U<F>D, UIs<F>DIs, DId<F>Dd			
Ca	Upwind		n/a	n/a	Li	Upwind		0.40	0.30
		Intershrub	n/a	n/a			Intershrub	0.42	0.29
		Undershrub	n/a	n/a			Undershrub	0.32	0.19
	Field		n/a	n/a		Field		0.42	0.29
	Downwind		n/a	n/a		Downwind		0.38	0.44
		Intershrub	n/a	n/a			Intershrub	0.39	0.44
		Undershrub	n/a	n/a			Undershrub	0.28	0.27
		Dune	n/a	n/a			Dune	0.33	0.41
		Interdune	n/a	n/a			Interdune	0.43	0.43
t-test results	n/a				*t-test results*	none			

Table 1b: Means and t-test results for all populations and sub-populations for Jornada samples
All numbers reported as µg/g soil. For t-test Results: **U**=Upwind, **UIs**=Upwind-intershrub, **S**=Scraped Site, **D**=Downwind, **DIs**=Downwind-intershrub

Species	Population	Sub-population	Mean	C.V.	Species	Population	Sub-population	Mean	C.V.
N_{Avail}	Upwind		35.7	0.94	Cl	Upwind		2.72	0.90
		Intershrub	29.9	1.34			Intershrub	2.44	0.94
		Undershrub	68.7	0.77			Undershrub	4.27	0.74
	Scraped Site		4.07	0.41		Scraped Site		3.80	1.34
	Downwind		6.59	0.98		Downwind		2.34	0.61
		Intershrub	5.83	0.97			Intershrub	2.37	0.62
		Undershrub	11.5	1.12			Undershrub	2.09	0.66
t-test results	D<U>S				*t-test results*	none			
Bicarb P	Upwind		11.2	0.63	SO_4	Upwind		9.05	1.84
		Intershrub	10.6	0.66			Intershrub	8.61	2.06
		Undershrub	14.8	0.50			Undershrub	11.5	0.92
	Scraped Site		2.36	0.66		Scraped Site		2.51	0.45
	Downwind		4.29	0.57		Downwind		1.67	0.69
		Intershrub	3.98	0.43			Intershrub	1.54	0.61
		Undershrub	6.32	0.83			Undershrub	2.57	0.81
t-test results	U>D>S				*t-test results*	S>D			
PO_4	Upwind		9.12	0.95	Na	Upwind		n/a	n/a
		Intershrub	7.94	0.89			Intershrub	n/a	n/a
		Undershrub	15.8	0.97			Undershrub	n/a	n/a
	Scraped Site		0.70	2.71		Scraped Site		n/a	n/a
	Downwind		3.05	0.46		Downwind		n/a	n/a
		Intershrub	2.85	0.35			Intershrub	n/a	n/a
		Undershrub	4.38	0.65			Undershrub	n/a	n/a
t-test results	U>D>S				*t-test results*	n/a			
Mg	Upwind		120	0.35	K	Upwind		337	0.22
		Intershrub	119	0.37			Intershrub	327	0.20
		Undershrub	122	0.20			Undershrub	390	0.30
	Scraped Site		98.7	0.34		Scraped Site		235	0.25
	Downwind		76.8	0.47		Downwind		215	0.28
		Intershrub	77.5	0.49			Intershrub	209	0.25
		Undershrub	72.6	0.40			Undershrub	251	0.40
t-test results	U>D				*t-test results*	D<U>S			
Ca	Upwind		1214.7	0.32	Li	Upwind		0.41	0.27
		Intershrub	1232	0.31			Intershrub	0.42	0.27
		Undershrub	1117	0.43			Undershrub	0.38	0.34
	Scraped Site		1490.4	0.34		Scraped Site		0.47	0.23
	Downwind		768	0.30		Downwind		0.33	0.27
		Intershrub	780	0.30			Intershrub	0.34	0.25
		Undershrub	694	0.25			Undershrub	0.31	0.43
t-test results	S>D<U				*t-test results*	S>D<U			

nitrogen content on the abandoned field indicates the application and persistence of chemical fertiliser to the site.

Plant available P and N_{Avail} are significantly correlated in both the upwind and downwind transects. They are not correlated in the field transect.

5.2.2. Jornada

Results from *t*-test comparisons of means indicate that N_{Avail} is lowest on the scraped site, intermediate for the downwind transect, and highest on the upwind transect. Mean N_{Avail}

Table 2a: Significant (α=0.01) Pearson Product-Moment Correlation Coefficients for all species, by transect, for Manix samples.

		N_{Aval}	Bicarb P	PO_4	Mg^{2+}	Ca^{2+}	Cl^-	SO_4^{2-}	Na^+	K^+	Li^+
All Samples	N_{Aval}	1.00									
	Bicarb P	0.36	1.00								
	PO_4	0.35	0.68	1.00							
	Mg^{2+}	n/a	n/a	n/a	n/a						
	Ca^{2+}	n/a	n/a	n/a	n/a	n/a					
	Cl^-	0.65	--	--	n/a	n/a	1.00				
	SO_4^{2-}	0.73	--	--	n/a	n/a	0.80	1.00			
	Na^+	--	--	--	n/a	n/a	0.39	0.36	1.00		
	K^+	--	--	--	n/a	n/a	--	0.35	--	1.00	
	Li^+	--	--	-0.37	n/a	n/a	--	--	0.47	--	1.00
Upwind Transect	N_{Aval}	1.00									
	Bicarb P	--	1.00								
	PO_4	0.63	--	1.00							
	Mg^{2+}	n/a	n/a	n/a	n/a						
	Ca^{2+}	n/a	n/a	n/a	n/a	n/a					
	Cl^-	--	--	--	n/a	n/a	1.00				
	SO_4^{2-}	0.48	--	--	n/a	n/a	--	1.00			
	Na^+	--	--	--	n/a	n/a	--	--	1.00		
	K^+	0.41	--	--	n/a	n/a	--	0.62	--	1.00	
	Li^+	--	--	--	n/a	n/a	--	--	--	0.65	1.00
Field Transect	N_{Aval}	1.00									
	Bicarb P	--	1.00								
	PO_4	--	--	1.00							
	Mg^{2+}	n/a	n/a	n/a	n/a						
	Ca^{2+}	n/a	n/a	n/a	n/a	n/a					
	Cl^-	0.77	--	--	n/a	n/a	1.00				
	SO_4^{2-}	0.74	--	--	n/a	n/a	0.82	1.00			
	Na^+	0.65	--	--	n/a	n/a	0.63	0.69	1.00		
	K^+	--	--	0.69	n/a	n/a	--	--	--	1.00	
	Li^+	--	--	-0.53	n/a	n/a	--	--	0.64	--	1.00
Downwind Transect	N_{Aval}	1.00									
	Bicarb P	0.73	1.00								
	PO_4	0.85	0.79	1.00							
	Mg^{2+}	n/a	n/a	n/a	n/a						
	Ca^{2+}	n/a	n/a	n/a	n/a	n/a					
	Cl^-	--	--	--	n/a	n/a	1.00				
	SO_4^{2-}	0.56	0.61	0.58	n/a	n/a	--	1.00			
	Na^+	--	--	--	n/a	n/a	--	--	1.00		
	K^+	-0.44	--	--	n/a	n/a	--	--	0.50	1.00	
	Li^+	-0.39	--	--	n/a	n/a	--	--	0.53	0.58	1.00

concentrations from Jornada samples are much higher than those from the Manix Basin despite the fact that these soils have never been fertilised.

Plant available P and N_{Avail} are significantly correlated in both the upwind and downwind transects. They are not correlated in the scraped site transect.

5.3. PHOSPHOROUS

5.3.1. Manix

Concentrations of bicarbonate-extractable PO_4 and water-extractable PO_4 are significantly correlated (ρ=0.68, significant at α=0.01) for all samples. The regression of water-extractable PO_4 against bicarbonate-extractable PO_4 for all samples yields:

$$PO_4 = -0.8 + 1.0 * (\text{bicarbonate-extractable } PO_4)$$

with r^2= 0.47 and the slope of the regression line significant at α=0.01. The near-zero intercept and slope of unity for this regression suggest that water extraction of PO_4 provides a reliable measure of plant-available P in these soils.

Results from t-test comparisons of means indicate that plant-available P is more concentrated downwind, especially in dunes, than on the field itself, and that it is least concentrated in the upwind site. That plant-available P is more concentrated in the intershrub spaces in the downwind transect relative to those in the upwind transect reflects the fact that relatively small shrub canopies can give rise to relatively large phosphorous-enriched dunes. The elevated phosphorous content on the abandoned field indicates the application and persistence of P-containing chemical fertiliser to the site.

Table 2b: Significant (α=0.01) Pearson Product-Moment Correlation Coefficients for all species, by transect, for Jornada samples.

		N_{Avail}	Bicarb P	PO_4	Mg^{2+}	Ca^{2+}	Cl	SO_4	Na^+	K^+	Li^+
All Samples	N_{Avail}	1.00									
	Bicarb P	0.72	1.00								
	PO_4	0.84	0.83	1.00							
	Mg^{2+}	0.35	0.47	--	1.00						
	Ca^{2+}	--	--	--	0.66	1.00					
	Cl	--	--	--	--	--	1.00				
	SO_4	0.56	0.75	0.76	--	--	--	1.00			
	Na^+	n/a	n/a	n/a	n/a	n/a	n/a	n/a	n/a		
	K^+	--	0.46	--	0.75	0.65	--	--	n/a	1.00	
	Li^+	--	--	--	0.64	0.73	--	--	n/a	0.42	1.00
Upwind Transect	N_{Avail}	1.00									
	Bicarb P	0.67	1.00								
	PO_4	0.82	0.87	1.00							
	Mg^{2+}	--	--	--	1.00						
	Ca^{2+}	--	--	--	--	1.00					
	Cl	--	--	--	--	--	1.00				
	SO_4	--	0.87	0.82	--	--	--	1.00			
	Na^+	n/a	n/a	n/a	n/a	n/a	n/a	n/a	n/a		
	K^+	--	0.70	0.57	--	--	--	--	n/a	1.00	
	Li^+	--	--	--	0.68	0.66	--	--	n/a	--	1.00
Scraped Site Transect	N_{Avail}	1.00									
	Bicarb P	--	1.00								
	PO_4	--	0.98	1.00							
	Mg^{2+}	0.73	--	--	1.00						
	Ca^{2+}	--	--	--	--	1.00					
	Cl	--	--	--	--	--	1.00				
	SO_4	--	--	--	--	--	--	1.00			
	Na^+	n/a	n/a	n/a	n/a	n/a	n/a	n/a	n/a		
	K^+	0.61	--	--	--	0.60	--	--	n/a	1.00	
	Li^+	--	--	--	--	0.84	--	--	n/a	--	1.00
Downwind Transect	N_{Avail}	1.00									
	Bicarb P	0.86	1.00								
	PO_4	0.57	0.57	1.00							
	Mg^{2+}	0.68	0.74	--	1.00						
	Ca^{2+}	0.63	0.59	--	0.83	1.00					
	Cl	--	--	--	--	--	1.00				
	SO_4	0.92	0.85	0.68	0.70	0.67	--	1.00			
	Na^+	n/a	n/a	n/a	n/a	n/a	n/a	n/a	n/a		
	K^+	0.85	0.89	0.59	0.78	0.73	--	0.87	n/a	1.00	
	Li^+	0.48	0.53	--	0.55	0.60	--	0.53	n/a	0.55	1.00

5.3.2. Jornada

Concentrations of bicarbonate-extractable PO_4 and water-extractable PO_4 are significantly correlated (ρ=0.83, significant at α=0.01) for all samples. The regression of water-extractable PO_4 against bicarbonate-extractable PO_4 for all samples yields:

$$\text{water-extractable } PO_4 = -1.25 + 0.86^* \text{ (bicarbonate-extractable } PO_4)$$

with r^2= 0.69 and the slope of the regression line significant at α=0.01. Since the slope of this regression line is significantly below 1.0, water extraction of PO_4 in these soils does not provide a reliable means of measuring plant-available P.

Results from t-test comparisons of means indicate that plant-available P is lowest on the scraped site, intermediate for the downwind transect, and highest on the upwind transect. Mean plant-available P concentrations from Jornada samples are much higher than those from the Manix Basin.

5.4. OTHER SPECIES: Cl, SO_4, Mg, Ca, K, Na, Li

5.4.1. Manix

Mean Cl, SO_4, Na, and K concentrations were all significantly higher in the field transect than in the other two transects. The differences of the means from samples taken in the field and those from the upwind transect were 2.72, 13.2, 18.9 and 65 µg/g for Cl, SO_4, Na and K, respectively. Mean concentrations of these ions were also significantly higher in samples taken from the field transect than those taken from the dunes in the downwind transect or from the intershrub areas in the upwind transect. Mean Na and SO_4 concentrations are higher in the downwind transect compared with the upwind transect, with 95% confidence.

Chloride and SO_4 concentrations are correlated with 99% confidence in samples from the field transect and not correlated at this confidence level for samples from the other two transects. Chloride and SO_4 concentrations are also correlated with 95% confidence in samples from the downwind transect ($\rho=0.45$). Sodium concentrations are significantly correlated with Cl and SO_4 in samples from the field, but not significantly correlated in the other transects. For all three ions, concentrations in samples from intershrub areas in either the upwind or downwind transects (or both) are lower than those from the abandoned field. Chloride, SO_4 and Na concentrations are also significantly correlated with N_{Avail} in samples from the abandoned field.

Potassium concentrations are higher in the abandoned field than in the other transects. Concentrations from the abandoned field are also higher than in intershrub areas of the upwind and downwind transects. Potassium concentrations are significantly correlated with Na concentrations in the downwind transect.

Lithium concentrations were very low in all samples, and t tests for lithium did not yield any significant relationships at the 99% confidence level. Significant correlations of Li with Na and K were obtained in samples from the downwind transect.

5.4.2. Jornada

There is no obvious pattern of enrichment of Cl, SO_4, Mg, Ca, K and Li in samples from any one transect. Mean Ca, SO_4 and Li concentrations are significantly higher in the scraped site relative to the downwind transect, possibly indicating that deflation of the scraped site has excavated $CaCO_3$ or $CaSO_4 \cdot 2H_2O$ illuvial horizons in these soils. The mean concentration of SO_4 in the upwind transect is higher than in the scraped site and downwind with 95% confidence. Mean concentrations of Mg, Ca and K are lower in the downwind transect than the upwind transect.

Chloride and SO_4 concentrations are not significantly correlated with any other species in the scraped site, and Cl concentrations are not significantly correlated with any other species in any other transect. This result is expected since the scraped site has never been irrigated.

In samples from the downwind transect, every measured species except PO_4 and Cl is significantly correlated with every other species.

6. Discussion

6.1. SALINISATION

The elevated concentrations of Cl, SO_4, Na and K in the abandoned field in the Manix Basin compared with the upwind transect strongly suggest that salt has accumulated in the upper 5 cm of soil due to evaporation of irrigation water from the field. That these ions are also significantly correlated with NO_3 and N_{Avail} in the field suggests that these chemical perturbations are related, namely through cultivation.

Assuming that the estimate from a series of Landsat MSS images of the basin at the time that this field was cultivated (seven years total) is correct, and that concentrations of these species in the upwind transect represent concentrations in the area which was to become the field, Cl, SO_4, Na and K accumulated at average rates of 0.4, 1.9, 2.7 and 9.3 μg/g per year, or approximately 9.9, 30, 29 and 4.9 % per year, respectively. This is a dramatic addition of ions to the soil, which may limit the usefulness of these areas for extended agriculture.

6.2. FERTILISATION

The elevated concentration of available nitrogen in samples taken from the abandoned agricultural field at Manix is a clear indication that fertiliser was applied to this field while it was in active use. Plant-available P concentrations are also significantly higher in the field than in the upwind samples, indicating the addition of P-containing fertiliser. Concentrations of these nutrients are elevated after approximately 10 years of disuse, indicating the persistence of plant-available nutrients in this ecosystem.

Despite elevated N and P in the abandoned agricultural field at Manix, the absence of shrubs indicates that recolonisation of fields by native shrubs after their abandonment is not simply related to nutrient content of the soils. There are, indeed, other abandoned fields in the Manix Basin which were also presumably fertilised and which have total plant covers (*A. polycarpa*) higher than in undisturbed desert.

In an experiment aimed at restoring Mojave Desert farmland by seeding native plants in order to reduce dust emissions, Grantz *et al.* (1998) found that by furrowing across the wind and direct seeding, they could establish *Atriplex canescens* in areas without deep sand. However,

> this revegetation was achieved in an anomalous year with above average and late rainfall that eliminated early competition from annual species and later fostered abundant shrub growth. This success was not reproducible in more normal years, when minimal disturbance protocols such as broadcasting of seed on the untilled soil surface were as effective and less costly (Grantz *et al.* 1998: 1209).

Thus, natural germination and regrowth of native perennial vegetation on abandoned fields may be rare, explaining the lack of cover on the field studied here. Bowers (1987) has suggested that allogenic factors affecting germination are also responsible for compositional dynamics of winter annuals in the Mojave Desert. The importance of germination conditions, and in particular the timing and duration of precipitation, highlight the dramatic role of inter-annual variability and long-term regional climate on the response of this ecosystem to human disturbance.

6.3. MATERIAL TRANSPORT

In the samples from the Jornada Basin, the downwind soils were all derived from the scraped site as is evident from the AVIRIS image in Plate 1B. In addition, the old pre-disturbance surface is apparent by digging through the mantle of sand derived from the scraped site. In contrast to Manix, where dunes tend to be discreet objects in the lee of shrubs separated by islands of the original surface, at Jornada, material blown from the scraped site on to the adjacent downwind area has completely blanketed the area. Results indicate that surface soils upwind are richer in N_{Avail} and plant-available P than those from downwind, indicating approximately a five-fold net loss of N_{Avail} and a three-fold net loss of plant-available P from the soils blown off the scraped site. In addition, the scraped site itself lost nearly 94% of its N_{Avail} and nearly 79% of its plant-available P if the soils from the upwind transect are considered to be representative of the original condition throughout the study site. The loss of N and P from the surface soils dramatically affects germination conditions for seeds in this area, and combined with other local environmental conditions - principally sand blasting from the scraped site - will reduce the establishment of new shrubs or grasses in these areas.

On the abandoned field in the Manix Basin, the addition of fertiliser for agriculture as well as other ions by irrigation provides an opportunity to probe the fate of material removed from the field by wind. Field relations and remote sensing images (Plate 1) suggest that sand has blown off the agricultural field on to adjacent, downwind areas.

Significantly elevated concentrations of N_{Avail} (α=0.05) and plant-available P (α=0.01) in the field and the downwind transect relative to the upwind transect, support this claim. Fertiliser, spread originally on the field for agriculture, has been broadcast across the landscape as the surface soil from the field has been eroded and transported by wind. Significantly elevated concentrations of Na and SO_4 (α=0.05) in the downwind transect at Manix relative to the upwind transect, indicate likewise that salt species, added inadvertently during irrigation, have been spread with the moving sands.

Anthropogenic additions of soil nutrients and salt species to the abandoned field at Manix provide us with the ability to examine the fate of soil removed from the field in terms of its subsequent nutrient content. Inferences may also be made about the texture of the mobilised soil. At the scraped site in the Jornada Basin, there have been no significant additions of chemical tracers (*i.e.* fertiliser or salts) to the soil. Thus, the issues of whether soil blown off the scraped site is depleted of nutrients by winnowing, and whether there is preferential removal of plant nutrients on dust particles available for long-range transport by wind suspension, may be addressed. These signals are not apparent in the N_{Avail} and plant-available P data from Manix, possibly because the natural nitrogen and phosphorous concentrations were overwhelmed by fertiliser additions.

Leys and McTainsh (1994) and Larney *et al.* (1998) have suggested that soil nutrients and other chemical species are concentrated on small dust-sized particles in soils, and that wind erosion preferentially removes these particles, thereby depleting the soil of nutrients. The removal of material from the field to downwind areas therefore may alter not only the soil chemistry, but the soil texture as well. In surface soils with a significant contribution of material eroded from a nearby disturbance, reduced clay content is thus expected due to this process of winnowing. At Jornada, mean Mg, Ca, K and Li concentrations upwind of the scraped site are higher than downwind concentrations. Samples from the abandoned agricultural field at Manix also exhibit patterns suggestive of winnowing: Cl, SO_4 and K display increased concentrations in the field relative to the material in the dunes. As long as all the material in the dunes originated from the field, the observed differences in concentrations indicate a net loss of these species, and the particles which carry them, by emission processes.

7. Implications for Biogeochemical Desertification in Arid Shrublands

A decade ago, Schlesinger *et al.* (1990) suggested a model for the degradation of semiarid grasslands. In their view, long-term grazing of grasslands leads to spatial and temporal heterogeneity of soil resources which promotes the invasion of shrubs. The presence of shrubs, in turn, enforces the localisation of soil resources in islands of fertility by concentrating organic material below shrub canopies, while intershrub spaces lose fertility through erosion and gaseous emissions.

The model hypothesis presented here indicates that in arid shrublands, direct anthropogenic disturbance which results in destruction of soil crusts and vegetation cover, can lead to indirect disturbance of adjacent areas by initiating the disintegration of islands of fertility. The model hypothesis of degradation in arid shrublands states that aeolian mobilisation of dust, sand and litter triggered by anthropogenic disturbance contributes to the destruction of islands of fertility by killing shrubs through burial and abrasion. This interrupts nutrient-accumulation processes and allows the loss of soil resources by abiotic transport processes. The resulting reduction of vegetation cover, in turn, increases runoff and wind transport, reduces latent heat flux through evapotranspiration, and results in increased surface temperatures. These feedbacks can result in continuing reduction of vegetation cover and may contribute to regional climate change. The degradation process places arid shrublands in a landscape and process continuum between semiarid grasslands and Sahara-like hyper-arid barren lands.

The results presented here strongly support the model hypothesis of the progressive degradation of arid shrublands associated with direct disturbance of the stable surface. In particular, they provide direct evidence that soil nutrients are depleted by wind erosion and redeposition. Removing the vegetation from the scraped site at Jornada has initiated a dramatic loss of nutrients from this ecosystem.

In addition, the scraped site provides an excellent example of the physical effects of degradation as ongoing saltation and aerosol fluxes remain higher here than anywhere else in the basin (Gillette, pers. comm. 1999): mesquite bushes growing near the scraped site exhibit severe branch abrasion and leaf stripping by wind, and grasses, once abundant, are now entirely absent. Regrowth of vegetation in the scraped site itself as well as downwind is therefore limited by both the relatively low nutrient concentrations and the physical effects of blowing sand.

The results of this study also highlight the roles and interactions of climate and physical processes in initiating and propagating desertification in areas adjacent to direct disturbances. Elevated concentrations of available N and P on and adjacent to the abandoned fields in the Manix Basin have not contributed to vegetation regrowth. Indeed, sand blowing from the abandoned field studied here has severely damaged the native vegetation and killed many individual plants. Human additions of N and P to this environment may not become significant until a period of increased available moisture. In the meantime, sand blowing off the field will continue to denude vegetation downwind.

The removal of soil nutrients by wind from disturbed areas in deserts has important implications for human activities in arid regions. Areas adjacent to severe anthropogenic disturbances which remove vegetation cover, as well as the direct disturbances themselves, therefore, can be expected to become less fertile with time, leading to progressive, expanding degradation and decreases in vegetation cover. Desertification of this sort can have serious environmental, agricultural, industrial, health and safety consequences in arid lands as blowing sand destroys machinery and crops, covers roads, contributes to dust storms, and liberates airborne allergens and even pathogens. When informed by process models, remote monitoring tools may be used to identify areas at risk of runaway degradation before large areas are adversely affected.

Landscape processes necessarily exist in the context of regional climate and can either be supported or hindered by climatic conditions and changes, a fact which makes ultimate control of the extent of degradation climate-related. However, cover reduction can cause regional warming while aerosol production due to dust emission may contribute to global cooling. As a result, anthropogenic landscape change can contribute to regional climate changes, and human activities in deserts are simultaneously subject to regional climate conditions and impacts upon them.

The vulnerability of arid lands to degradation argues for development of linked degradation process models and monitoring strategies in order to minimise environmental damage and to promote sustainable management of human activities in arid lands. This is particularly true in the face of global and regional socioeconomic factors which may compromise remediation strategies or promote inexpensive, but damaging, land-use practices.

8. Acknowledgments

We would like to thank Dale A. Gillette of the United Stated Environmental Protection Agency in Raleigh, North Carolina for his many insights on the role and magnitude of wind and sediment transport in the Jornada Basin. Our discussions have been most helpful to the completion of this study.

9. References

Bach, A. J. (1998) Assessing conditions leading to severe wind erosion in the Antelope Valley, California, 1990-1991, *Professional Geographer* **50**, 87-97.

Bowden, L. W., Huning, L. R., Hutchinson, C. F., and Johnson, C. W. (1974) Satellite photograph presents first comprehensive view of local wind: the Santa Ana, *Science* **184**, 1077-1078.

Bowers, M. A. (1987) Precipitation and the relative abundances of desert winter annuals: a 6-year study in the northern Mohave Desert, *Journal of Arid Environments* **12**, 141-149.

Bulloch, H. E., Jr. and Neher, R. E. (1977) *Soil Survey of Dona Ana County Area, New Mexico*, Soil Conservation Service, United States Department of Agriculture, Washington, D. C.

Campbell, C. E. (1972) Some environmental effects of rural subdividing in an arid area: a case study in Arizona, *The Journal of Geography* **71**, 147-154.

Clements, T., Mann, J. F., Stone, R. O. and Eymann, J. L. (1963) *A study of windborne sand and dust in desert areas*, United States Army Natick Laboratories, Earth Sciences Division, Natick, Massachusetts.

Dohrenwend, J. C., Bull, W. B., McFadden, L. D., Smith, G. I., Smithe, R. S. U. and Wells, S. G. (1991) Quaternary geology of the Basin and Range province in California, in R.B. Morrison (ed.), *Quaternary Nonglacial Geology: Coterminous U.S.*, Vol. K-2, Geological Society of America, Boulder, Colorado, pp. 321-352.

Evans, J. R. (1962) Falling and climbing sand dunes in the Cronese ('Cat') Mountain Area, San Bernardino County, California, *Journal of Geology* **70**, 107-113.

Fryrear, D. W. (1981) Long-term effect of erosion and cropping on soil productivity, in T.L. Péwé (ed.), *Desert Dust: Origin, Characteristics, and Effect on Man*, Vol. Special Paper 186, Geological Society of America, Boulder, Colorado, pp. 253-259.

Grantz, D. A., Vaughn, D. L., Farber, R., Kim, B., Zeldin, M., Van Curen, T. and Campbell, R. (1998) Seeding native plants to restore desert farmland and mitigate fugitive dust and PM10, *Journal of Environmental*

Quality **27**, 1209-1218.

Hyers, A. D. and Marcus, M. G. (1981) Land use and desert dust hazards in central Arizona, in T.L. Péwé (ed.), *Desert Dust: Origin, Characteristics, and Effect on Man*, Vol. Special Paper 186, Geological Society of America, Boulder, Colorado.

Khalaf, F. I. and Al-Ajmi, D. (1993) Aeolian processes and sand encroachment problems in Kuwait, *Geomorphology* **9**, 111-134.

Larney, F., Bullock, M., Janzen, H., Ellert, B. and Olson, E. (1998) Wind erosion effects on nutrient redistribution and soil productivity, *Journal of Soil and Water Conservation* **53**, 133-140.

Leathers, C. R. (1981) Plant components of desert dust in Arizona and their significance for man, in T.L. Péwé (ed.), *Desert Dust: Origin, Characteristics, and Effect on Man*, Vol. Special Paper 186, Geological Society of America, Boulder, Colorado, pp. 191-206.

Leys, J. and McTainsh, G. (1994) Soil loss and nutrient decline by wind erosion - cause for concern, *Australian Journal of Soil and Water Conservation* **7**, 30-35.

Meek, N. (1989) Geomorphic and hydrologic implications of the rapid incision of Afton Canyon, Mojave Desert, California, *Geology* **17**, 7-10.

Meek, N. (1990) Late Quaternary Geochronology and Geomorphology of the Manix Basin, San Bernardino County, California, Ph.D., University of California at Los Angeles.

Meek, N. (1997) Paleoclimatic Implications of the Mojave River System, *The Association of American Geographers, 93rd Annual Meeting*, Fort Worth, Texas, p. 175.

Morrison, R. B. (1991) Quaternary geology of the southern Basin and Range Province, in R.B. Morrison (ed.), *Quaternary Nonglacial Geology: Coterminous U.S.*, Vol. K-2, Geological Society of America, Boulder, Colorado, pp. 283-320.

National Climate Data Center (1993) Solar and Meteorological Surface Observation Network: 1961-1990, Volume III: Western United States, [CD-ROM], update: September 1993.

National Climate Data Center (1997) U.S. Precipitation by State, California, [HTML Document], http://www.ncdc.noaa.dov/ol/climate/online/coop-precip.html, update: January 1, 1999.

Ray, T. W. (1995) Remote Monitoring of Land Degradation in Arid/Semiarid Regions, Ph.D., California Institute of Technology.

Schlesinger, W. H., Raikes, J. A., Hartley, A. E. and Cross, A. F. (1996) On the spatial pattern of soil nutrients in desert ecosystems, *Ecology* **77**, 364-374.

Schlesinger, W. H., Reynolds, J. F., Cunningham, G. L., Huenneke, L. F., Jarrell, W. M., Virginia, R. A. and Whitford, W. G. (1990) Biological feedbacks in global desertification, *Science* **247**, 1043-1048.

Sharp, R. P. (1966) Kelso Dunes, Mojave Desert, California, *Geological Society of America Bulletin* **77**, 1045-1074.

Sokal, R. R. and Rohlf, F. J. (1981) *Biometry: Principles and Practice of Statistics in Biological Research,* W.H. Freeman and Company, New York.

Spitzer, H. A. (1993) Antelope Valley emergency soil erosion control, *Land and Water*, March 1993, 20-24.

Tugel, A. J. and Woodruff, G. A. (1978) *Soil Survey of San Bernardino County, California: Mojave River Area*, Soil Conservation Service, United States Department of Agriculture, Washington, D. C.

Wilshire, H. G. (1980) Human causes of accelerated wind erosion in California's deserts, in D.R. Coates and J.D. Vitek (eds.), *Thresholds in Geomorphology*, George, Allen and Unwin, London, pp. 415-433.

AUTHORS

GREGORY S. OKIN[1], BRUCE MURRAY[1] and WILLIAM H. SCHLESINGER[2]

Corresponding author Greg Okin. Phone: (626) 395-6465. Fax: (626) 585-1917.

[1] Division of Geological and Planetary Sciences 100-23, California Institute of Technology, Pasadena, CA, 91125. okin@gps.caltech.edu, bcm@caltech.edu.

[2] Department of Botany, Duke University, Durham, NC, 27708-340. schlesin@duke.edu

CHAPTER 5

THE POTENTIAL FOR DEGRADATION OF LANDSCAPE FUNCTION AND CULTURAL VALUES FOLLOWING THE EXTINCTION OF *MITIKA* (*BETTONGIA LESUEUR*) IN CENTRAL AUSTRALIA

J.C. NOBLE, J. GILLEN, G. JACOBSON, W.A. LOW, C. MILLER
and the *MUṮITJULU* COMMUNITY

1. Abstract

This paper canvasses the implications of declining biodiversity in the context of landscape degradation in *Uluṟu-Kata Tjuṯa* National Park, Northern Territory, following the regional extinction of *mitika*. Field studies undertaken to gain more detailed insights into the landscape ecology of these mesomarsupials are described. These studies were based on the hypothesis that the original distribution, approximate densities and habitat preferences of *mitika* could be delineated by mapping relict warrens, most of which had subsequently been usurped by rabbits. Data collected included locations, dimensions, entrances (active and passive), habitat geology, and local vegetation. Landscape function analyses demonstrated that relict *mitika* warrens continue to play a significant role as major obstruction elements capable of trapping and retaining rainfall and soil nutrients. The results illustrate how these high fertility patches contribute significantly to mesoscale landscape heterogeneity and primary productivity. Traditional owners are also concerned about degradation of cultural values in these landscapes following the demise of *mitika*. Local knowledge of these animals, still extant in *Uluṟu-Kata Tjuṯa* up until the early 1950s, was therefore recorded during discussions in the field with senior *Aṉangu*. The possibility of eventually re-introducing this culturally significant species back into appropriate landscapes in *Uluṟu-Kata Tjuṯa* is briefly discussed.

"Is an exhausted and threatened land a symbol of what we have become? The issue is, to my mind, a religious one. To learn humility."

Tony Kelly (1990)

2. Introduction

Although classed as one of the 'megadiverse' countries, largely because of its diverse marsupial fauna, many of Australia's endemic species are now totally extinct following European settlement. The World Conservation Monitoring Centre claimed in 1992 that this loss represented 30% of global mammalian extinctions. Much of this attrition occurred within a specific group, the medium-sized marsupials ranging in size from 1.5 to 8 kg (Noble 1996), generally corresponding to the 'critical weight range' (CWR) of 35 g to 5.5

71

A.J. Conacher (ed.), Land Degradation, 71–89.

kg for vulnerable species as defined by Burbidge and McKenzie (1989). Some are now totally extinct, some are extinct on the mainland, while others are restricted to small, isolated populations having disappeared from much of their original distribution (Morton and Baynes 1985; Low 1986; Burbidge and McKenzie 1989; Johnson *et al.* 1989; Baynes and Baird 1992).

One of these mesomarsupials was the *mitika*, a *Pitjantjatjara* term used by related language groups (Hobson 1990) when referring to the burrowing bettong (*Bettongia lesueur*) throughout the central desert regions of Australia (Plate 1). The generic title *Bettongia* was originally coined by the British taxonomist J.L. Gray in 1837 from the Western Australian Aboriginal word *bettong* or small wallaby (Strahan 1981). The specific epithet was proposed by Quoy and Gaimard (1824), both medical officers aboard the corvettes *Uranie* and *Physicienne* during the 1817-1820 expedition commanded by Louis de Freycinet (Strahan 1981), to commemorate the artist and naturalist Charles-Alexandre Lesueur (1778-1846). Lesueur collected widely during the voyages of the *Géographe* and *Naturaliste* (1800-1804) under the command of Nicholas-Thomas Baudin (1754-1803). His detailed and delicate paintings and sketches have provided an invaluable artistic and scientific legacy with more than 5000 documents comprising the Lesueur Collection now held in Fort de Tourneville above Le Havre (Hunt 1999).

Plate 1. *Mitika* or burrowing bettong (*Bettongia lesueur*). (Photo: G. Robertson).

Commonly referred to by early European settlers as rat-kangaroos, most mesomarsupials were particularly vulnerable to changes in habitat imposed during early pastoral settlement (Noble 1993,1997). Prior to European settlement, *mitika* had one of the widest mainland distributions for any native mammal (Finlayson 1958, Burbidge 1988). Ultimately, regional extinction followed as habitat quality deteriorated through prolonged overgrazing by introduced herbivores such as domestic livestock and the European rabbit (*Oryctolagus cuniculus*). Increased predation by the fox (*Vulpes vulpes*) (Short and Smith 1994), together with bounty hunting and poisoned grain laid out for rabbit control, accelerated the ultimate demise of these 'native vermin' (Noble 1993, Short 1998, Baker and Noble 1999). Eventually the only surviving *mitika* populations were, until recently, located on three offshore islands in the Indian Ocean (Short and Turner 1993).

The comprehensive definition of desertification developed at the 1977 United Nations Conference on Desertification (UNCOD) recognised, in part, that any diminution of biological diversity had the potential to contribute towards landscape degradation (Mainguet 1994). The relationship between biodiversity and sustainable land use is now receiving increasing attention since Australia became a signatory in 1993 to the 1992 International Convention on Biological Diversity. Despite the controversy and misunderstanding often associated with community perceptions of biodiversity, Walker and Nix (1993) claim there are essentially three different kinds of values, or reasons, justifying its conservation. The first relates to direct economic values including products based on food and fibre; tourism; and genes for genetic engineering. The second group involves 'ecosystem services' whereby biological diversity plays a fundamental role in regulating basic ecosystem functions such as nutrient cycling and water redistribution. Finally, the third group incorporates most of the aesthetic factors characterising 'quality of life' values – analogous to such cultural values as fine art and music but of particular importance to indigenous communities. The extinction of *mitika* throughout most of its former range, for example, has also created a deep sense of loss amongst traditional landowners.

It is the second group, though, which is currently attracting the attention of many ecologists, including rangeland ecologists, as the fundamental roles of biologically-mediated processes in the functioning of semi-arid savanna ecosystems become more widely recognised. Repercussions arising from the reduction in biological diversity following the extinction of many mesomarsupials are still to be adequately evaluated, particularly in terms of any diminution of landscape function and resultant loss of soil fertility in these arid ecosystems. Although Kelly (1990) used the term humility in a religious context in the quotation at the beginning of this paper, its etymology has strong ecological undertones based on the Latin *humus* = soil.

Clearly, much of the mesoscale heterogeneity (Holling 1992) found in functional landscapes throughout arid Australia is based on surface-soil biopedturbations (*sensu* Whitford and Kay 1999) derived from past fossorial, i.e. burrowing, activity of soil-resident fauna. Studies of the size, distribution and abundance of *mitika* warrens in various landscapes of western New South Wales led to preliminary estimates of potential population sizes, as well as formulation of specific geo-ecological hypotheses (Noble 1996). It was postulated that as the abundance of larger fossorial species such as *mitika* declined, fundamental landscape processes such as nutrient cycling, dispersal of seeds and spores of mycorrhizal fungi, and regulation of shrub populations, may have become progressively impaired as their abundance declined (Noble 1993, 1996).

This paper reviews what is currently known about the ecology of *mitika* in semi-arid and arid ecosystems of Australia, particularly in the context of potential degradation of landscape function (Ludwig *et al.* 1997) following their regional extinction. The results of field studies into the landscape ecology and habitat geology of *mitika* undertaken during 1995 and 1996 at *Uluṟu-Kata Tjuṯa* National Park (Fig. 1), *c*. 450 km southwest of Alice Springs, Northern Territory, are presented. In particular, the paper examines the hypothesis that relict warrens continue to influence mesoscale landscape heterogeneity and patch dynamics in key *mitika* habitats. The possibility of reintroducing this culturally significant species back into selected landscapes is also briefly discussed.

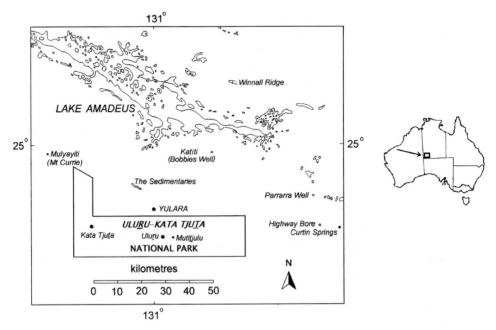

Figure 1. Locations of *Uluṟu-Kata Tjuṯa* National Park and *Katiti* (Bobbies Well) south of Lake Amadeus.

3. The Habitat Geology and Landscape Ecology of *Mitika*

Mitika construct substantial warrens in diverse habitats. A powerful burrower, it was able to build conspicuous, and long-lived, gallery systems or warrens in rocky terrain or 'hard red' landscapes on the mainland (Noble 1999), these sites being normally inhospitable to smaller fossorial animals. Because of their structural integrity, these warrens provided valuable refugia during protracted droughts (Morton 1990) even though the restriction of *mitika* to these sites probably exposed them to increased risk of predation by the dingo (*Canis familiaris dingo*). Relict *mitika* warrens on the mainland are frequently characterised by a horseshoe-shaped perimeter mound up to 30 m in diameter surrounding a calcrete 'lens' around 30 cm thick, partially exposed in the central depression. In large warren systems, several entrances may commonly be found wherever the calcrete was sufficiently soft to enable excavation and deposition of secondary calcrete mounds overlying the original perimeter mound (Noble 1993).

These 'lenses' of pedogenic calcrete frequently occur in mid-catenary positions (Plate 2a) overlying extensive fluvial/sheetflow calcrete (English 1998) and owe their origin to wetting and drying cycles over geological time. On a regional scale, *Uluṟu-Kata Tjuṯa* National Park is underlain by sedimentary rocks of the Amadeus Basin. Prominent formations include the Mount Currie Conglomerate, which constitutes both the *Uluṟu* and *Kata Tjuṯa* monoliths (Wells *et al.* 1970). Soil carbonate is dissolved during wet periods by groundwater shed from monolith catchments. As it moves slowly through the soil towards adjacent drainage lines, evaporation eventually becomes dominant and the carbonate is precipitated within the soil matrix (Plate 2b). The terms vadose and phreatic are commonly used in the calcrete and karst literature (e.g. Jacobson and Arakel 1987), with vadose seepage referring to rainwater or soil water percolating downwards through fissures and pores. As the soil profile becomes progressively drier, the carbonate is eventually precipitated out of the soil solution to form vadose calcrete. Phreatic calcrete, however, is formed in the phreatic zone below the groundwater table where all substrate cavities remain permanently saturated.

In sandplain landscapes, *mitika* warrens are more subdued due to partial burial by more mobile substrates. Nevertheless, vestiges of the original subcircular perimeter mound constructed by *mitika* can often still be recognised in these sites, even where warrens have been extensively re-worked by rabbits. Rabbit entrances generally follow the outline of the original perimeter mound where digging has been facilitated by softer substrates. The presence of pre-fabricated habitation provided by the original *mitika* populations, particularly in otherwise unattractive landscapes dominated by shallow soils and rocky substrates, proved to be a major legacy for rabbits invading arid and semi-arid ecosystems last century. Not only did *mitika* warrens facilitate rapid invasion, they also provided stable refuges enabling rabbits to survive extended droughts by recycling dung and concentrating their urine (Myers and Parker 1965).

A successful rabbit control program based on a combined strategy of ripping (Plate 3a and b) followed by fumigation of re-activated warrens, has been underway for the past 10 years in *Uluṟu-Kata Tjuṯa* National Park (Low *et al.* 1999a). Maps prepared from 1:25 000 air photographs and differential GPS locations (Low *et al.* 1999b) illustrate distinct patterning of warrens relative to local relief (Fig. 2a & b), with highest warren densities generally located midslope along calcareous interfluves or *puḻi*, a *Pitjantjatjara* term for fans and alluvium (Baker and Muṯitjulu Community 1992). The frequent presence of pedogenic calcrete, either within these warrens or in close proximity, suggested that the majority of warrens had originally been constructed by *mitika*.

4. Field Studies of *Mitika* Sites

4.1 *ULUṞU-KATA TJUṮA*

An arid landscape functioning efficiently has the capacity to capture, and retain, soil-water following episodic rainfall events (Ludwig *et al.* 1997). Rainfall redistribution is fundamentally important in transporting organic matter, mineral soil and constituent nutrients to fertile patches or 'islands' where much of the runoff and subsequent herbage production is concentrated (Noble *et al.* 1998). Dysfunctional landscapes become 'leaky' once there has been a reduction in the density of obstruction elements such as perennial

(a)

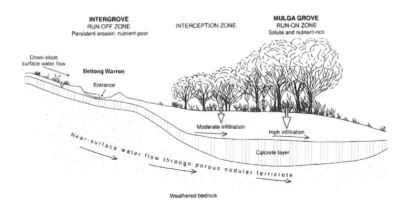

(b)

Plate 2. (a) Topographic section of a typical banded mulga (*Acacia aneura*) community growing on a massive, Quaternary red earth (Qr) catena. Warrens constructed by *mitika* are generally found in the mid-slope regions of soil catenae (adapted from Tongway and Ludwig 1997 and English 1998). (b) Close-up view of an entrance beneath the calcrete lens of a relict *mitika* warren examined near *Katiti* (Bobbies Well) south of Lake Amadeus. (Photo: J. Noble).

(a)

(b)

Plate 3. (a) Aerial view of a ripped warren on *Uluṟu-Kata Tjuṯa* National Park clearly showing the semi-circular outline on the left-hand side of the original *mitika* perimeter mound. The ripped area is 30 m in diameter. (Photo: W. Low). (b) A large *mitika* warren and central calcrete lens prior to ripping. (Photo: C. Smith).

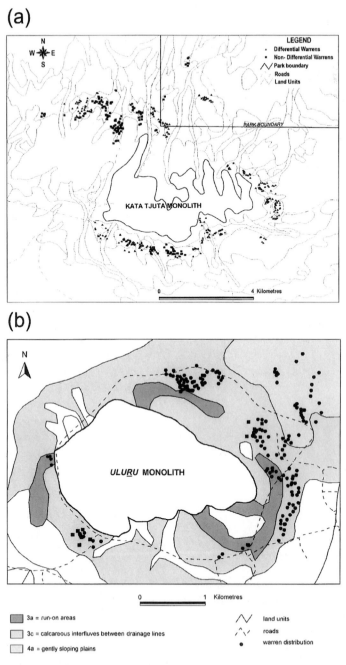

Figure 2. (a) Map of *Uluru-Kata Tjuta* National Park illustrating the concentrations of rabbit/*mitika* warrens in the calcareous land units and pediments surrounding both monoliths in relation to major land units as defined by Allan (1984) (after Low *et al*. 1999b). (b) Map showing the disposition of warrens around *Uluru* in relation to major land units (after Low *et al*. 1999b).

grass tussocks following excessive grazing pressures (Tongway and Ludwig 1997). It is further postulated that the degree of dysfunction has been exacerbated by the demise of 'ecosystem engineers' such as *mitika* whose warrens constituted substantial obstruction elements. In order to test this hypothesis, landscape function analysis (LFA) (Tongway 1991) was undertaken at replicate *mitika* sites within *Uluru* and *Kata Tjuta* National Park (Fig. 1). After recording their positions using a global positioning system (GPS), relevant features of their habitat geology and associated vegetation (dominant ground and overstorey species) were noted. Topographic profiles were constructed from data collected using a dumpy level (± 1 cm). Duplicate transects (140-220 m in length), one running downslope through the middle of a selected warren, the other a control transect about 20 m away and parallel with the warren transect, were surveyed at each site. The position and size of individual obstruction elements, i.e. long-lived features which obstruct or divert water flow and/or collect/filter out material from runoff (Tongway and Hindley 1995), were recorded along each transect. In addition, details of the nature, size and distribution of fetch elements, where fetch is the distance between successive obstruction features (Tongway and Hindley 1995), were also noted. Finally, an index of surface roughness (m) was calculated for slopes above and below individual *mitika* warrens as well as for adjacent control (non-warren) transects using a boundary analysis technique (Ludwig and Cornelius 1987; Ludwig and Tongway 1995).

Warren profiles indicated that the *mitika* warrens constituted major obstruction elements capable of trapping significant quantities of overland flow together with any transported organic matter and surface soil. The presence of a relict mitika warren resulted in a marked increase in surface roughness (Fig. 3a). Topographic profiles also indicated that warrens were generally located where there was a distinct break in slope. Adjacent control profiles showed a similar, though less pronounced, change in slope at the same relative position (Fig. 3b), suggesting a change in the underlying geology possibly related to the deposition of pedogenic calcrete.

The results obtained from a landscape function analysis undertaken at one *mitika* site near *Kata Tjuta* indicate major differences in diversity of obstruction elements above and below the warren. In contrast, there was little difference in the nature of the fetch elements. Such differences are, however, probably site-specific and largely controlled by the steepness of gradient above (Fig. 4a) and below (Fig. 4b) major elements like *mitika* warrens.

Soil samples were also taken at two depths (0-2 and 2-4 cm) at various points along the transect running through the *mitika* warren to observe spatial variations in nutritional status. Subsequent determinations were made of pH (1:5 water) and electrical conductivity (Loveday 1974), organic carbon (Colwell 1969), and available nitrogen (Gianello and Bremner 1986). There were marked contrasts between internal and external *mitika* warren microsites in terms of all soil chemical parameters measured (Fig. 5).

As expected, the surface soil inside the warren was considerably more alkaline due to inversion of the underlying calcrete during warren construction. Similarly, the highest values for electrical conductivity and organic carbon content were those obtained from samples taken at the lowest point inside the warren. Whilst total nitrogen values were also high inside the warren, the maximum concentration was recorded outside the warren in the surface 2 cm immediately upslope of the perimeter mound.

4.2 *Katiti*

Quaternary calcrete of late Pleistocene origin (22 000-75 000 years) (Jacobson *et al.* 1988) outcrops in the Yulara Creek- *Katiti* (Bobbies Well) area northeast of *Uluru* (Fig. 1) where it forms a carapace (up to 12 m thick) between the sand dunes. Because of its groundwater

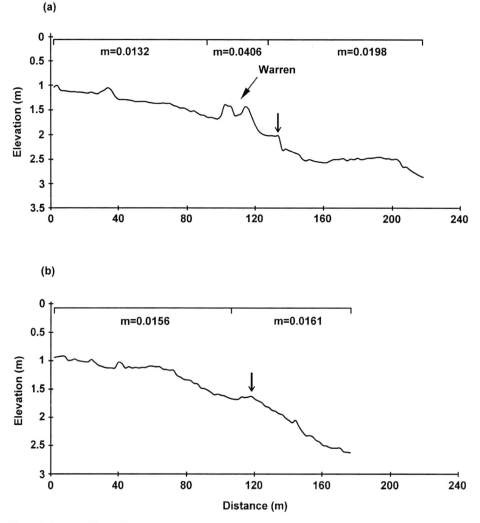

Figure 3. Topographic profiles for (a) the warren and (b) adjacent control transects recorded at a *mitika* site near *Kata Tjuta*. Indices of surface obstruction (m) for different segments of each transect are also shown. The arrows indicate the break in slope possibly associated with pedogenic calcrete deposition.

origin, the surface is mounded due to continual crystallisation at the base. During June 1996, an aerial survey was undertaken by flying along a north-south transect between Lake Amadeus and *Uluru* in proximity to a palaeodrainage system. This transect was chosen

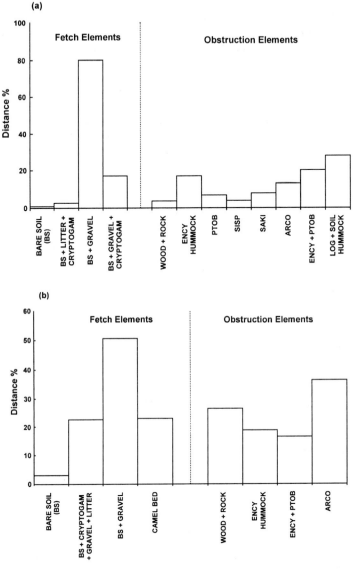

Figure 4. Results of landscape function analysis undertaken at a *mitika* site near *Kata Tjuṯa* (vegetation site no. 8 - UTM 675 242, 7204 930) illustrating changes in diversity of obstruction elements (a) above the warren and (b) below the warren. Key to plant species acronyms along the X-axis: ENCY = *Enneapogon cylindricus*; PTOB = *Ptilotus obovatus*; SISP = *Sida* spp.; and ARCO = *Aristida contorta*.

primarily because it covered several geological habitats along a vadose-phreatic calcrete sequence, thereby enabling validation of the hypothesis that well-preserved *mitika* warrens are closely associated with calcrete-rich landscapes known to exist in the region (Wells *et al.* 1970). This transect also happened to approximate the route followed by the Horn expedition in 1894 during its journey south to inspect *Uluṟu*, when they kept mounting "...

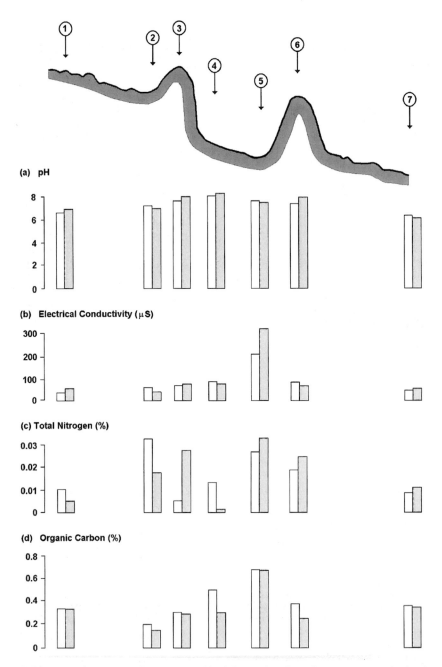

Figure 5. Diagrammatic representation (not to scale) of changes in soil nutrient status at two depths along a transect passing through a *mitika* warren near *Kata Tjuṯa* (UTM 678 113, 7208 226). Key: open columns = 0-2 cm; hatched = 2-4 cm.

(a)

(b)

Plate 4. (a) Aerial and (b) ground views of the *mitika* warren (UTM 723 158, 7230 872) near *Katiti* (Bobbies Well). Warren dimensions were 27 x 30 m. The spinifex ridge to the east of the warren is visible in the top left hand corner of (a). (Photos: J. Noble).

one sandhill after another, all covered with tussocks of Porcupine grass, amongst which the kangaroo-rats, *Bettongia lesueuri*, kept dodging in and out with remarkable agility" (Spencer 1896: 84).

Because this aerial reconnaissance was undertaken by helicopter, it was possible to undertake further landscape analyses by landing close to chosen sites. One of these was a well-formed *mitika* warren situated in a landscape dominated by Quaternary calcrete (evaporational calcrete overlying depositional calcrete) near *Katiti* (Plate 4 a & b). As at *Uluṟu-Kata Tjuṯa*, a topographic profile was constructed for this site. Herbage density and composition were recorded along the same transect using a point-centred quarter technique (Mueller-Dombois and Ellenberg 1974). Herb density was highest in the depression at the base of the dune and immediately adjacent to the outside edge of the perimeter mound (Fig. 6). The evidence suggests that both microsites were acting not only as 'sinks' for water and nutrients but also as 'safe sites' for germination and seedling establishment. The high plant

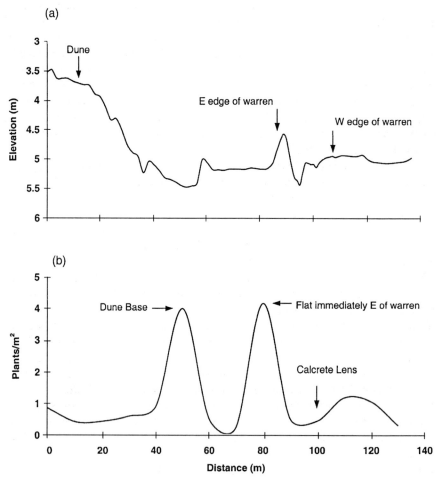

Figure 6. Topographic profile of the *mitika* transect established near *Katiti*. Data on herbage density collected along the same transect illustrate high density zones at the dune base and immediately adjacent to the warren perimeter mound.

density to the east of the warren reflects a peak in soil-water resulting from rainfall redistributed from the spinifex dune, whereas the high density value recorded inside the warren, where the fetch is considerably less, is more a 'water + nutrients peak'.

Mitika warrens were examined at various other sites including the Sedimentaries, northwest of Yulara, and Curtin Springs Station, east of *Uluṟu*. Overall, information detailing position (determined using a global positioning system), class, diameters, total and active rabbit entrances, habitat geology and local vegetation were collected for a total of 38 putative *mitika* sites at *Uluṟu-Kata Tjuṯa* and regional sites with results recorded elsewhere (Noble *et al.* 1996).

5. Discussion

While population estimates of *mitika* must necessarily remain conservative, they can be derived in those communities where the density and size distributions of relict warrens can still be recorded. Even then however, the influence of climatic variability on *mitika* survival is uncertain. Finlayson (1958: 242) found that numbers ".... fluctuate greatly and its occurrence is local and discontinuous and not uniform. Warrens housing a big population during one season may be found quite deserted the next". Although it is extremely difficult to determine at what exact time these warrens were first constructed, it appears unlikely that they were constructed earlier than the late Pleistocene/early Holocene. Initial radiocarbon dating of calcrete samples collected at the nearby Sedimentaries indicates precipitation of calcrete occurred on the pediment over a considerable time period ranging from 6000-12 000 years BP (Jacobson, unpubl. data).

The regional extinction of *mitika* populations in central Australia is now recognised as an important symptom of desertification in these landscapes (Noble *et al.* 1999) because it is apparent that they were involved in various ecosystem processes, especially those relating to patch dynamics. *Mitika* warrens clearly behave as major obstruction elements in the landscape, resulting in the formation of fertile patches immediately upslope of the perimeter mound as well as inside the central depression. Increased fertility in these microsites, in turn, leads to higher densities of herbaceous plants. Landscape function analysis indicated that the density of minor obstruction elements, such as wood residues and grass hummocks, was generally higher above the warren although obstruction density overall was site-specific and strongly related to slope.

Prior to pastoral settlement, light grazing by native herbivores such as large kangaroos (*Macropus* spp.) enabled sufficient fuel to accumulate which, in turn, resulted in more frequent fire. In the event of minimal post-fire rainfall occurring, regenerating shrub coppices were browsed by medium-sized marsupials including the formerly widespread burrowing bettong (*Bettongia lesueur*) and brush-tailed bettong (*B. penicillata*). Even shrub seedlings surviving in unburnt patches were vulnerable to mesomarsupial browsing, thereby promoting the maintenance of complex vegetation mosaics, or multiple stable states, throughout much of the semi-arid woodlands (Noble 1996, 1997; Noble *et al.* 1996). Today, woody plants commonly establish inside central depressions of *mitika* warrens which act as 'safe sites' for germination and seedling establishment.

Some aspects of Aboriginal knowledge in relation to *mitika* ecology have been summarised in an earlier report (Noble *et al.* 1996). Senior *Anangu*, a term used by Western

Desert language speakers when referring to themselves (Baker and Mutitjulu Community 1992), recall *mitika* living as large families within a single warren with individuals known to forage a long distance away from their home warrens, often sheltering underneath spinifex hummocks (Spencer 1896). Senior *Anangu* also recalled *mitika* burying seeds of the hemi-parasitic quandong (*Santalum acuminatum*), as well as utilising seeds and roots of pituri (*Duboisia hopwoodii*). However, the ecological significance of these associations has yet to be determined.

Another, but less obvious, manifestation of landscape degradation following the decline in *mitika* numbers is the resultant reduction in cultural diversity. One of the long-term goals contained in the 1991 Plan of Management for *Uluru-Kata Tjuta* National Park is the desirability of re-establishing extinct fauna of cultural significance. Subsequent discussions have clearly indicated that *Anangu* rate *mitika* as the most important of the *Tjukurpa* species they wish to see restored to their traditional lands. *Tjukurpa* is a *Pitjantjatjara* term generally used to refer to the religious and social Law governing all aspects of Aboriginal life (Baker and Mutitjulu Community 1992).

Appropriate strategies for conserving these rare and endangered populations, based on their translocation back into areas where they were once believed to have been present, are currently being developed elsewhere in Australia (Beckmann 1990; Christensen and Burrows 1994; Nelson *et al.* 1992; Short and Smith 1994; Short *et al.* 1989, 1992). To be successful, however, not only must exotic predators and competitors be controlled effectively, but re-introductions of viable populations (Sinclair *et al.* 1998) must also be made into preferred habitats, where such information is available. Fortunately in many landscapes, *mitika* habitat preferences have been clearly imprinted in the form of relict warrens which have since been invaded by rabbits. Given the importance attached by the *Anangu* to the reintroduction of *Tjukurpa* fauna into *Uluru-Kata Tjuta*, it seems only a matter of time before such activities are given the necessary scientific and funding support. In the interim, it is important that *mitika* warrens which have not been ripped be preserved as future reintroduction foci. Where possible, alternative non-destructive options such as fumigation should be used for future rabbit control.

Although the studies outlined in this paper are by no means definitive, there is now a basic understanding emerging in regard to *mitika* habitat ecology. Such information, when applied with the experience gained from reintroduction projects currently underway (for example, Short and Smith 1994), and the application of appropriate predator-prey models (Sinclair *et al.* 1998), suggest appropriate strategies can now be designed for testing at a realistic, regional, scale (Wilson *et al.* 1992). Subsequent exploitation for either commercial (for example ecotourism) or cultural (such as *Tjukurpa*, traditional food) purposes, is a fundamental issue still to be resolved in dialogue with the *Anangu* to ensure self-sustaining *mitika* populations can be established successfully.

6. Acknowledgments

Norman Tjakaljiri, Billy Wara, Johnny Tjingo and Edith Richards provided invaluable insights into Aboriginal knowledge of *mitika* during field trips that included tape-recorded interviews at key sites. The tapes were later translated by Greg Snowdon (Uncharted Journeys). We wish to also acknowledge the support and encouragement provided by the following people: Bob and Pat Barber, Mt Skinner Station; Dennis Matthews, Finke Gorge National Park; Bob and Marie Purvis, Atartinga Station; and Peter Severin and staff, Curtin Springs Station, for allowing access and assisting in the location of *mitika* sites; Anne Grattidge, Low Ecological Services, for early

GPS and GIS mapping of rabbit warrens; Graham Griffin and Craig James, CSIRO Alice Springs, for helpful discussions and logistical support; and finally Norman Hindley (soil chemical analysis), Gil Pfitzner (preparation of figures), and Liz Poon (photographs), all at CSIRO Wildlife and Ecology, Canberra. Valuable comments on earlier drafts of this paper were kindly provided by Isabel McBryde, Arthur Conacher, David Spratt, David Tongway and two anonymous referees.

7. References

Allan, G.E. (1984) Uluru (Ayers Rock-Mount Olga) National Park Revised Land Unit Map, scale 1:250 000, in E.C. Saxon (ed.), *Anticipating the Inevitable: A Patch-burn Strategy for Fire Management at Uluru (Ayers Rock-Mount Olga) National Park*, CSIRO, Melbourne.

Baker, B.W. and Noble, J.C. (1999) People, vermin, and loss of biodiversity: prairie dogs in North America and burrowing bettongs in Australia, in D. Eldridge and D. Freudenberger (eds.), *People and Rangelands – Building the Future. Proceedings of VI International Rangeland Congress, Townsville*, VI International Rangeland Congress, Inc., Aitkenvale, vol. 2, pp. 647-648.

Baker, L.M. and Mutitjulu Community (1992) Comparing two views of the landscape: Aboriginal traditional ecological knowledge and modern scientific knowledge, *Rangeland Journal* 14, 174-189.

Baynes, A. and Baird, R.F. (1992) The original mammal fauna and some information on the original bird fauna of Uluru National Park, Northern Territory, *Rangeland Journal* 14, 65-91.

Beckmann, R. (1990) Bringing back the bettongs, *Ecos* 65, 32.

Burbidge, A.A. (1988) Burrowing bettong *Bettongia lesueur*, in R. Strahan (ed.), *The Australian Museum Complete Book of Australian Mammals*, Angus and Robertson, Sydney, pp. 187-189.

Burbidge, A.A. and McKenzie, N.L. (1989) Patterns in the modern decline of Western Australia's vertebrate fauna: causes and conservation implications, *Biological Conservation* 50, 143-198.

Christensen, P. and Burrows, N. (1994). Project desert dreaming: experimental reintroduction of mammals to the Gibson Desert, Western Australia, in M. Serena (ed.), *Reintroduction Biology of Australian and New Zealand Fauna*, Surrey Beatty & Sons, Chipping Norton, pp. 199-207.

Colwell, J.D. (1969) Autoanalyser procedure for organic carbon analysis of soil, CSIRO Australia Division of Soils. National Soil Fertility Project Circular No. 5.

English, P. (1998) Palaeodrainage at Uluru-Kata Tjuta National Park and implications for water resources, *Rangeland Journal* 20, 255-274.

Finlayson, H.H. (1958) On central Australian mammals (with notice of related species from adjacent tracts). Part III - the Potoroinae, *Records of the South Australian Museum* 13, 235-302.

Gianello, C. and Bremner, J.M. (1986) Comparisons of chemical methods of assessing potentially available organic nitrogen in soil, *Communications in Soil Science and Plant Analysis* 17, 215-36.

Gray, J.L. (1837) Description of some new or little known Mammalia, principally in the British Museum Collection, *Magazine of Natural History and Journal of Zoology, Botany, Mineralogy, Geology, and Meteorology* 1, 577-87.

Hobson, J. (1990) *Current Distribution of Central Australian Languages*, Institute for Aboriginal Development, Alice Springs.

Holling, C.S. (1992) Cross-scale morphology, geometry, and dynamics of ecosystems, *Ecological Monographs* 62, 447-502.

Hunt, S. (1999) Paris, Le Havre, Sydney, in S. Hunt and P. Carter (eds.), *Terre Napoléon: Australia Through French Eyes 1800-1804*, Historic Houses Trust of New South Wales, Sydney, pp. 7-19.

Jacobson, G. and Arakel, A.V. (1987) Calcrete aquifers in the Australian arid zone, in *Proceedings of International Conference on Groundwater Systems Under Stress*, Brisbane, 1986, Australian Water Resources Council Conference Series, Australian Government Publishing Service, Canberra, vol. 13, pp. 515-523.

Jacobson, G., Arakel, A.V. and Chen, Y. (1988) The central Australian groundwater discharge zone: evolution of associated calcrete and gypcrete deposits, *Australian Journal of Earth Sciences* 35, 549-566.

Johnson, K.A., Burbidge, A.A. and McKenzie, N.L. (1989) Australian macropods: status, causes of decline, and future research and management, in I. Hulme, G. Grigg and P. Jarman (eds.), *Kangaroos, Wallabies and Rat-Kangaroos*, Surrey Beatty & Sons, Chipping Norton, pp. 641-657.

Kelly, A. (1990) *A New Imagining: Towards an Australian Spirituality*, Collins Dove, Melbourne.

Loveday, J. (1974) *Methods for Analysis of Irrigated Soils*, Technical Communication No. 54, Bureau of Soils, Commonwealth Agricultural Bureau, Canberra.

Low, W.A. (1986) The disappearance of mammal species from pastoral areas of central Australia: a land management problem? In P.J. Joss, P.W. Lynch and O.B. Williams (eds.), *Rangelands: A Resource Under Siege*, Australian Academy of Science, Canberra/Cambridge University Press, Sydney, pp. 551-552.

Low, W.A., Miller, C., Baker, L., Dobbie, W., Grattidge, A. and Gillen, J. (1999a) A review of the rabbit control program at Uluru-Kata Tjuta National Park, February 1989 to December 1998. Unpublished completion report (Project DN69), Parks Australia, Environment Australia.

Low, W.A., Miller, C., Grattidge, A., Allan, G., Gillen, J. and Dobbie, W. (1999b) The distribution and abundance of rabbit warrens at Uluru-Kata Tjuta National Park. Unpublished completion report (Project DN69), Parks Australia, Environment Australia.

Ludwig, J.A. and Cornelius, J.M. (1987) Locating discontinuities along ecological gradients, *Ecology* **68**, 448-450.

Ludwig, J.A. and Tongway, D.J. (1995) Spatial organisation of landscapes and its function in semi-arid woodlands, Australia, *Landscape Ecology* **10**, 51-63.

Ludwig, J., Tongway, D., Freudenberger, D., Noble, J. and Hodgkinson, K. (eds.) (1997) *Landscape Ecology, Function and Management: Principles from Australia's Rangelands*, CSIRO Publishing, Melbourne.

Mainguet, M. (1994) *Desertification: Natural Background and Human Mismanagement*, Springer-Verlag, Berlin.

Morton, S.R. (1990) The impact of European settlement on the vertebrate animals of arid Australia: a conceptual model, *Proceedings of the Ecological Society of Australia* **16**, 201-213.

Morton, S.R. and Baynes, A. (1985) Small mammal assemblages in arid Australia: a reappraisal, *Australian Mammalogy* **8**, 159-169.

Mueller-Dombois, D. and Ellenberg, H. (1974) *Aims and Methods of Vegetation Ecology*, John Wiley & Sons, New York.

Myers, K. and Parker, B.S. (1965) A study of the biology of the wild rabbit in climatically different regions in eastern Australia, *CSIRO Wildlife Research* **10**, 1-32.

Nelson, L.S., Storr, R.F. and Robinson, A.C. (1992) *Plan of Management for the Brush-Tailed Bettong,* Bettongia penicillata Gray, 1837 *(Marsupialia, Potoroidae) in South Australia*, National Parks and Wildlife Service, Department of Environment and Planning, South Australia.

Noble, J.C. (1993) Relict surface-soil features in semi-arid mulga (*Acacia aneura*) woodlands, *Rangeland Journal* **15**, 48-70.

Noble, J.C. (1996) Mesomarsupial ecology in Australian rangelands: burrows, bettongs (*Bettongia* spp.) and biocontrol of shrubs, in N.E. West (ed.), *Rangelands in a Sustainable Biosphere*. Proceedings of the Vth International Rangeland Congress, Salt Lake City, Utah, Society for Range Management, Denver, pp. 395-396.

Noble, J.C. (1997) *The Delicate and Noxious Scrub*, CSIRO Wildlife and Ecology, Canberra.

Noble, J.C. (1999) Fossil features of mulga (*Acacia aneura*) landscapes: possible imprinting by extinct Pleistocene fauna, *Australian Zoologist* **31**, 396-402.

Noble, J.C., Greene, R.S.B. and Müller, W.J. (1998) Herbage production following rainfall redistribution in a semi-arid mulga (*Acacia aneura*) woodland in western New South Wales, *Rangeland Journal* **20**, 206-225.

Noble, J.C., Detling, J., Hik, J, and Whitford, W.G. (1999) Soil biodiversity and desertification in *Acacia aneura* woodlands, in D. Eldridge and D. Freudenberger (eds.), *People and Rangelands – Building the Future. Proceedings of VI International Rangeland Congress, Townsville*, VI International Rangeland Congress, Inc., Aitkenvale, vol. 1, pp. 108-109.

Noble, J., Low, W., Jacobson, G. and Gillen, J. (1996) Distribution and abundance of relict warrens constructed by the burrowing bettong (*Bettongia lesueur*) at Uluṟu-Kata Tjuṯa National Park. Unpublished report to the Australian Nature Conservation Agency, Canberra (Project DN70).

Quoy, J.R. and Gaimard, J.P. (1824) Zoologie, in M. Louis de Freycinet (ed.), *Voyage Autour du Monde: Enterpris Par Ordre du Roi ... Execute Sur les Corvettes de S.M. l'Uranie et la Physicienne, Pendant les Annees 1817, 1818, 1819 et 1820 ...* Chez Pillet Aine, Imprimeur-Libraire, 1824-1839, Paris.

Short, J. (1998) The extinction of rat-kangaroos (Marsupialia: Potoroidae) in New South Wales, Australia, *Biological Conservation* **86**, 365-377.

Short, J. and Turner, B. (1993) The distribution and abundance of the burrowing bettong (Marsupialia: Macropodoidea), *Wildlife Research* **20**, 525-534.

Short, J. and Smith, A. (1994) Mammal decline and recovery in Australia, *Journal of Mammalogy* **75**, 288-297.

Short, J., Turner, B. and Majors, C. (1989) The distribution, relative abundance, and habitat preferences of rare macropods and bandicoots on Barrow, Boodie, Bernier and Dorre Islands. Final Report to the National Kangaroo Monitoring Unit, Australian National Parks and Wildlife Service, CSIRO, Midland.

Short, J., Bradshaw, S.D., Giles, J., Prince, R.I.T. and Wilson, G.R. (1992) Reintroduction of macropods (Marsupialia: Macropodoidea) in Australia - a review, *Biological Conservation* **62**, 189-204.

Sinclair, A.R.E., Pech, R.P., Hickman, D., Mahon, P. and Newsome, A.E. (1998) Predicting the effects of predation on conservation of endangered prey, *Conservation Biology* **12**, 564-575.

Spencer, B. (Ed.) (1896) *Report on the Work of the Horn Scientific Expedition to Central Australia, Parts 1-4*, Melville, Mullen and Slade, Melbourne.

Strahan, R. (1981) *A Dictionary of Australian Mammal Names*, Angus & Robertson, Sydney.

Tongway, D.J. (1991) Functional analysis of degraded rangelands as a means of defining appropriate restoration techniques, in A. Gaston, M. Kernick and H. Le Houérou (eds.), *Proceedings of the 4ᵗʰ International Rangeland Congress,* Congress, 22-26 April, 1991, Montpellier, France, Service Central d'Information Scientifique et Technique, Montpellier, vol. 1, pp. 166-168.

Tongway, D. and Hindley, N. (1995) *Manual for Soil Condition Assessment of Tropical Grasslands*, CSIRO Division of Wildlife and Ecology, Canberra.

Tongway, D.J. and Ludwig, J.A. (1997) The conservation of water and nutrients within landscapes, in J. Ludwig, D. Tongway, D. Freudenberger, J. Noble and K. Hodgkinson (eds.), *Landscape Ecology, Function and Management: Principles from Australia's Rangelands*, CSIRO Publishing, Melbourne, pp. 13-22.

Walker, B. and Nix, H. (1993) Managing Australia's biological diversity, *Search* **24**, 173-178.

Wells, A.T., Forman, D.J., Ranford, L.C. and Cook, P.J. (1970) *Geology of the Amadeus Basin, Central Australia,* Bureau of Mineral Resources, Australia, Bulletin 100.

Whitford, W.G. and Kay, F.R. (1999) Biopedturbation by mammals in deserts: a review, *Journal of Arid Environments* **41**, 203-230.

Wilson, G., McNee, A. and Platts, P. (1992) *Wild Animal Resources: Their Use by Aboriginal Communities,* Bureau of Rural Resources, Canberra.

World Conservation Monitoring Centre (1992) *Global Biodiversity: Status of Earth's Living Resources,* Chapman and Hall, London.

AUTHORS

J.C. NOBLE[1], J. GILLEN[2], G. JACOBSON[3], W.A. LOW[4], C. MILLER[4] AND THE *MUṮITJULU* COMMUNITY[5]

[1] CSIRO Wildlife and Ecology, GPO Box 284, Canberra, ACT 2601
Corresponding author; email: jim.noble@dwe.csiro.au
[2] *Uluṟu-Kata Tjuṯa* National Park, PO Box 119, Yulara, NT 0872
[3] Bureau of Rural Sciences, PO Box E11, Kingston, ACT 2604
[4] W.A. Low Ecological Services, PO Box 3130, Alice Springs, NT 0871
[5] CMA Ininti Store, Ayers Rock, NT 0872

Part Two

Land Degradation as a Response to Land Use Practices and Land Use Change

The theme of the conference from which the papers in this book have been selected was 'Agriculture, Land Degradation and Desertification'. This reflects the fact that in terms of spatial extent, in most regions agricultural practices, or changes in agricultural land use (including pastoralism), are the most important causes of land degradation.

The four chapters in this part illustrate the point well, from widely dispersed parts of the world. The first, by Donald Davidson, reviews the data on soil erosion in Scotland, a cool temperate region not usually high on the list of regions normally associated with land degradation. But Davidson shows the problem is real and that there is a complex array of historical and more contemporary land use practices, including grazing, trampling by humans and cultivation, which are responsible for the degradation. The following chapter by Guðrún Gísladóttir describes severe land degradation in an even colder climate, and serves as a reminder that 'desertification' can be a significant problem in cold as well as hot arid environments. Here, too, grazing by introduced stock is a significant cause of land degradation. The idea and reality of 'patchiness' features in this paper, and the concept reappears in a number of other contributions.

D'Angelo, Enne, Madrau and Zucca look at a much warmer part of the world (Sardinia), relating changes in land use (largely clearing of the forests for pastoralism) to increased sensitivity of the land to degradation. The final paper by Tadashi Kato, Xiaoju Wang and Shinichi Tokuta is much more specific. They show how the heavy applications of nitrogen to Japanese tea fields is being reflected in severe soil acidification, retarded root growth, and water contamination in ponds, rivers and groundwater, with fish deaths in some ponds. They state that whilst tea fields cover marginally over 1% of the total arable land in Japan, the export of nitrogen from the tea fields almost equals that from all other arable lands in the country. It would be interesting to know how land managers and policy makers are responding to the problem - or whether they even recognise that the problem exists.

CHAPTER 6

AN ASSESSMENT OF SOIL EROSION BY WATER IN SCOTLAND

DONALD A. DAVIDSON, IAN C. GRIEVE and ANDREW N.
TYLER

1. Abstract

This paper critically assesses the evidence for soil erosion by water in a cool temperate country, Scotland. In the Scottish uplands, peat erosion is the most extensive form of erosion but there are marked regional variations in extent and severity. Potential influencing factors are discussed, but it is concluded that no simple causal relationships can be proposed; instead peat erosion is linked to a combination of such factors as grazing by sheep and deer, increases in rainfall and resultant sub-surface pipe enlargement, and natural cycles of peat growth and decay. For upland mineral soils, data are provided to demonstrate the loss of organic matter as a result of human trampling pressure. In the lowlands, the incidence of soil erosion is demonstrated to be highly variable in space and time. Fields with light-textured soils and which have been ploughed or sown with autumn cereals are the most susceptible to gullying and soil wash as a result of particular weather sequences. Though the impacts of individual erosional events may be low, concern is expressed about the cumulative impacts. For one site, the cumulative impacts over a c. 40 year timescale are demonstrated by the application of the ^{137}Cs technique and the results indicated that erosion rates were ≤ 2.00 mm yr^{-1}; the corresponding deposition rates were ≤ 2.4 mm yr^{-1}. In this field, the main impact was damage to postholes from a Neolithic (ca. 3000 BC) timber structure. The effect of soil erosion on loss of the archaeological record is thus demonstrated. The overall conclusion is that soil erosion is an issue in a cool temperate country such as Scotland with different impacts in the uplands and lowlands. New approaches to modelling and predicting soil erosion are required for such environments to reflect relationships between erosion and hydrology.

2. Introduction

Since 1980 there has been a growing appreciation that soil erosion in Europe is not limited to Mediterranean regions, but also occurs to significant extents in more northerly and temperate areas. Data on soil erosion have been acquired by a range of approaches including aerial photograph analysis (Watson and Evans 1991), field measurement of erosion features following specific hydrological events (Wade and Kirkbride 1998), caesium-137 budgets (Walling and Quine 1991), plot or field monitoring (Chambers *et al.* 1992) , sedimentation rates in reservoirs (Duck and McManus 1990) and monitoring

93

A.J. Conacher (ed.), Land Degradation, 93–108.

at the catchment scale (Ferguson *et al.* 1991). Initiatives have been taken by the EU to assess soil erosion risk, for example through the development of the EUROSEM model (Morgan *et al.*1997). In temperate Europe, specific research projects have dealt with measuring, assessing and predicting erosion in such countries as France, Germany, Sweden, Norway, England and Scotland (Oldeman 1994). From such work, a number of key issues in soil erosion in temperate parts of Europe can be summarised:

i)   marked spatial and temporal variability in erosion rates (McEwen and Werritty 1988; Wade and Kirkbride 1998);

ii)   rates well above accepted soil loss tolerance values for specific rainfall events (Duck and McManus 1987);

iii)   extensive erosion in specific regions depending upon particular soil and land use conditions (Chambers *et al.* 1992);

iv)   soil tillage translocation is an important component of soil erosion (Quine *et al.* 1999);

v)   increases in winter rainfall resulting in greater soil erosion when the soil surface is unprotected (Harrison 1993; Kirkbride and Reeves 1993; Davidson and Harrison 1995);

vi)   difficulties in differentiating accelerated erosion from year-to-year variability in background rates (Ferguson *et al.* 1991), and

vii)   controversies over the significance of erosion (Frost and Speirs 1996).

The overall risk of soil erosion in Scotland has been evaluated recently at a scale of 1:250 000 on the basis of slope, topsoil and runoff classes (Lilly *et al.* 1999). Soils were assumed to be free of vegetation. Through the use of a rule-based approach, 100 m grid cells (1 ha) were classified as subject to high, moderate or low geomorphological risk of soil erosion; no account was taken of land cover, land management or rainfall attributes. Separate assessments were made for mineral soils and for soils with organic surface layers. The results for Scotland as a whole indicate that 53.4% of the country is at moderate risk of erosion by overland flow and a further 32.5% at high risk. In practice, much of Scotland has a protective land cover, and the actual incidence of erosion is much less than these figures might suggest. Nevertheless, the results from this study by Lilly *et al.* (1999) highlight the potential susceptibility of soils in a cool temperate maritime environment to erosion if there is no vegetation or crop cover.

The aim of this paper is to review and evaluate these issues with reference to recent investigations of soil erosion by the authors in both the Scottish uplands and lowlands. Data from much of this work have been published elsewhere (Davidson and Harrison 1992; Grieve *et al.* 1995; Davidson *et al.* 1998; Tyler *et al.* 1998; Grieve 2000). This paper integrates the results from these studies and others in Scotland to provide an overall assessment of the significance of soil erosion in a humid temperate country with a wide diversity of land use intensity and population density.

3. Erosion in the Uplands

3.1. EROSION SURVEY OF UPLAND SCOTLAND

In recent years, the conservation agency in Scotland, Scottish Natural Heritage (SNH), has become increasingly aware of erosion issues in the uplands. These issues derived from concern over climate and land use changes and increased trampling pressures due to outdoor recreation. Given the lack of baseline data, a key need was for an overall assessment of erosion types and distribution; such a survey was undertaken based on interpretation of aerial photographs (Grieve *et al.* 1994; Grieve *et al.* 1995). One hundred and forty four sampling squares, 5 km by 5 km, were selected in the Scottish uplands by stratified random sampling. The sample represented approximately 20% of the upland area. Erosion features were classified by photo interpretation and marked on overlays. Features were digitised and their areas or lengths were measured by image analysis (Table 1). Eroded peat was the most extensive form of erosion, affecting 219 km^2 (6% of the area). Gullied slopes on mineral soils occupied a further 170 km^2. Areas of peat erosion

Table 1. Areas, lengths and percentages of different erosion types in upland Scotland (Grieve *et al.* 1994)

Erosion type or land management	Area (km^2)or length (km)	Percentage or m km^{-2} of area examined
Dissected peat	219 km^2	6.02%
Gullied areas	170 km^2	4.69%
Debris flow/cone	22.2 km^2	0.61%
Landslide	2.9 km^2	0.08%
Sheet erosion	22.0 km^2	0.61%
Linear gullies	297 km unvegetated 115 vegetated 182	80 m km^{-2}
Footpaths	18 km	5 m km^{-2}
Vehicle tracks	146 km	40 m km^{-2}
Forestry	422 km^2 Standing 331 Felled/ploughed 91	11.6%
Upland management	329 km^2 Muirburn 167 Drained 81 Heavily grazed 81	9.1%

are usually striking features on both black and white, and colour photographs. The banks of bare peat and the eroded gullies show as very dark tones, contrasting with the lighter, often slightly mottled tones of the surrounding vegetation. The confidence in mapping these areas is therefore high and in the study was estimated at over 90%.

In contrast to the area affected by peat erosion, the evidence of erosion of footpaths in mountain areas triggered by trampling was much more limited. Table 1 shows there is a total of 5 m km^{-2} of footpath erosion. However, within popular mountain areas such as the Cairngorms (Fig. 1), the incidence of footpath erosion was much greater (72 m km^{-2}). Other features such as debris flows or scree also affected relatively small areas.

In an attempt to gain information on the extent and severity of peat erosion, and also of the sensitivity of areas to erosion, three broad categories were mapped:

P1 *Peat erosion with narrow gullies* - areas with narrow, well-defined gullies which are still actively eroding into peat; the gullies usually form a characteristic dendritic pattern according to the slope.

P2. *Peat erosion with broad gullies* - the gullies are often eroded down to the underlying mineral soil which may be partially revegetated and flanked by bare peat banks; they are frequently more than 3 m wide, and are often used by animals for shelter; a pattern is usually discernible on the air-photo, especially on steeper slopes, but the overall appearance on air-photos is not as striking as for the narrow gullies.

P3. *Remnant peat blocks* - these stand out as isolated 'tables' on the hillside and it is assumed they are the remnants of a formerly much more extensive peat cover; the vegetation on the surrounding land is usually a grass-dominated replacement community and can appear as a lighter tone.

Figure 2 shows the regional variations in the percentage area affected and its distribution among the three categories of severity. The sub-region with the greatest area affected by peat erosion (20%) was the Monadhliath, south west of Inverness. However, almost all of the erosion in this sub-region was in the least severely eroded (P1) class. In the east of the country much of the peat erosion occurred in the more severely eroded classes. In the eastern Southern Uplands and eastern Grampians, more than 80% of the peat erosion was in classes P2 or P3. These areas had the greatest evidence of upland management with more than 30% affected by burning or increased grazing by sheep and deer, and the greatest lengths of hill tracks, up to 200 m km^{-2}. There is also evidence of substantially increased numbers and densities of grazing animals in the last few decades (Grieve and Hipkin 1996).

These findings from this erosion survey need to be set within a wider context before their significance can be discussed. Calculations based on the mapping and analysis of the British Soil Surveys indicate that some 75% of the total organic carbon in British soils is stored in Scottish peat (Howard *et al.* 1995). Blanket peat covers large areas of upland Scotland and much has been and continues to be eroded. The spatial extent of eroded peat can be assessed from the maps and publications of the Soil Survey of Scotland, since eroded peat is categorised as a separate phase in their mapping programs. In eastern Scotland as defined by the 1:250 000 scale Soil Survey Sheet 5 map, Walker *et al.* (1982) indicate that blanket peat occupies 1790 km^2 or 7.2% of the land area. In the same area, the eroded phase covers 821 km^2 or almost half the total area of blanket peat. These authors also cite instances of isolated tabular areas of peat usually about 1 m thick and about 10 to 20 m wide on many exposed summits, such as in the Monadhliath Mountains and the Ladder Hills. These remnant blocks are completely surrounded by subalpine and alpine soils and are almost certainly the remnants of a much more extensive cover of peat.

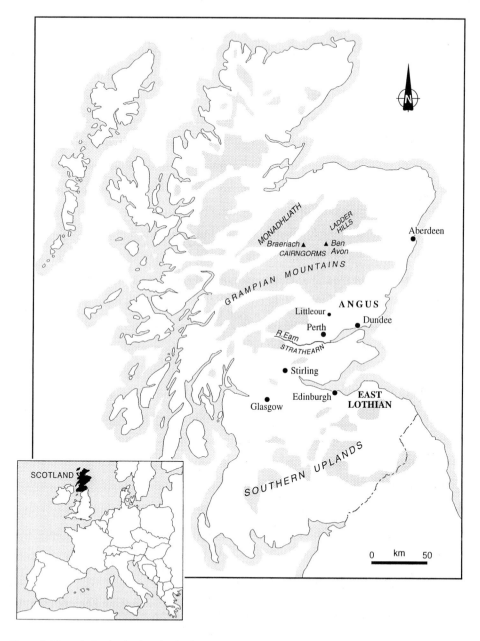

Figure 1. Places in Scotland referred to in the text.

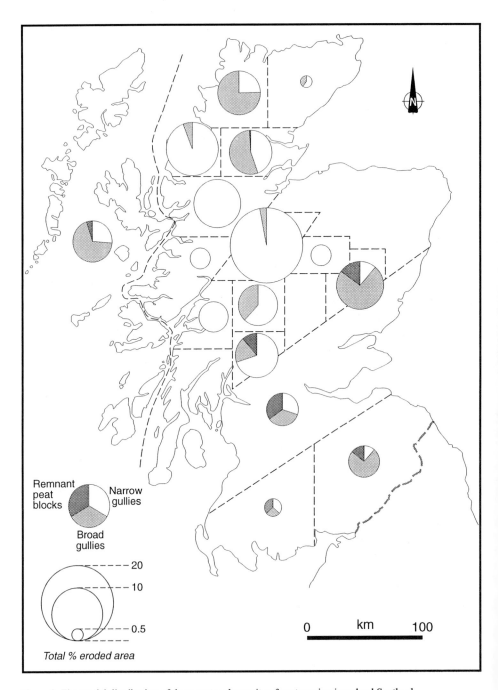

Figure 2. The spatial distribution of the extent and severity of peat erosion in upland Scotland.

3.2. PEAT EROSION IN THE U.K.

Peat erosion in the UK as a whole has been the subject of much speculation and study for some time (Bower 1962). In the southern Pennine uplands, Tallis (1985) reported that about three-quarters of the total peat area of some 300 km^2 is seriously eroded. Early work (Tallis 1965) showed that peat erosion had proceeded in two distinct stages: a slow, headward extension of streams into the peat along pre-established lines of weakness (the pre-glacial stream channels) beginning c. 3000 B.C. and continuing into historic times, and a very rapid extension of gullying after c. A.D. 1770. Tallis (1985), using pollen and macrofossil analyses, interpreted the onset of peat erosion on Featherbed Moss 1000-1200 years ago as natural. A series of bog slides and bursts around the margin of the peat blanket had led to drainage gullies extending back into the peat, downslope slumping of peat and drainage of the water pools near the watershed (interfluve). It was suggested that this sequence of events was initiated by an early spread of peat over flatter ground to its topographic limits, followed by a rapid build-up of largely unhumified peat peat over more consolidated peat and resulting in the development of an unstable peat mass. A wet Sphagnum-dominated bog surface developed after 1100 A.D. and partial recolonisation of some erosion features may have occurred. Renewed peat erosion, still in progress, is attributed to the death of Sphagnum from air pollution some 200-300 years ago. Tallis (1985) also pointed out that some components of the erosion system are not explained by these findings - several large areas of massive peat erosion, still active, were known to have been caused by catastrophic accidental fires 40 years previously. Erosion at the peat margin is clearly accentuated, if not caused, by intensive sheep grazing. He concluded that peat erosion may have been a 'natural' component of the uplands for many centuries. The upland peat blanket is an inherently unstable system and its break-up can be triggered in a number of ways. Stevenson et al. (1990), using a palaeolimnological approach, concluded that the main phase of erosion in Scotland was initiated between AD 1500 and 1700 either by climatic change associated with the Little Ice Age and/or by an increase in the intensity of burning.

Other research has concentrated on the rates and processes of erosion. Francis (1990) showed that maximum recession of eroding peat surfaces occurred in summer, with up to 80% being due to shrinkage. The highest rates of loss from peat faces due to rain wash occurred in autumn and early winter, with the sediment supply becoming limited as winter progressed. Mackay and Tallis (1996) concluded that the acceleration of blanket bog erosion in the Forest of Bowland, Lancashire, England during the early 1900s was due to a combination of below average rainfall, exceptional drought in one year (1921), a decline in management through fewer gamekeepers and one catastrophic fire. Birnie (1993) used erosion pins to estimate erosion rates on bare peat surfaces in Shetland and concluded that average annual rates of loss are 1- 4 cm yr^{-1}, similar to rates found on bare peat at higher altitudes in the Pennines by other researchers. Biological sustainability of upland management systems on peatland vegetation grazed by sheep was considered by Birnie and Hulme (1990), using ecological studies from the UK to evaluate the situation in Shetland. Their findings suggested that the present system, which is based on numbers of sheep, has set stocking rates at levels which are not sustainable, resulting in overgrazing and in some areas peat erosion.

3.3. LOSS OF ORGANIC MATTER FROM UPLAND MINERAL SOILS: THE ROLE OF HUMAN TRAMPLING

Loss of organic matter from upland peat soils is very clearly expressed in the erosional features as assessed from the aerial photograph survey. This survey suggested that the spatial extent of footpath erosion was small nationally, and essentially confined to popular mountain areas such as the Cairngorms. However, the upland soils affected are vulnerable alpine or sub-alpine podzols and loss of organic matter in these fragile upland environments is a topic of conservation importance. In an investigation of soil degradation on the Cairngorm plateau in north-eastern Scotland, 43 soil profiles were sampled on granitic parent material at altitudes ranging from 970 to 1300 m (Grieve 2000). Many of the soils under undisturbed vegetation were alpine podzols with distinctive Bh or Bs horizons. Approximately 50% of the sites sampled had no or partial vegetation cover, due either to human activity (trampling) or active geomorphic processes such as cryoturbation on patterned ground. Exposed cols and summit ridges were most heavily affected by trampling, with extensive areas of path development and some erosion. In these areas most soil profiles which had surface vegetation disturbance were truncated (Fig. 3). The Ben Avon profiles were located at 980 m in a heavily trampled col. At the vegetated site, organic matter was about 8% in the surface horizon but there was a second maximum at 50 cm. The partially vegetated site, located 1 m away, lacked the surface organic horizon but also showed an increase in organic matter content at 40 cm in the subsoil. The Braeriach profiles were located at 1135 m in another col, with complete and disturbed vegetation cover respectively. Again both profiles have evidence of organic matter eluviation, but the surface organic horizon is absent from the disturbed profile and the horizon of illuvial organic matter was slightly shallower. Overall, the sites with disturbed vegetation had less organic matter than those with undisturbed vegetation. Analysis of variance showed that there was a significant difference in total organic matter between disturbed and undisturbed profiles. In the disturbed soils total organic matter was reduced to less than half that in those with undisturbed vegetated. This magnitude of difference was similar for sites in the cols, where vegetation disturbance was more strongly influenced by human trampling, and those in areas such as snow-patch hollows where natural cryoturbation processes were likely to be more significant.

4. Soil Erosion in the Lowlands

Soil erosion on arable areas is also a topic of increasing concern in Scotland but there have been few studies on the subject. Previous research has included studies of sediment load in rivers (e.g. Al-Ansari et al. 1977) and reservoirs (Duck and McManus 1990). Catchments in central Scotland which are dominated by arable farming have greater sediment yields than those of open moorlands, but the greatest yields are associated with afforested catchments. However, reservoir sedimentation rates are not necessarily related to erosion on fields or hillslopes. Some information is also available from earlier field surveys of erosion. Speirs and Frost (1987) described an increase in the incidence of water erosion events on arable land in eastern Scotland between 1969 and 1985. Frost and Speirs (1984) and Duck and McManus (1987) described individual events, and Watson and Evans (1991) reported an

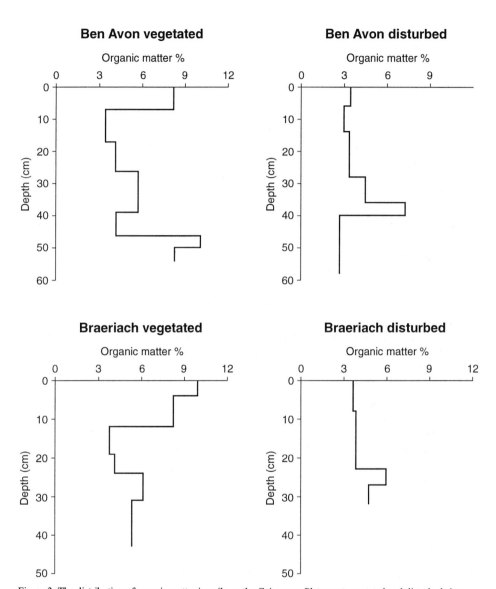

Figure 3. The distribution of organic matter in soils on the Cairngorm Plateau at vegetated and disturbed sites.

erosion survey in north-east Scotland, estimated eroded volumes from ground photographs, and suggested that Scottish soils are more erodible than English. Kirkbride and Reeves (1993) reported severe erosional damage to fields in Angus following a low-intensity rainfall event of prolonged duration. Also for Eastern Scotland, Wade and Kirkbride (1998) demonstrated the role of snow distribution, thaw rate and the amount and timing of rainfall during the thaw phase in initiating erosion. Conventional use of rainfall erosivity values does not help in such circumstances for the explanation or prediction of erosion. Frost and Speirs (1996) also provided a note of caution, based on results from a survey following a severe rainfall event in an area susceptible to erosion in East Lothian, Scotland; they found that only 10 fields out of 265 had significant erosion.

4.1. EROSION SURVEY – STRATH EARN

An area of 40 km^2, located to the immediate west of Perth in central Scotland, was selected for survey following excessive rainfall and snowfall in January, 1993 (Davidson and Harrison 1995). The area is mantled with thick, reddish brown till derived from Lower Old Red Sandstone. During the first 18 days of January, 1993, there was a succession of frontal systems which brought very wet conditions. Rain or snow was recorded on every day at most stations in the Tay and Earn catchments. On 11-13th January there were strong winds and heavy snow showers in central Scotland. By the 14th a slow thaw had occurred which resulted in flooding by the 15th. The weather situation further deteriorated on the 16th with heavy rainfall and a marked rise in temperature. The results of the rapid thaw combined with heavy rainfall were expressed in a peak discharge on the River Earn at Forteviot gauging station of 415 m^{3}s^{-1} at 5.45 a.m. on the 17th. This discharge had an estimated return period of 80 years. Over the two months following the storm, an erosion survey provided an inventory of the nature and occurrence of erosion on arable land. The survey had to be undertaken rapidly before farmers began to repair damage to fields.

A total of 208 fields was surveyed and 76 of these exhibited erosional features such as sheetwash, rill erosion, tramline erosion and gullies. These were further subdivided and land cover type was also recorded (Table 2). Erosion was found to be most frequent on autumn-sown cereal fields and ploughed land, and virtually absent on pasture and stubble fields.

Table 2. Incidence of erosion according to land cover in the survey area in Strath Earn (Davidson and Harrison 1995)

Land cover type	Number of eroded fields	Number of uneroded fields
pasture	0	102
stubble	1	51
autumn sown cereal	43	12
ploughed land	29	36
fodder crops & oilseed rape	3	7

Seventy eight percent of autumn-sown cereal fields had evidence of erosion and the corresponding figure for ploughed land was 45%. The most common erosional features were single ephemeral gullies along topographic hollows. Sheetwash between cereal rows or plough furrows and gullies along end furrows were also common. The dominant cause was concentration of runoff within fields as a result of topographic form. Next in importance was runoff generated in fields or woodland upslope of the damaged fields, and the effect of cereal planting or ploughing in concentrating runoff. Although less common, erosion due to flow from a drain, ditch or road and erosion from runoff concentration in end furrows was also important.

The conclusion from this and similar surveys (Kirkbride and Reeves 1993; Wade and Kirkbride 1998) is that substantial erosion can occur on arable land as a result of particular circumstances – unprotected soil (autumn-sown cereals or ploughed land), susceptible soils and a distinctive sequence of weather conditions (fields at field capacity and then further precipitation, or sudden thaw of snow along with further precipitation or a few days of extreme intensity rainfall). Overall, erosion as a result of one storm event as occurred in Strath Earn has no major on-site or off-site impact. However, the impact must be assessed on the basis of cumulative effects and if there is an increasing frequency of such events as a result of changes in rainfall, there will be degradation of the soil resource base in the long term. Any increase in winter rainfall will thus result in the more frequent occurrence of soil erosion. At present, incidences of substantial erosion on arable land in central and eastern Scotland seems to occur on average every third winter. The site at Littleour offers the opportunity to discuss the cumulative effects of erosion over the last c. 40 years.

4.2. EROSION RATES AT ONE SITE – LITTLEOUR

Erosion can cause significant damage to and ultimate loss of subsurface archaeological features. Much of the archaeology of eastern Britain, an area of intensive arable agricultural activity, appears only as cropmarks. At such sites surface features have been ploughed flat but pits, postholes and ditches lying immediately below the topsoil can be detected from the air. The cropmarks are evident as areas of crop stress due to soil moisture deficits over the underlying archaeological features. The excavation of a site at Littleour in Perthshire, Scotland, provided an opportunity to test a new methodology for assessing erosion risk (Davidson et al. 1998; Tyler et al. 1998). The site is an enclosure measuring 22 m long by 7.5 m wide, defined by deep pits which had held oak posts, probably supporting a fence. Within the enclosure was a further massive axial post. Radiocarbon dating places the use of the enclosure around 3000 BC (calibrated).

The site is located on a gently sloping bench in a 6 ha field (Fig. 4). The soil is a humus-iron podzol of the Corby series derived from fluvioglacial sands and gravels. The Ap horizon incorporates the original Ah and Ea horizons and lies directly above the Bs. The soil texture is loamy sand or sand and the Ap horizon varies in thickness downslope from 20 to more than 50 cm. The ^{137}Cs technique for estimating soil erosion was chosen because it is a well established and validated technique for quantifying erosion rates over the period extending from the early 1950s (Kachanoski 1993; Quine 1995). For the Littleour site, 24 sampling points were selected including two reference sites in unploughed grassland within the field; cores of known surface area were extracted and sectioned at defined intervals. Wet bulk density, moisture content and dry bulk density were determined. ^{137}Cs activity concentrations of subsamples were determined by standard laboratory based gamma spectrometry (Davidson et al. 1999; Tyler 1999). Total erosion or deposition for the profile during the 43 years since the maximum deposition of ^{137}Cs were calculated from the difference between total activity and that at the reference sites by the directly proportional method (Walling and Quine 1991).

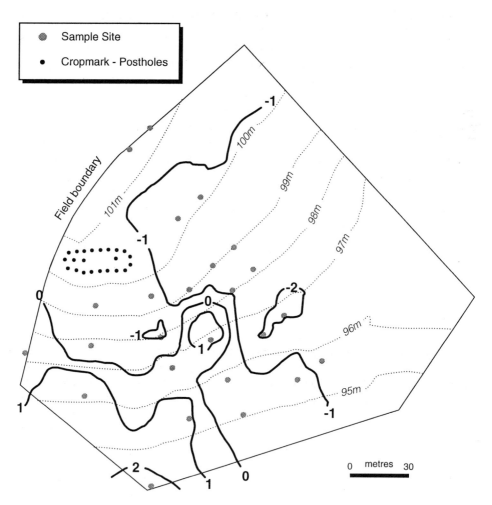

Figure 4. Erosion survey field at Littleour, Blairgowrie, Scotland, showing the Neolithic cropmark site, sample sites for ^{137}Cs determination, contours and erosion/deposition rate isopleths (Davidson *et al.* 1998; Tyler *et al.* 1998).

Mean rates of soil redistribution, as mm a^{-1} over 43 years, ranged from net erosion of 2.0 mm yr^{-1} to net deposition of 2.4 mm yr^{-1}. Figure 4 shows the spatial pattern of soil redistribution as an isopleth map. Maximum deposition occurs in the southern corner of the field and maximum erosion on the steeper midslope sections, but the erosion rate isopleths are not parallel to the ground surface contours. By simple interpolation from the spatial pattern, a mean net erosion rate of 0.5 mm yr^{-1} can be proposed for the cropmark since c.1953 though this may be an underestimate. Sample sites located at comparable slope positions to the crop mark have annual losses of 0.9 and 1.9 mm yr^{-1}. Sites close to the upper break of slope are the most vulnerable to erosion and the erosion rate experienced over the cropmark may be nearer to 1 mm yr^{-1} than to 0.5 mm yr^{-1}. An erosion rate of 0.5 to 1 mm yr^{-1} is markedly greater than the soil loss tolerance value of 0.1 mm yr^{-1} which has been proposed for the UK (Kirkby 1980). The immediate implication is that, if the average erosion rate which has been present for the last 43 years persists, there will be a progressive thinning of the topsoil.

Further ploughing at the site will thus penetrate to an increasing depth into the Bs horizon, resulting in damage to, and ultimately loss of, the archaeological features cut into it.

The thickness of the topsoil on the excavated site ranges from 20 to 30 cm. In the lower part of the field where deposition is dominant, the topsoil (Ap and A horizons) is between 50 and 60 cm thick. A simple calculation suggests that a loss of 15 cm from the area of the cropmark could account for the increase in thickness in the lower part of the field. On the basis of the estimated erosion rate of between 0.5 mm yr^{-1} and 1.0 mm yr^{-1}, the accumulation of soil in the lower part of the field could have been achieved over a period of 150-300 years. It seems likely, however, that an acceleration in erosion rate has occurred, with greater erosion rates following the introduction of new cultivation and cropping techniques since 1945. Although the postholes of the Littleour structure are between 0.5 and 1.05 m deep, and therefore at little immediate risk of complete loss, the pit containing the pottery is only 0.3 m deep. Many cropmark sites are made up entirely of features of this or less depth, and could be at risk of damage or complete loss within a few decades.

5. Discussion and Conclusions

5.1. UPLAND AREAS

The upland survey has demonstrated for the first time the considerable areal extent and potential significance of a range of erosional processes in the Scottish uplands. When all forms of erosion are taken into account (Table 1), the overall total of almost 12% of the area affected is close to the estimated average of 11.3% affected by water erosion in Europe (Oldeman 1994). Such figures do not take into account lowering of organic matter contents in the upper horizons of mineral soils. The loss of vegetation cover from trampling is crucial as recovery of vegetation is extremely slow on upland plateaux given cryoturbation processes and extreme exposure. The erosion of peat has impacts in terms of landscape quality, nature conservation and loss of organic carbon. Blanket and basin peat cover 30.2 and 0.7% respectively of the land area of Scotland and contain an estimated 75% of the total soil carbon in Great Britain (Howard et al. 1995). In the Monadhliath region, 82% of the area of blanket peat was eroded, 80% in the least severe category. In the eastern Grampians, 74% of the area mapped as blanket peat was eroded, 65% in categories P2 and P3. The 6% of the area affected by peat erosion could thus represent a very significant loss of the soil carbon store.

Although the extent of blanket peat erosion in upland Scotland is now known and there are indications of declines in organic carbon in mineral soils, there are considerable scientific uncertainties regarding the processes and causes of such degradation. It is clear that the growth and decay of blanket bogs is a natural phenomenon, and the isolation of any causal anthropogenic factor is extremely difficult. Furthermore, subtle changes in spatial and temporal rainfall patterns may contribute, but a dearth of upland rain gauges is a major constraint to any analysis. There has been an increase in annual precipitation over western and northern Scotland over the last 30 years, much of which can be attributed to increases during the winter (Smith 1995). Annual totals in these areas are up to 40% greater than during the drier years 1970-71. Since 1987/88, such changes in winter precipitation over

Scotland have been associated with an increasing frequency of maritime westerly airstreams and a decrease in cold continental easterly airstreams (Mayes 1996). The greater winter rainfall has led to many soils being at field capacity for most of the winter. Precipitation has fallen less frequently as snow, and the characteristic variability inherent in Atlantic weather systems has meant that snow cover has been more ephemeral (Harrison 1993). In the absence of the protection afforded by surface freezing or by a snow cover, soils have been more vulnerable to erosion. Therefore, such climatic variability may well contribute to increased erosion risk both in the uplands and lowlands. However, these changes in rainfall pattern cannot account for the increases in peat erosion as determined for the drier eastern areas of Scotland, for example, the Grampian region. In such localities there may well be an interplay of increased grazing pressure from sheep and deer, with subtle changes in upland peatland systems which have become more vulnerable to sudden and substantial change through erosion. Another possible explanation is the extension or enlargement of sub-surface drainage systems in blanket peats; no research has been done in Scotland on pipe networks and their role in initiating erosion.

5.2. LOWLAND AREAS

Much debate surrounds the significance of soil erosion on arable land in Scotland. To most farmers it is an occasional nuisance factor. Data on off-site impacts do not exist though there is an appreciation of significant effects on blockage of field drains and on particulate movement of phosphates to lakes. Though guidelines have been issued on good management practices to reduce soil erosion risk, there has been no co-ordinated approach to foster soil conservation. Historic Scotland is the only government agency to accept the potential impacts, especially on crop mark sites.

The processes of accelerated erosion in the lowlands are rather better understood than in the uplands and land use practices are of crucial importance in determining the likelihood of erosion. As in the uplands, the major uncertainty in predicting future patterns of erosion is in relation to rainfall patterns, with possibilities of changes in the amount, seasonal distribution and intensity. The results from soil erosion surveys on arable land indicate the inapplicability of rainfall erosivity measures for predicting erosion; instead there is growing awareness of the complexity of erosion/runoff relationships for rain and snow events which are of particular relevance to temperate regions. Soil losses of up to 5 mm have been determined from a short duration rainfall event (over three days) when antecedent conditions exacerbated erosion. In addition, estimation of mean annual erosion rates seems of little value given the considerable spatial and inter-annual variability. The implication is that the assessment of soil erosion risk for a temperate area such as Scotland needs to be probability-based and to focus on predicting the consequences of particular weather conditions for soil erosion under a range of land use, soil and topographic conditions.

6. References

Al Ansari, N.A., Al Jabbari, M. and McManus, J. (1977) The effect of farming upon solid transport in the River Almond, Scotland, in *Erosion and Solid Matter Transport in Inland Waters - Symposium.* Proeedings of the Paris Symposium, July 1977. IAHS-AISH Publication No. 122, 118-125.
Birnie, R.V. (1993) Erosion rates on bare peat surfaces in Shetland, *Scottish Geographical Magazine* **109**, 12-17.

Birnie, R.V. and Hulme, P.D. (1990) Overgrazing of peatland vegetation in Shetland, *Scottish Geographical Magazine* **106**, 28-36.

Bower, M.M. (1962) The cause of erosion in blanket peat bogs: a review of recent work in the Pennines, *Scottish Geographical Magazine* **78**, 33-43.

Chambers, B.J., Davies, D.B., and Holmes, S. (1992) Monitoring of water erosion on arable farms in England and Wales, 1989-90, *Soil Use and Management* **8**, 163-170.

Davidson, D.A., Grieve, I.C., Tyler, A.N., Barclay, G.J. and Maxwell, G.S. (1998) Archaeological sites: assessment of erosion risk, *Journal of Archaeological Science* **25**, 857-860.

Davidson, D.A. and Harrison, D.J. (1995) Water erosion on arable land in Scotland: results of an erosion survey, *Soil Use and Management* **11**, 63-68.

Duck, R.W. and McManus, J. (1987) Soil erosion near Barry, Angus, *Scottish Geographical Magazine* **103**, 44-46.

Duck, R.W. and McManus, J. (1990) Relationships between catchment characteristics, land use and sediment yield in the Midland Valley of Scotland, in J. Boardman, I.D.L. Foster and J.A. Dearing (eds.), *Soil Erosion of Arable Land*, Wiley, Chichester, pp. 285-299.

Ferguson, R.I., Grieve, I.C. and Harrison, D.J. (1991) Disentangling land use effects on sediment yield from year-to-year climatic variability, *International Association of Hydrological Sciences Publication* **203**, 13-20.

Francis, I.S. (1990) Blanket peat erosion in a mid-Wales catchment during two drought years, *Earth Surface Processes and Landforms* **15**, 445-456.

Frost, C.A. and Speirs, R.B. (1984) Water erosion of soils in south-east Scotland - a case study, *Research and Development in Agriculture* **1**, 145-152.

Frost, C.A. and Speirs, R.B. (1996) Soil erosion from a single rainstorm over an area in East Lothinan, Scotland, *Soil Use and Management* **12**, 8-12.

Grieve, I,C. (2000) Effects of human disturbance and cryoturbation on soil iron and organic matter distributions and on carbon storage at high elevations in the Cairngorm Mountains, Scotland, *Geoderma* **95**, 1-14.

Grieve, I.C., Davidson, D.A. and Gordon. J.E. (1995) Nature, extent and severity of soil erosion in upland Scotland, *Land Degradation and Rehabilitation* **6**, 41-55.

Grieve, I.C. and Hipkin, J.A. (1996) Soil erosion and sustainability, in A.G. Taylor, J.E. Gordon and M.B. Usher (eds.), *Soils, Sustainability and the Natural Heritage*, HMSO, Edinburgh, pp. 236-248.

Grieve, I.C., Hipkin, J.A. and Davidson, D.A. (1994) *Soil erosion sensitivity in upland Scotland*, Scottish Natural Heritage Research Survey and Monitoring Report No 24.

Harrison, S.J. (1993) Differences in snow cover on Scottish ski slopes between cold and warm winters, *Scottish Geographical Magazine* **109**, 37-44.

Howard, P.J.A., Loveland, P.J., Bradley, R.I., Dry, F.T., Howard, D.M. and Howard D.C. (1995) The carbon content of soil and its geographical distribution in Great Britain, *Soil Use and Management* **11**, 9-15.

Kachanoski, R.G. (1993) Estimating soil loss from the changes in soil caesium-137, *Canadian Journal of Soil Science* **73**, 629-632.

Kirkbride, M.P. and Reeves, A.D. (1993) Soil erosion caused by low-intensity rainfall in Angus, Scotland, *Applied Geography* **13**, 299-311.

Kirkby, M.J. (1980) The problem, in M.J. Kirkby and R.P.C. Morgan (eds.), *Soil Erosion*, Wiley, Chichester, pp.1-16.

Lilly, A., Hudson, G., Birnie, R.V and Horne, P.L. (1999) The inherent geomorphological risk of soil erosion by overland flow in Scotland, Report submitted to Scottish Natural Heritage, Macaulay Land Use Research Institute, Aberdeen.

Mackay, A.W. and Tallis, J.H. (1996) Summit-type blanket mire erosion in the Forest of Bowland, Lancashire, UK: predisposing factors and implications for conservation, *Biological Conservation* **76**, 31-44.

Mayes, J.C. (1996) Spatial and temporal fluctuations of monthly rainfall in the British Isles and variations in the mid-latitude westerly circulation, *International Journal of Climatology* **16**, 585-596.

McEwen, L.J. and Werritty, A. (1988) The hydrological and geomorphological significance of a flash flood in the Cairngorm Mountains, Scotland, *Catena* **15**, 361-377.

Morgan, R.P.C., Quinton, J.N., Smith., R.E., Govers, G., Poesen., J.W.A., Auerswald, K., Chisci, G., Torri, D. and Styczen, M.E. (1998) The European soil erosion model (EUROSEM): a dynamic approach for predicting sediment transport from fields and small catchments, *Earth Surface Processes and Landforms* **23**, 527-544

Oldeman, L.R. (1994) The global extent of soil degradation, in D.J. Greenland and I. Zabolcs (eds.), *Soil Resilience and Sustainable Land Use*, CAB International, Wallingford, pp. 99-118.

Quine, T.A. (1995) Estimation of erosion rates from Caesium-137 data: the calibration question, in I.D.L. Foster, A.M. Gurnell and B.W. Webb (eds.), *Sediment and Water Quality in River Catchments*, John Wiley and Sons, Chichester, pp. 307-329.

Quine, T.A., Govers, G., Poesen, J., Walling, D., van Wesemel, B. and Martinez-Fernandez, J. (1999) Fine-earth translocation by tillage in stony soils in the Guadalentine, South East Spain: an investigation using ^{137}Cs, *Soil and Tillage Research* **51**, 270-301.

Speirs, R.B. and Frost, C.A. (1987) Soil water erosion on arable land in the UK, *Research and Development in Agriculture* **4**, 1-11.

Smith, K. (1995) Precipitation over Scotland, 1757-1992; some aspects of temporal variability, *International Journal of Climatology* **15**, 543-556.

Stevenson, A.C., Jones, V.J. and Battarbee, R.W. (1990) The cause of peat erosion: a palaeolimnological approach, *New Phytologist* **114**, 727-735.

Tallis, J.H. (1965) Studies on Southern Pennine peats IV. Evidence of recent erosion, *Journal of Ecology* **57**, 509-520.

Tallis, J.H. (1985) Mass movement and erosion of a Southern Pennine blanket peat, *Journal of Ecology* **73**, 283-315.

Tyler, A.N. (1999) Monitoring anthropogenic radioactivity in salt marsh environments through *in situ* gamma ray spectrometry, *Journal of Environmental Radioactivity* **45**, 235-252.

Tyler, A.N., Davidson, D.A. and Grieve, I.C. (1998) Estimating soil loss from cropmark sites: using the Caesium 137 methodology at Littleour, in G.J. Barclay and G.S. Maxwell (eds.), *The Cleaven Dyke and Littleour: Monuments in the Neolithic of Tayside*, Society of Antiquaries of Scotland Monograph Series Number 13, pp. 83-91.

Wade, R.J. and Kirkbride, M.P. (1998) Snowmelt-generated runoff and soil erosion in Fife, Scotland, *Earth Surface Processes and Landforms* **23**, 123-132.

Walling, D.E. and Quine, T.A. (1991) The use of caesium-137 measurements to investigate soil erosion on arable fields in the UK: potential applications and limitations, *Journal of Soil Science* **42**, 147-165.

Walker, A.D., Campbell, C.G.B., Heslop, R.E.F., Gauld, J.H., Laing, D., Shipley, B.M. and Wright, G.G. (1982) *Soil and Land Capability for Agriculture, Eastern Scotland*, Soil Survey of Scotland Monograph, Macaulay Institute for Soil Research, Aberdeen.

Watson, A. and Evans, R. (1991) A comparison of estimates of soil erosion made in the field and from photographs, *Soil and Tillage Research* **19**, 17-27.

AUTHORS

DONALD A DAVIDSON, IAN C GRIEVE and ANDREW N TYLER

Department of Environmental Science, University of Stirling, Stirling, Scotland FK9 4LA
d.a.davidson@stir.ac.uk

CHAPTER 7

ECOLOGICAL DISTURBANCE AND SOIL EROSION ON GRAZING LAND IN SOUTHWEST ICELAND

GUÐRÚN GÍSLADÓTTIR

1. Abstract

Severe land degradation and desertification are considered the most serious environmental problems in Iceland. The Krísuvíkurheiði heathland in southwestern Iceland is one of the areas where degradation, prompted by overgrazing, has had a serious effect on vegetation, causing severe soil erosion.

The disturbance on the structure and function of the ecosystem will be discussed, as well as the effect of spatial patches and the heterogeneity of the ecosystem during soil erosion. For this purpose various ecological conditions have been taken into account, including plant species composition, distribution of plant communities and erosion forms, and the effect of spatial ecological patterns on land degradation.

At Krísuvíkurheiði, long-term land use has led to changes in the vegetation communities and species composition as well as changed micro-scale patterns of the plant species. The overall pattern of degradation within various plant communities indicates that the susceptibility to erosion is variable. This variability plays an important role in the severity of soil erosion of the grazing land. The heterogeneous moss heaths, dwarf shrub heaths and grass heaths are highly sensitive to erosion, whereas the homogeneous grassland is much more resistant. The study showed that 86% of Krísuvíkurheiði is either characterised by erosion or runs the risk of being so characterised.

2. Introduction

Land degradation has been a serious environmental problem for the last few centuries in Iceland. It has mainly been explained as the direct or indirect effect of anthropogenic activity, in combination with highly erodible soils and a severe climate. Together, these factors have greatly influenced the ecological conditions, resulting in reduced vegetative cover, a changed botanical composition, and a plant community distribution which contributed to increased soil erosion (examples: Bjarnason 1942; Þórarinsson 1961; Bjarnason 1979; Kristinsson 1979; Þorsteinsson 1980a, 1980b, 1986; Arnalds 1987; Arnalds et al. 1997; Gísladóttir 1993, 1998).

The objective of this paper is to evaluate the effect of land-use on land degradation and soil erosion in Krísuvíkurheiði, southwest Iceland. Present species cover in plant communities, the ratio of unpalatable:palatable species and the micro-scale pattern of species abundance are used as indicators of the effects of long-term grazing and for estimating the susceptibility of various plant communities to soil erosion. Furthermore, the role of patches in the expansion of deserts will be examined.

109

A.J. Conacher (ed.), Land Degradation, 109–126.

The term land degradation is used for vegetation where plant species composition has changed, with the result of that the plant community's capacity to withstand soil erosion is reduced, leading to a loss of vegetation cover, which in turn contributes to erosion and desertification. The term soil erosion is used for areas where land degradation has resulted in the depletion of vegetation followed by active erosion.

3. The Krísuvíkurheiði Heathland

The study area, Krísuvíkurheiði, extends over approximately 24 km² of the Reykjanes Peninsula in south-western Iceland (Fig. 1). The substratum is made up of a gently sloping basaltic lava field, covered mainly by dryland andosols and to a lesser extent by histosols (Gísladóttir 1998). The climate is cold-temperate oceanic with small annual and daily variations in temperature, and the mean growing season (mean daily temperature above +4°C) was 190 days during the 30 year period 1961-1990 (unpublished data from the Icelandic Meteorological Office, Reykjavik, Iceland). According to Einarsson (1988), mean annual precipitation was 1400-2000 mm during the years1931-1960.

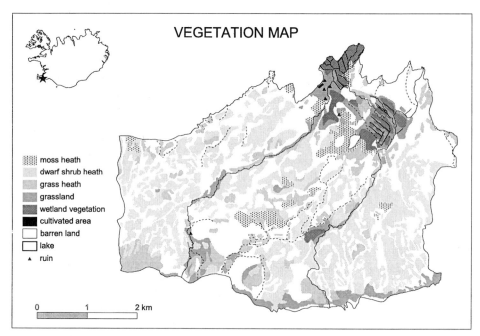

Figure 1. Vegetation map of Krísuvíkurheiði. (From Gísladóttir 1998).

Krísuvíkurheiði belongs to the birch woodland region of the boreal vegetation zone (Nordiska Ministerrådet 1984). Birch (*Betula pubescens*) and willows (*Salix sp.*) did grow there in the beginning of the 18th century when people living in the area used birch for charcoal production (Magnússon and Vídalín 1923-1924). Low and creeping *Betula pubescens* is currently found only locally on Krísuvíkurheiði. Dwarf birch (*Betula nana*) and small and fragile willow shrubs (*Salix* sp.) have also been identified, showing that birch and willow could still thrive in the area. Although birch woodland is absent, heathland is common (Fig. 1).

Krísuvíkurheiði was a valuable common grazing land for livestock for centuries. It has been estimated that serious degradation commenced there in the beginning of the 19th century (Gísladóttir 1993). The common is at present characterised by discontinuous vegetative cover and marked soil erosion (Gísladóttir 1993, 1998; Arnalds *et al.* 1997; Magnússon 1998). Farmers living in the area until the middle of the 20th century used it as grazing land for their livestock, mainly sheep. Furthermore, farmers living in neighbouring communities had access to the common as well. The sheep population reached its peak in 1930, when approximately 12 000 winter-fed sheep from 10 communities grazed on Krísuvíkurheiði and the surrounding lands, covering an area of 300 km². Up to around 1960 sheep grazed in the area from early May until November, when they were rounded up and taken to grazing land near the farms. Winter grazing was practised up to 1960 (Gísladóttir 1993). The area has now lost its former importance as grazing land due to diminished farming in the area in general. It is still used for sheep grazing during the summer from late May until early October; however, this is only for sheep belonging to urban dwellers for whom farming is a part of their lifestyle (Gísladóttir 1998). In 1995, 1000 winter-fed ewes as well as their lambs grazed on the common (unpublished data from the Agricultural Society of Iceland).

4. Methods

Because of the role of vegetation in land degradation, the vegetation communities of the Krísuvíkurheiði heathland were mapped. When evaluating land degradation and soil erosion, the effect of long-term land use on vegetation will be taken into consideration as one of the factors affecting the outcome. Moreover, considering the importance of plant species composition for grazing and its role in recognising thresholds of different ecological stages, the species composition and cover of the plant communities of Krísuvíkurheiði will be discussed, as well as the ratio of unpalatable:palatable species.

For inventories of species cover in dryland plant communities, 281 sample plots, each 0.25 m² in size, were measured in late summer 1996. The measurements were carried out using Þorsteinsson's method (1980a). At each site within the community, the sample circle (0.25 m²) was repeatedly thrown at random. The cover of each physiognomic layer was estimated visually, with combined cover of 100%. The bottom layer consisted of mosses and lichens and the field layer contained graminoids, herbs and dwarf shrubs. The micro-patterns of species abundance of the various plant communities were studied with respect to their ability to withstand soil erosion.

The mapping of plant communities was done by visual interpretation of CIR aerial photographs with high geometrical and spectral resolution. The photographs (on a scale of 1:25 000) were interpreted with a Zeiss Jena Interpretoscope with a magnification of 2-16x, based on the existing method by Ihse (1978). Because of the patchy vegetation, detailed mapping was difficult. The map was therefore generalised and the boundaries drawn around vegetated areas and patches that were at least 25 m in extent and separated by barren land. The vegetation map was digitised and imported to a Geographic information system (GIS) database and geometrically rectified by the rubber sheeting method of the Environmental Systems Research Institute (1991).

Since Krísuvíkurheiði is characterised by patches of vegetation merged with barren areas, it is important to understand the role of spatial patterns and scales on the function and development of the ecosystem and soil erosion. Therefore, the spatial distribution of patches and fragments of vegetation and deserts was mapped in detail. For this purpose a map was drawn showing vegetation and barren land with a pixel resolution of 1 m² (Fig. 2). The map was based on orthophotos from CIR aerial photographs, which enabled aerial measurements (Gísladóttir, 1998).

The exact proportions of various plant communities were measured by laying the vegetation map (Fig. 1) over the map showing the vegetation patches and barren land (Fig. 2), in order to evaluate land degradation and soil erosion. These measurements are important due to the different susceptibility of various plant communities to soil erosion.

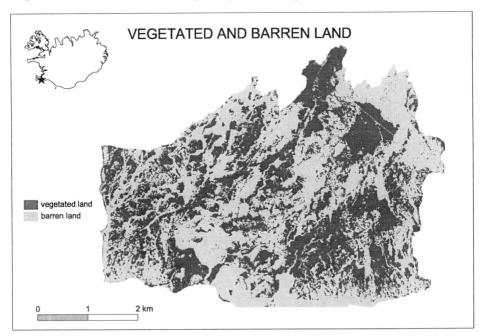

Figure 2. Vegetated and barren land in Krísuvíkurheiði. Pixel size is 1 m². (From Gísladóttir 1998).

5. Results and Discussion

5.1. STRESS, DYNAMICS AND THRESHOLDS OF THE ECOSYSTEMS

Dryland plant communities dominate the vegetation in Krísuvíkurheiði heathland (Figs. 1 and 3). Since wetland vegetation is not affected by land degradation in the area (Gísladóttir 1998), it will not been discussed further in this paper.

5.1.1. Disturbance Of The Woodland Ecosystem
Large areas of barren land and vegetation patchiness reflect ecological disturbances of the grazing land (Figs. 2 and 3). The meaning of disturbance is, however, not always obvious because ecological systems are probably always in a non–equilibrium state at

some spatial and temporal scale, as they continuously adjust to fluctuating abiotic and biotic conditions (Archer and Stokes 2000). When changes to vegetation are caused by human interference, it often leads to dramatic changes in the erosion rate (Viles 1990).

Archer and Stokes (2000) suggest that anthropic activities may reinforce or accelerate changes in an ecosystem triggered by natural events. Gísladóttir (1993) has argued that the near absence of woodland in the area is a response to intensive use for charcoal production and fuel. Additionally, leaves were harvested to add to the hay crop and intensive grazing took place in earlier times during spring and winter. Sheep are very selective in grazing, and during winter sheep selected birch and willows (Þorsteinsson and Ólafsson 1967; Þorsteinsson 1964, 1980a). Over-defoliation causes winter killing (Heady and Child 1994), and concurrent grazing and intensive use of woodland on Krísuvíkurheiði and its surroundings have accelerated the effect of severe climate spells on disturbance of the ecosystem (Gísladóttir 1993). The result is a change in the vegetation by the encroachment of unpalatable, evergreen dwarf shrubs into the woodlands, the loss of woodland and herb communities, and the succession of heathland communities which dominate the vegetation in the area. This conclusion is in agreement with the loss of woodlands in general in Iceland (for example, Arnalds 1987; Bjarnason 1979). The heathland may have resulted from long-term grazing of woodlands which originally grew on homogeneous and fertile soils, but which are becoming increasingly heterogeneous spatially and temporally, promoting the invasion of shrubs (Schlesinger et al. 1990). This also agrees with the model, proposed by Aradóttir et al. (1992), which describes the shift in vegetation succession in Iceland due to intensive livestock grazing from a continuous cover of palatable deciduous shrubs, grasses and broad-leaved herbs to less productive heathland dominated by unpalatable evergreen dwarf shrubs and narrow-leaved herbs.

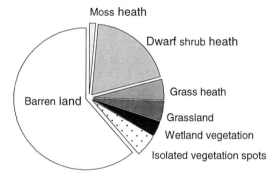

Figure 3. Proportion of different land cover types on Krísuvíkurheiði. (From Gísladóttir 1998).

Noy-Meir and Walker (1986) have explained the lag effects influencing species changes observed in the changes in the ratio of unpalatable (U) to palatable (P) species which occur in response to changes in grazing pressure. One explanation involves changes in competitive abilities related to tuft size and relative defoliation. If the changes in species conform to the dynamics associated with a single stable equilibrium, removing grazing pressure should return the species composition to its original state. Taking into account the reduced grazing pressure in Krísuvíkurheiði since 1930 (Gísladóttir 1993), a vegetation succession towards woodland should have taken place if the explanation of a single stable equilibrium were applied. However, this has not occurred, suggesting that

despite quite substantial reductions in stocking rates there has been no significant
increase in the proportion of palatable plants or increase in the areas covered by birch
and willows. This indicates a multi-equilibrium ecological system as explained by Noy-
Meir and Walker (1986), where heathland vegetation represents a hysteresis effect of
the unpalatable: palatable (U:P) species ratio in response to change in grazing pressure
at another equilibrium stage. The intensive land use on Krísuvíkurheiði has probably
passed the threshold of the woodland ecosystem and it may be irreversible for some
time. It has evidently lost its resilience and changed into another state which does not
overlap the previous one, and will remain so for a long time. According to Noy-Meir
and Walker (1986), such catastrophic changes in composition sometimes occur in
grazing land when shrubs take over. Wissel (1984) has emphasised that the
characteristic return time to equilibrium increases when a threshold is approached, and
dramatic changes to thresholds in multiple stable ecosystems may be irreversible if
caused by human actions.

5.1.2. The Formation Of Present Plant Communities
Constant removal through browsing on Krísuvíkurheiði, especially during harsh
weather, has impeded regeneration. In earlier times, horses and cattle were grazed in the
study area in addition to sheep, thus greatly adding to the pressure from browsing
(Gísladóttir 1993). Gísladóttir (1998) has argued that the species composition of the
present plant communities on Krísuvíkurheiði (Fig. 4), with an extensive cover of dwarf

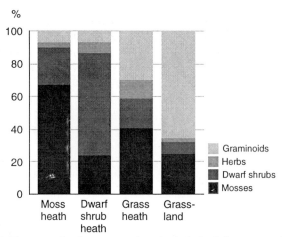

Figure 4. The cover of various groups of species in dryland plant communities on Krísuvíkurheiði.

shrubs and mosses and a small cover of graminoids and herbs, is a result of long-term
grazing. Given that birch woodland with grassland and broad leaved herbs dominated
Krísuvíkurheiði, the ratio of U:P might be used to reflect the change of the plant
communities in response to grazing pressure indicating a multi-state equilibrium. The
groups of species which are unpalatable on Krísuvíkurheiði are mosses and dwarf
shrubs, while graminoids and herbs are palatable (Þorsteinsson 1980a).
    Figure 5 shows the ratio of U:P species by plant communities in response to changes
in grazing pressure on Krísuvíkurheiði. Grazing pressure can be measured on various
scales such as time-scale (short- and long term), seasonal scale (spring and winter

grazing and/or summer grazing), and grazing intensity (herbivores per m²). During warm periods the carrying capacity of the grazing land is higher than during cold periods, and the carrying capacity of the grazing land is lower during spring and winter than during mid summer. Grazing land in good condition can sustain a high stocking rate without having a destructive effect on the vegetation, while a low stocking rate can have a devastating effect on already degraded grazing land.

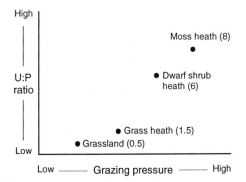

Figure 5. Dynamics of proportional composition of unpalatable and palatable species by plant communities in response to changes in grazing pressure on Krísuvíkurheiði. Figures in brackets show the U:P ratio. (Figure adapted from Noy-Meir and Walker 1986).

Is it possible to use the U:P ratio of the various plant communities an indicator of the dynamics of the ecosystem and its response to intensive grazing? The highest U:P ratio within this region is in moss heath which covers 1% of the area. This plant community is probably the youngest plant community on Krísuvíkurheiði, as it reclaims barren land. Bjarnason (1991) has found that *Racomitrium* sp. is able to remain monospecific for a long period of time, even though small ruptures in the moss carpet might enable some vascular species to become established, such as *Empertum nigrum* and *Salicx herbacea*. It is therefore unlikely that the moss heath is a result of increased disturbance after having passed the threshold of dwarf shrub heath, because when passing the threshold of dwarf shrub heath, the ecosystem will turn into a desert, which later can be reclaimed by moss heath (Gísladóttir 1998). Figure 5 shows that intensive disturbance created by high grazing pressure can change the ecosystem dramatically, to unpalatable dwarf shrub heath. Krísuvíkurheiði is covered to 20% by dwarf shrub heath (Figs. 1 and 3), which indicates that the ecosystem has passed the threshold of woodland. This unpalatable state is an effect of the shift from woodland to dwarf shrub heath, and any reversal of the change, according to Noy-Meir and Walker (1986), should be extremely difficult to achieve. The much lower U:P ratio and area covered (5%) by grass heath is a response to less grazing pressure. This allows another stage level of grass heath, where the vegetation grows on more humid soils, with a smaller proportion of dwarf shrubs and mosses (Figs. 3 and 4). The lowest U:P ratio in grassland (Fig. 5) indicates that this plant community is more resistant to intensive grazing than other dryland plant communities on Krísuvíkurheiði and more stable than the other plant communities (see Noy-Meir and Walker 1986), even though it covers only 4% of Krísuvíkurheiði (Fig. 3). These various ecological stages might indicate that the susceptibility of the plant communities to land degradation and soil erosion should increase from grassland – grass heath – dwarf heath to moss heath.

5.1.3. The Effect Of The Micro-scale Pattern Of Species Abundance On Soil Erosion
The plant communities consist to a great extent of the same vascular species, but their micro-scale patterns differ. The effect of land-use on species composition has not only been considerable on Krísuvíkurheiði but it has also affected the pattern of species abundance. Here it will be argued that the pattern plays an important role in the plant communities' susceptibility to soil erosion. Kolasa (1989) developed a model of habitat, either homogeneous or heterogeneous depending on the resolution with which species see their microhabitat. The species operate on a broad range in homogeneous habitat, followed by intermediate species in a more heterogeneous habitat to a narrow range of species which operate in an heterogeneous habitat. These patterns can be seen in the plant communities on Krísuvíkurheiði (Plate 1). The graminoids and pleurocarpous moss species in the grassland plant community are evenly distributed, making a homogeneous pattern of species (Plate 1a) compatible with Kolasa's (1989) model. Here the grassland is composed of generalists as described by Kolasa (1989), or species that contrast with the specialists, such as the dwarf shrub heath, which use small subdivisions of the hierarchical structure of the habitat. The large and evenly distributed cover of graminoids with a dense root mat in the grassland makes it resistant to trampling.

The pattern of species abundance in grass heath fall within Kolasa's (1989) slightly more heterogeneous habitat than grassland characterised by intermediate range species (Plate 1b). The pleurocarpous species are evenly distributed in the field layer covered by the graminoids and herbs. In the drier parts *Racomitrium lanigunosum* and dwarf shrubs have built up a pattern of small isolated patches where the species occupy their own space within the more homogeneous pattern of graminoids, herbs and pleurocarpous species. Grass heath is more susceptible to soil erosion than grassland because of the pattern caused by the species in the drier parts of the plant community.

Dwarf shrub heath is the most heterogeneous habitat (Plate 1c), which shows that species operating in a narrow range of values of temporal and spatial variables may involve either morphological or physiological specialisation and/or restrictions imposed by the habitat. Grazing and trampling herbivores, competitors and unfavourable physical conditions such as the hummocks, which are unstable environments (Arnalds 1994), or poor resource availability might cause these restrictions. The dwarf shrub heath is distinguished by a hummocky surface, where the top of the hummocks are covered by drought-resistant species like *Racomitrium lanigunosum* and *Empetrum nigrum* which tolerate this harsh environment, while graminoids and herbs seek the depressions. This results in a micro-scale mosaic of patches built up of various shrubs, covering their own spaces. In between are spaces occupied by *R. lanigunosum*. The graminoids and herbs are small and lacking in vigour; they cover only a small part of the plant community and share the same physical space in the hollows between the hummocks or mix with the shrubs and mosses. The prediction that can be made, based on this model, and which is important in explaining the function of the ecosystem and the susceptibility of the dwarf shrub heath in Krísuvíkurheiði to erosion, is that groups of species are clustered by similarities in their ecological range and abundance and that they should be vulnerable to disturbance. Hence, dwarf shrub heath is more susceptible to soil erosion than the previously mentioned plant communities as *Racomitrium lanigunosum* is the most susceptible species to soil erosion (Bjarnason 1991; Gísladóttir 1998).

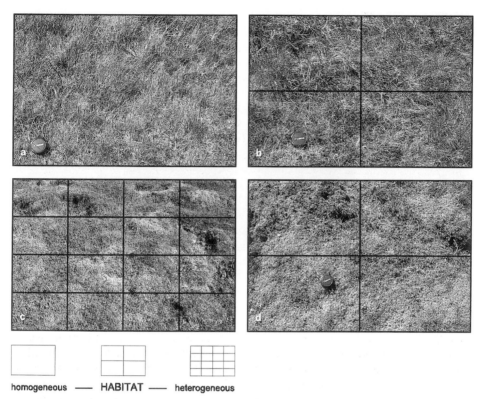

homogeneous ——— **HABITAT** ——— heterogeneous

Plate 1. Micro-scale patterns of species abundance in various plant communities, in which habitas homogeneity or heterogeneity depend on species microhabitats (Kolasa 1989). 1a Anhomogeneous grassland of broad range species. 1b Grass heath of intermediate range species. 1c A heterogeneous dwarf shrub heath of narrow range species. 1d Moss heath of intermediate range species. (Photos: Guðrún Gísladóttir).

Where the moss heath has reached its highest developed stages in Krísuvíkurheiði, the pattern of species abundance can be compared with grass heath belonging to the intermediate range of species (see Kolasa 1989): less heterogeneous than dwarf shrub heath and less homogeneous than grassland (Plate 1d). *Racomitrium lanigunosum* comprises the largest patches in space, dwarf shrubs occupy smaller spaces, and the small proportion of graminoids and herbs mix with the shrubs or mosses. Where *Racometrium* sp. are still reclaiming land they are homogeneous in accordance with Kolasa's model (1989). Even though the moss heath is not as heterogeneous a community as dwarf shrub heath, it is the plant community that is the most susceptible to soil erosion because of the mosses' sensitivity to trampling (Gísladóttir, unpublished data). In this vegetation type mosses cover large areas without any protection by other species more resistant to trampling than the mosses. However, the moss heath community has a limited distribution in Krísuvíkurheiði (Figs. 1 and 3).

5.2. SOIL EROSION

Soil erosion in Krísuvíkurheiði is extensive and severe and occurs on both vegetated and barren land. The major forms of erosion on vegetated areas are erosion spots and erosion escarpments (Plate 2). Other forms are fluvial channels and desert formations such as scree slopes, gravel and bare soil or andosol remnants, as described by Arnalds (1999).

Plate 2. Erosion escarpment at the edges of grassland. The height of the escarpment is over 2 m. (Photo: Guðrún Gísladóttir).

5.2.1. Development of Erosion Spots

The most common erosion form within the vegetated areas is erosion spots (Plate 3), measuring from 0.5 m up to 8 m in diameter, frequently found in dwarf shrub heath and to a lesser degree in grass heath (Gísladóttir 1998). Erosion spots are characterised by a bare surface, sometimes with exposed roots of dwarf shrubs and herbs. Dwarf shrub heath is the plant community most susceptible to spot erosion. The hummocks are susceptible to damage because of the unstable environment caused by freezing and thawing in andosols, as discussed by Arnalds (1994). By virtue of its loose structure *Racomitrium lanigunosum* is particularly sensitive to trampling (Bjarnason 1991, Gísladóttir 1998). When combined with the hummocks, freezing and thawing results in an extremely sensitive location. When the moss has been damaged, the soil becomes exposed and open to natural erosion processes. The drought resistant dwarf shrubs are also sensitive to soil erosion as they cluster adjacent to the mosses and/or on the tops of the hummocks. Because of the limited cover of graminoids, the plant communities lack the dense root mat of grasses that impedes soil erosion. A small exposure of soil on the hummocks is often the beginning of an erosion spot, the growth of which is induced by trampling which in turn causes additional detachment from the surface. Soil erosion is enhanced and further accelerated by natural processes (Gísladóttir 1998). During winter the temperature fluctuates just above or below freezing point and the limited snow cover increases the possibility of frost and damage to the vegetation on top of the hummocks, especially when windy. Erosion processes such as fluvial erosion accelerate the erosion, enlarging the spots, which then coalesce, leaving a large area without vegetation (Plates 4 and 5). The length of the eroded perimeter increases and can develop into a distinct escarpment where the soil is thick (Plates 2 and 5), or a serrated escarpment, caused by the erosion spots developed throughout the vegetated area, which further increase the length of the erosion front (Plates 4 and 6). Where the soil is thin, the borders between the vegetated areas and bare land are low in height and serrated, and become successively diffuse until the vegetative cover completely loses its coherence and dissolves into fragments which successively deplete and give way to barren land (Plates 4 and 5) (Gísladóttir 1998). Large areas within the dwarf shrub heath erode and divide the vegetation into patches (Fig. 2). This development results in a system of active erosion, both within the plant communities and along the borders of the plant communities along the escarpments.

Plate 3. Erosion spot in dwarf shrub heath. The spot in the foreground is approximately 1 m in diameter. Note the trampling by sheep. (Photo:Guðrún Gísladóttir)

Plate 4. Erosion spots in dwarf shrub heath in the foreground. The vegetation cover has successively lost its coherence and given way to desert (gravel). To the right of the gravel the active erosion front of dwarf shrub heath can be seen. Active erosion exists also along the escarpments on the hills. (Photo: Guðrún Gísladóttir)

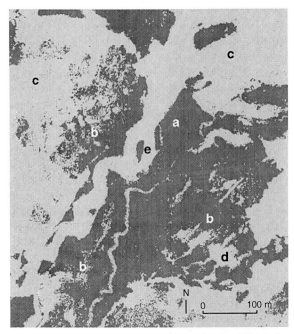

Plate 5. Erosion forms on flatland connected to grassland (a), dwarf shrub heath (b) and deserts (c). The escarpment connected to grassland is about 2 m high with fluvial erosion active along the escarpment. Erosion spots of various sizes are connected to dwarf shrub heath. They can coalesce (d), become dense or form more or less serrated edges. Desert forms are gravels (c) and andosol remnants (e). (From Gísladóttir 1998).

Plate 6. Escarpments on the edges of dwarf shrub heath patches with active erosion. Active erosion also exists in erosion spots within the borders of the plant community. (Photo: Guðrún Gísladóttir).

Erosion spots frequently occur in grass heath for the same reason as in dwarf shrub heath, even though they are not as common. This reduced susceptibility to erosion is because of the micro-scale pattern of species, the species cover and the fact that the hummocks in grass heath are not as high and frequent as in dwarf shrub heath. This makes grass heath more resistant to soil erosion (Gísladóttir 1998). However, the small areas of individual grass heaths and their spatial distribution adjacent to barren land and dwarf shrub heath increase their susceptibility to soil erosion.

5.2.2. Escarpment Development

As escarpments form at the edges between vegetated and desert land, the edges of practically all vegetated areas could be classified as escarpments (Plates 5 and 7). The size of the vegetated area marked by an escarpment varies from a few metres to about 1300 m in diameter (Fig. 2). Their heights range from less than 250 mm to more than 3 m (Gísladóttir 1998). Andosols, which lack cohesion (Arnalds et al. 1995), make up the thickest part of the formation and are susceptible to erosion when exposed, while the underlying materials as well as the root mat on the top are more resistant (Arnalds 2000a). Many complex processes are active along the escarpments, as have been studied by Arnalds (2000a). Water erosion, both rills along the escarpments and lateral rain impact during storms, needle-ice formation and the action of freezing and thawing, greatly reduce surface cohesion, making the soil particles easily detached. The root mat is eventually undermined, causing slumping (Plate 2). The effect of sheep trampling also increases detachment from the surface.

Plate 7. Desert formation. In the foreground is an andosol remnant. Behind it is an extensive area of gravel. The soil beneath the gravel surface ranged between 0.4 and 1 m in depth. Note the grazing sheep. (Photo: Guðrún Gísladóttir).

5.2.3. Desert Formation

As described by Arnalds (2000a, 2000b), deserts cover the largest part of the Krísuvíkurheiði (Fig. 2 and Plate 7). Deserts are the most serious erosion forms and represent a stage dominated by geophysical processes, which are likely to prevent restoration of land (Archer and Stokes 2000). Many of the presently barren areas are probably remnants of larger tracts of dwarf shrub heath, as indicated by vegetated spots close to the remaining dwarf shrub heaths. Despite the enormous soil erosion, mosses and dwarf shrubs have reclaimed barren land in certain places, as have dwarf shrubs in areas characterised by erosion spots. The overall extent of such reclamation is, however, negligible.

Wind erosion is not as active process as might be expected from the persistent winds. The presence of *Racomitrium lanigunosum* at the edges of plant communities indicates that wind-borne material does not accumulate (Gísladóttir 1998). This would have a negative effect on the mosses because of an increase in the decaying process (Bjarnason

1991). Furthermore, the mosses do not withstand abrasion by aeolian materials. However, when strong winds and high precipitation occur simultaneously, the wind adds to the effect of fluvial erosion.

Many of the reasons for the extensive soil erosion in Krísuvíkurheiði have been discussed above. Once depletion of vegetation has occurred the properties of andosols make them highly susceptible to soil erosion. The volcanic parent materials in Iceland have a large surface area and weather rapidly, both volcanic glass and crystalline material (Gíslason *et al.* 1996), to form clay minerals such as allophane, ferrihydrates and imogolite (Wada 1985; Wada *et al.* 1992). They almost entirely lack phyllosilicate minerals that provide cohesion (Wada *et al.* 1992). The soils are characterised by high infiltration rates and hydraulic conductivity, but also high wind erosion susceptibility (Arnalds 2000a; Gísladóttir 2000). Andosols have extremely high water retention, but a low plasticity index (Maeda *et al.* 1977). All these properties contribute to the high susceptibility of the soils to frost heaving, landsliding and transport by wind, rain-splash and running water (Arnalds 2000a).

Pressure from present land use, even though much less than that exerted by former land use (Gísladóttir 1993), still contributes to the degradation and erosion of the heathland plant communities. The grazing behaviour of Icelandic sheep presented by Thórhallsdóttir and Thorsteinsson (1993) indicates how intensively sheep cover the area. Their study showed that during the light nights in July sheep grazed throughout the 24 hour day with grazing bouts usually at an interval of ½ -2 hours. With the shortening of the day length the grazing time also shortened. Furthermore the sheep ranged over both heath and wetlands. This means that there is a trampling effect all over the grazing land even if the grazing value of dwarf shrub heath and grass heath is minimal (Gísladóttir 1998). The patchiness of the vegetation on Krísuvíkurheiði, however, contributes to the movement of sheep, which further increases the trampling effect.

5.2.4. The Importance of Scale in Detecting Soil Erosion Severity

Even though erosion is extensive the detection of erosion forms can be missed if the scale used is not detailed enough. This is especially important when mapping erosion spots, as emphasised by Gísladóttir (1998). These forms were detected in the field, but were easily missed when viewed from a distance, being hidden by hummocks and by their low form. They could also be detected when the CIR aerial photographs were viewed with a stereoscope, since bare ground is easily distinguished from vegetation by colour and form. The mapping of vegetation classes did not allow for a scale detailed enough to map features down to 1 m^2 in size. This was possible, however, when the same CIR aerial photographs were scanned and classified into only two classes, vegetated areas and barren lands, with a pixel resolution of 1 m^2 showing the patchy landscape on Krísuvíkurheiði in detail (Fig. 2).

Where the land is covered by remnants of andosols, their susceptibility to erosion is such that the environment is very unstable, which can delay any land reclamation for some time. This state is comparable to the 'disfunctional' landscape of Tongway and Ludwig (1997), which has lost its patchiness (vegetation and soil mosaic). They found that resources leaked out as the landscape had lost its ability to capture, store, conserve and recycle water and nutrients. On the deserts on Krísuvíkurheiði, isolated vegetation spots as small as 1 m^2 could be mapped on the CIR aerial photographs (Figs. 2 and 3 and Plate 5). These small vegetation spots down to 1 m^2 are more common on gravels where the soil remnants are thin and species have successively reclaimed land. These isolated vegetation spots can be important in increasing surface stability (Magnússon 1994) and capturing seeds from vegetation patches which are distributed in the area. This scale used is in accordance with Ludwig and Tongway (2000), who emphasised the importance of using fine-scale or micro-scales of less than 1 m² in patchy landscapes in order to understand the function of landscape processes in capturing scarce resources in favour of vegetation. In general, however, the deserts on Krísuvíkurheiði are more or less unstable because of the susceptibility of the andosols to erosion.

The model proposed by Aradóttir et al. (1992) emphasises the role of bare patches in the degradation and desertification process, which expose the andosol mantle (Archer and Stokes 2000), which is then removed by erosion processes (Gísladóttir 1998). It is therefore important in studying such areas to use a method that can show the pattern of spatial distribution of patches and fragments of vegetation. This was possible by using the CIR photographs. Because the spatial pattern of species within the plant communities affects the erosion severity, as well as the patchiness of the vegetation, it is important to know the actual proportions of the various communities. This was determined by laying the map showing plant communities (Fig. 1) over the map showing the detailed difference between the barren land and vegetated areas (Fig. 2), thus enabling quantification of the absolute proportion of the various plant communities (Fig. 3) (Gísladóttir 1998). The total vegetation cover was 39% and the remaining 61% was barren land. Due to the marked soil erosion occurring in the dwarf shrub heath and grass heath, 86% of Krísuvíkurheiði was either characterised by soil erosion or was highly susceptible to soil erosion.

The condition of Krísuvíkurheiði corresponds to an ecosystem as described by Aradóttir et al. (1992) which has passed a threshold such that further erosion cannot be prevented. The vegetation cannot be reclaimed unless economic inputs are greatly increased. The development of soil erosion in Krísuvíkurheiði has reached such an extent that the vegetative cover has lost coherence and dissolves into fragments which successively give way to barren land; compare the affect of heterogeneity as described by Schlesinger et al. (1990). The overall heterogeneity of fragmented vegetation and desert land on Krísuvíkurheiði has caused the redistribution of areas enhanced by detachment and transportation of materials mainly by water and to a lesser extent by wind. The intensive soil erosion further redistributes the valuable nutrients such as the N and organic C content of the soil, as shown by Magnússon (1998). Here, the mean C/N ratio decreased successively with intensified soil erosion. It ranged from 11.7 for gravel with hardly any soil, to 17 for vegetated land in a process affected by heterogeneity and patchiness as discussed by Schlesinger et al. (1990). Thornes (1990) has emphasised the competition between erosion and vegetation where vegetated and eroded areas are concomitant: the two types establish the conditions under which one or the other will dominate. In this respect the Krísuvíkurheiði heathland is dominated by erosion.

6. Conclusions

Long term land use has caused depletion of the woodland in Krísuvíkurheiði, which has lost its resilience, thus making the loss of woodland irreversible at present. The extensive distribution of dwarf shrub heath and grass heath is a result of the changed ecosystem from a woodland environment to a heathland environment, expressed by the increasing U:P ratio of the vegetation communities. Long-term grazing has changed the spatial pattern of species in the various plant communities, strongly affecting their susceptibility to soil erosion and the erosion feature formed. This implies the necessity of large-scale (small area) analysis of plant communities and resolution details when mapping the fragmentation of land. High-resolution CIR aerial photographs which can show eroded versus vegetated areas with 1 m² resolution are important to detect the spatial heterogeneity and patchiness of vegetation and barren land.

Long-term land use which has caused changes in the ecosystem of Krísuvíkurheiði underlies the various forms and extension of erosion. The outcome is a heavily degraded heathland with 86% either eroded or highly susceptible to soil erosion. Having already been greatly eroded, the dwarf shrub heaths and grass heaths offer little resistance to further degradation, while continued grazing and trampling exacerbate the erosion processes.

Species composition and cover in combination with the pattern of spatial heterogeneity and fragmentation play an important role in the development of the soil erosion processes. The effect of patches of vegetation and bare soil on the ecological functions of Krísuvíkurheiði has accelerated land degradation and desertification.

7. Acknowledgments

I am grateful for financial support from the following funds: The University of Iceland Research Fund, Student Innovation Fund (Iceland), Carl Mannerfelts Fund, Hans W:son Ahlmans Fund, Swedish Society for Anthropology and Geography, Andrée Fund, and the Lagrelius Fund. Financial support was received from Professor Margareta Ihse through funding from the Swedish Environmental Protection Agency of the project Remote Sensing and GIS in Landscape Ecological Studies - for monitoring of landscape changes and valuation of landscape diversity in an ecological and historical perspective. Thanks to Thomas Berggren for help with scanning the CIR aerial photographs, making the orthophotos and the preparation for image classification in Erdas Imagine. Thanks to two reviewers and Kristín Svavarsdóttir who made valuable comments on the paper and Terry Lacy and Richard Yeo for improving the English.

8. References

Aradóttir, Á. L., Arnalds, Ó. and Archer, S. (1992) Hnignun gróðurs og jarðvegs [Degradation of vegetation and soils. In Icelandic], *Græðum Ísland* **IV**, 73-82.

Archer, S. and Stokes, C. (2000) Stress, disturbance and change in rangeland ecosystems, in O. Arnalds and S. Archer (eds.), *Rangeland Desertification*, Kluwer Academic Publishers, Dordrecht, pp. 17-38.

Arnalds, A. (1987) Ecosystem Disturbance in Iceland. *Arctic and Alpine Research* **19**, 508-513.

Arnalds, Ó. (1994) Holklaki, þúfur og beit [Frost action, hummocks and grazing. In Icelandic], *Græðum Ísland* **V**, 115-120.

Arnalds, Ó. (1999) Soils and soil erosion in Iceland, in H. Ármannsson (ed.), *Geochemistry of the Earth's Surface*, A.A.Balkema, Rotterdam, pp. 135-138.

Arnalds Ó. (2000a) The Icelandic ´rofabard´ soil erosion features, *Earth Surface Processes and Landforms* **25**, 17-28.

Arnalds O. (2000b) Desertification: an appeal for a broader perspective, in O. Arnalds and S. Archer (eds.), *Rangeland Desertification*, Kluwer Academic Publishers, Dordrecht, pp 5-15.

Arnalds, Ó., Hallmark, C. T. and Wilding, L. P. (1995) Andisols from four different regions of Iceland, *Soil Science Society of America Journal* **59**, 161-169.

Arnalds, Ó., Þórarinsdóttir, E. F., Metúsalemsson, S., Jónsson, Á., Grétarsson, E. and Árnason, A. (1997) *Jarðvegsrof á Íslandi* [Soil erosion in Iceland. In Icelandic], The Soil Conservation Service and the Agricultural Research Institute, Reykjavík.

Bjarnason, Á. H. (1979) Beit og gróður [Grazing and vegetation. In Icelandic], *Ársrit Skógræktarfélags Íslands* **1979**, 40-47.

Bjarnason, Á. H. (1991) Vegetation on lava fields in the Hekla areaa, Iceland, *Acta Universitatis Upsaliensis, Acta Phytogeographica Suecia* **77**, Almquist & Wiksell International, Stockholm.

Bjarnason, H. (1942) Ábúð og örtröð [Farming and overstocking. In Icelandic with English summary], *Ársrit Skógræktarfélags Íslands* **1942**, 8-40.

Einarsson, M. Á. (1988) Precipitation in southwestern Iceland. *Jökull* **38**, 61-70.

Environmental Systems Research Institute, INC (1991) *Arc/Info Users' Guide. 6.0 Surface modelling with TIN^{TM}: Surface analysis and display.* XXP ESRI 2nd ed., Redlands, CA.

Gísladóttir, F. Ó. (2000) *Umhverfisbreytingar og vindrof sunnan Langjökuls* [Environmental change and eolian processes south of the Langjökull glacier. In Icelandic with English abstract], Unpublished M.Sc. thesis, University of Iceland, Reykjavík.

Gísladóttir, G. (1993) Geographical analysis of natural and cultural landscape. A methodological study in Southwestern Iceland, *Meddelanden från Naturgeografiska institutionen vid Stockholms universitet Nr.* **A 289**, Stockholm University, Stockholm.

Gísladóttir, G. (1998) *Environmental Characterisation and Change in South-western Iceland*, Department of Physical Geography, Stockholm University, Dissertation Series 10, Stockholm.

Gíslason, S.R., Arnórsson, S. and Ármannson, H. (1996) Chemical weathering of basalt in Southwest Iceland: effects of runoff, age of rocks and vegetative/glacial cover, *American Journal of Science* **296**, 837-907.

Heady, H.F. and Child, R.D. (1994) *Rangeland Ecology and Management*, Westview Press, Oxford.

Ihse, M. (1978) Flygbildstolkning av vegetation i syd och mellansvensk terräng - en metodstudie för översiktlig kartering [Aerial photo interpretation of vegetation in south and central Sweden - a methodological study of medium-scale mapping. In Swedish with English summary], *SNV PM* **1083**, Statens naturvårdsverk, Stockholm.

Kolasa, J. (1989) Ecological systems in the hierarchical perspective: Breaks in community structure and other consequences, *Ecology* **70**, 36-47.

Kristinsson, H. (1979) Gróður í beitarfriðuðum hólmum á Auðkúluheiði og í Svartárbugum [The vegetation on islands protected from grazing in the Icelandic highlands. In Icelandic with English summary] *Týli* **9**, 33-46.

Ludwig, J.A. and Tongway, D.J. (2000) Viewing rangelands as landscape systems, in O. Arnalds and S. Archer (eds.), *Rangeland Desertification*, Kluwer Academic Publishers, Dordrecht, pp. 39-52.

Maeda, T., Takenaka, H. and Warketin, B.P. (1977) Physical properties of allophane soils, in. N.C. Brady (ed.), *Advances in Agronomy*, Academic Press, New York, pp. 229-264.

Magnússon, Á. and Vídalín, P (1923-1924) *Jarðabók* **III** [Land and Cencus Register III. In Icelandic], Hið íslenska fræðafjelag, Copenhagen.

Magnússon, S.H. (1994) *Plant Colonization of Eroded Areas in Iceland*, Department of Ecology, Plant Ecology, Lund.

Magnússon, S.H. (1998) *Ástand lands í Krýsuvík sumarið 1997. Áætlun um uppgræðslu* [The condition of land in Krýsuvík in the summer 1997. Land reclamation plan. In Icelandic], unpublished report.

Nordiska Ministerrådet (1984) Naturgeografisk regionalindelning I Norden [The physiogeographical regions of the Nordic countries. In Swedish, Danish and Norwegian], Nordiska ministerrådet, Copenhagen.

Noy-Meir, I. and Walker, B.H. (1986) Stability and resilience in rangelands, in P.J. Joss, P.W. Lynch, and O.B. Williams (eds.), *Rangelands a Resource under Siege*, Cambridge University Press, Cambridge, pp. 21-25.

Schlesinger, W.H., Reynolds, J.F., Cunningham, G.L, Huenneke, L.F., Jarrell, W.M., Virginia, R.A., and Whitford, W.G. (1990). Biological feedbacks in Global Desertification, *Science* **247**, 1043-1048.

Thórhallsdóttir, A.G. and Thorsteinsson, I. (1993) Behaviour and plant selection, *Icelandic Agricultural Sciences* **7**, 59-77.

Thornes, J.B. (1990) The interaction of erosional and vegetational dynamics in land degradation: spatial outcomes, in J.B. Thornes (ed.), *Vegetation and Erosion. Processes and environments,* John Wiley & Sons, Chichester, pp. 41-53.

Thorsteinsson, I. (1986) The effect of grazing on stability and development of Northern rangelands: A case study from Iceland, in O. Gudmundsson (ed.), *Grazing Research at Northern Latitudes.* Plenum, New York, pp. 37-43.

Tongway, D.J. and Ludwig, J.A. (1997) The nature of landscape dysfunction in rangelands, in J.A. Ludwig, D.J. Tongway, D. Freudenberger, J. Noble, K. Hodgkinson (eds.), *Landscape Ecology. Function and Management. Principles from Australia's Rangelands*, CSIRO Publishing, Melbourne, pp. 49-61.

Viles, H.A. (1990) The agency of organic beings: a selective review of recent work in biogeomorphology, in J.B. Thornes (ed.), *Vegetation and erosion. Processes and environments,* John Wiley & Sons, Chichester, pp. 5-24.

Wada, K. (1985) The distinctive properties of andosols, *Advances in Soil Science* **2**, 173-229.

Wada, K., Arnalds, O., Kakuto, Y., Wilding, L.P. and Hallmark, C.T. (1992) Clay minerals in four soils formed in eolian and tehra materials in Iceland, *Geoderma* **52**, 351-365.

Walker, B. H. (1993) Rangeland ecology: Understanding and managing change, *Ambio* **22**, 80-87.

Wissel, C. (1984) A universal law of the characteristic return time near thresholds, *Oecologia* **65**, 101-107.

Þorsteinsson, I. (1964) Plöntuval sauðfjár. Rannsóknir á afréttarlöndum, [Plant selection by sheep - research on common grazing lands. In Icelandic], *Freyr* **60**, 194-201.

Þorsteinsson, I. (1980a) Gróðurskilyrði, gróðurfar, uppskera gróðurlenda og plöntuval búfjár [Environmental data, botanical composition and production of plant communities and the plant preference of sheep. In Icelandic with English abstract], *Journal of Agricultural Research in Iceland* **12**, 85-99.

Þorsteinsson, I. (1980b) Nýting úthaga–beitarþungi [Grazing intensity–proper use of rangelands. In Icelandic with English abstract], *Journal of Agricultural Research in Iceland* **12**, 113-122.

Þorsteinsson, I. and Ólafsson, G. (1967) Fjárbeit í skóglendi og úthaga [Sheep grazing in woodland and open grazing land. In Icelandic], *Ársrit Skógræktarfélags Íslands* **1967**, 6-14.

Þórarinsson, S. (1961) Uppblástur í ljósi öskulagarannsókna [Soil erosion in Iceland in the light of tephrochronological investigations. In Icelandic], *Ársrit Skógræktarfélags Íslands* **1960-1961**, 17-54.

AUTHOR

GUÐRÚN GÍSLADÓTTIR

Department of Geology and Geography, University of Iceland, IS 101- Reykjavík, Iceland, and Department of Physical Geography, Stockholm University, Sweden

E-mail: ggisla@hi.is

CHAPTER 8

LAND COVER CHANGES IN SARDINIA (ITALY): THE ROLE OF AGRICULTURAL POLICIES IN LAND DEGRADATION

M. D'ANGELO, G. ENNE, S. MADRAU and C. ZUCCA

1. Abstract

During recent decades in European Mediterranean Countries the shift from traditional to modern agricultural systems, favoured by national and common agricultural policies, has led in many cases to the overexploitation and degradation of natural resources. This is particularly true for Sardinia, where the growing demand for sheep cheese has caused the cultivation of unsuitable areas for forage production.

This paper evaluates the effects of past agricultural policies on current land degradation phenomena. With this aim in view, land cover changes were monitored in a test area subjected to severe anthropic pressure, located in central-eastern Sardinia.

Multi-temporal land cover maps were derived from the interpretation of aerial photographs and compared by using GIS techniques. Land cover dynamics in the period 1955-1996 highlighted that the shift from extensive to semi-extensive production systems caused deep changes in the landscape: in particular, a severe decrease in forest land, mainly due to the creation of artificial pastures, was observed. By matching the presence of artificial pastures and suitability classes derived from a previously developed model for the evaluation of Land Suitability, different degrees of environmental sensitivity to land degradation were highlighted.

2. Introduction

In the Mediterranean basin, animal breeding has always been one of the main agricultural activities (Papanastasis 1998). The scientific community has recognised that agropastoralism, when practised irrationally (overgrazing and irrational grazing systems), is one of the main causes of land degradation (Mairota et al. 1998; Narjisse 1998).

Different authors (Previtali 1996; Enne et al. in press) have highlighted that the impact of agropastoral activities on land degradation processes can be direct (animals' impact on the soil and vegetation) and indirect (irrational practices related to agropastoral activities, such as the use of fire and the creation of artificial pastures on unsuitable areas). These kinds of land degradation are particularly severe in Sardinia, a Mediterranean island with a total area of 2 408 000 ha and with a total number of 762 800 Livestock Units (1 LSU = 500 kg of body weight = 1 cattle = 10 goats = 10 sheep), and 4 297 700 sheep in particular. Table 1 shows the growth of the sheep numbers in Sardinia and Italy from 1881 to 1994. During the last 30 years in Sardinia there has been a remarkable increase in sheep,

127

mainly favoured by the rise in the price of sheep milk during the 1970s and the 1980s.

In order to meet the higher feeding requirements of animals, the regional government of Sardinia adopted policies aimed at increasing forage production, and also provided subsidies. Unfortunately, such policies were not accompanied by the necessary guidelines for their implementation and, in addition, the interventions were carried out on areas which were unsuitable due to morphologic and pedologic conditions. Extensive areas of Mediterranean maquis were cleared to create artificial pastures by using fire and deep mechanical tillage. This resulted in a severe impact on the environment by intensifying erosion processes. Furthermore, these artificial pastures have to be periodically cleared of shrubs and non-palatable invasive species, work generally done by using fire or heavy machinery. These interventions are carried out every three-four years at the end of the dry season, thus increasing soil vulnerability in relation to the first and intense autumn rainfalls (Seuffert 1992; Vacca et al. 1994); in situations of low biological potential the frequency of clearing interventions is higher, thus increasing soil degradation rates.

Table 1. Sheep numbers in Sardinia and Italy in the period 1881-1994 (Benedetto et al. 1995; ISTAT 1996).

Year	Sardinia (24 080 km²)	Italy (301 318 km²)
1881	845 000	8 863 000
1908	1 877 000	11 426 000
1930	2 054 000	10 268 000
1950	2 576 000	10 295 000
1970	2 558 000	7 948 000
1980	3 021 000	9 277 000
1985	3 822 000	11 451 000
1990	4 097 000	10 847 000
1994	4 297 700	10 651 500

The aim of this paper is to evaluate the effects of past agricultural policies on current land degradation phenomena by monitoring land cover changes and their consequences on land due to the aforementioned dynamics, over an area located in Central-Eastern Sardinia.

3. Materials and Methods

The test area is located between 40°20' and 40°31' latitude north and between 9°34' and 9°50' longitude east (municipalities of Irgoli, Onifai, Orosei), covering about 20 900 ha. The climate is typically dry sub-humid Mediterranean, with a mean annual temperature of about 17°C and mean annual precipitation ranging from 500 to 700 mm; local climate varies with elevation which ranges from sea level to 862 m a.s.l. (Monte Senes). Geology is quite heterogeneous: Palaeozoic granites and metamorphic rocks; Jurassic crystalline limestones; Plio-Pleistocene basalts; old and recent alluvial deposits. The landscape is therefore highly complex (Fig. 1).

With reference to land use patterns, the agricultural area amounts to 11 045 ha, 63% of which is represented by meadows and pastures (ISTAT 1992). According to official data, wooded areas cover about 6000 ha mainly represented by Mediterranean maquis; the limited extent of forest (coppice of *Quercus ilex* L.) is the result of severe anthropic pressures during the 20th century. There were major tourist developments in the coastal areas in the 1990s, although agropastoral practices are still the most important economic activity.

In order to understand the impact of agrosilvopastoral systems, all available information was acquired. Due to the limited availability of cartographic information on land cover changes, aerial photographs taken over the study area in 1955 (flight IGMI - Istituto Geografico Militare Italiano, black and white, at a scale of 1:33 000) and in 1989 (flight IGMI, black and white, at a scale of 1:33 000) were obtained. Traditional photo-interpretation procedures were then applied to derive the pedological map and the multi-temporal land cover maps (1955 and 1996) at a scale of 1:25 000. The 1996 land cover map was derived by updating the information acquired from the 1989 aerial photographs using SPOT-HRV satellite data (IGMI 1993) and field surveys.

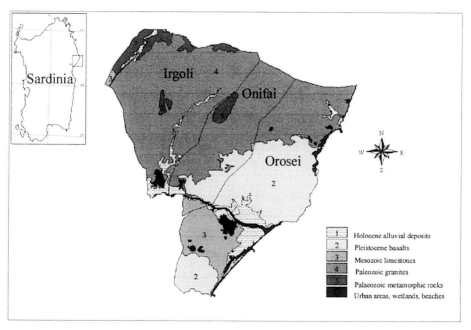

Figure 1. Location of the study area.

With reference to land cover mapping, the CORINE (1992) legend was used and adapted to the Sardinian environment. The different classes were then grouped into the following main typologies: agricultural areas (arable land, permanent crops, heterogeneous agricultural areas); pasture lands as natural pastures and artificial pastures, and shrublands, Mediterranean maquis, woodlands and forest plantations.

Land cover maps were digitised and integrated into a GIS implemented via ARC-INFO software. Statistics of land cover classes in 1955 and 1996 were obtained and land cover changes were highlighted. The complex land cover dynamics were derived by overlaying the two maps. Emphasis was placed on the changes from naturally vegetated areas to agricultural land (crops and pastures).

With particular reference to artificial pastures, a model for the evaluation of "Land Suitability for the creation of artificial pastures", previously developed and tested (d'Angelo *et al.* 2000), was applied to the area, in order to match land cover change and real land suitability. This comparison highlighted the increased degradation risk related to land use changes.

4. Results

Land cover changes in the period 1955-1996 are the result of a general marked intensification of anthropic activities in the study area. By simply comparing the two land cover maps it is evident that the 1996 pattern is more complex than the 1955 one (Fig. 2). A major change was the shift from extensive to semi-extensive production systems. Different trends were observed between coastal and hilly areas, towards tourist development and a marked intensification of agropastoral systems, respectively.

In 1955 the area was characterised by extensive land utilisation (Table 2). Pastures and forests (woodlands, Mediterranean maquis and forest plantations) almost covered the whole area (85.2%) whereas urban areas were very limited. Agricultural areas covered only 11.7% of the land, and they were mainly located on the Orosei plain and along the river Cedrino, areas which have been cultivated for centuries (Le Lannou 1979). Perennial crops were very limited and generally mixed with horticultural crops (1047 ha, 5% of the total area), whereas annual crops (meadows) covered 1406 ha (6.7%). Grasslands (natural pastures, artificial pastures and shrublands) constituted 44.3% of the total area. Artificial pastures were only 18% of the total (1711 ha), mainly located on flat areas on basaltic

Table 2. Land cover changes in the study area (1955-1996).

Land cover classes	Surface				Difference	
	1955		1996		(1955-96)	
	(ha)	*(%)*	*(ha)*	*(%)*	*(ha)*	*(%)*
Agricultural areas	2 454	10.7	2 932	12.8	+ 478	+19.5
Natural pastures	3 334	14.6	2 803	12.2	- 531	-15.9
Artificial pastures	1 711	7.5	3 388	14.8	+ 1 677	+98.0
Shrublands	4 216	18.5	5 462	23.9	+ 1 246	+29.5
Woodlands	1 625	7.1	1 517	6.6	- 108	- 6.7
Mediterranean maquis	6 240	27.3	2 409	10.5	- 3 831	-61.4
Forest plantations	687	3.0	1 582	6.9	+ 895	+130.3
Other	627	2.7	796	3.5	+ 169	+ 26.8

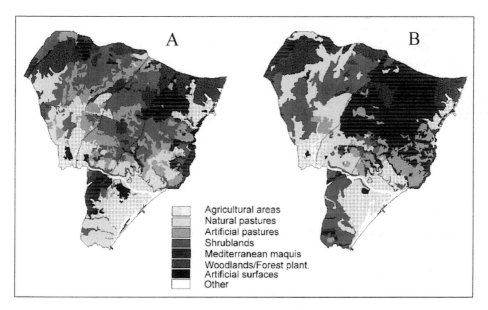

Figure 2. Land cover maps; A. 1996; B. 1955. The legend is purposely simplified with respect to the original one.

landscapes. Hilly areas were mainly covered with shrublands whose boundaries with Mediterranean maquis were seldom clear, particularly in the areas with uneven morphology. It is worth highlighting that natural pastures and shrublands covered 63% of the total area of Irgoli municipality (Table 3). Forest areas were mainly constituted by woodlands (1625 ha, 7.1%) and above all by Mediterranean maquis (6241 ha, 73%), mainly located in areas with uneven morphology.

In general, the distribution of the different kinds of pasture lands was linked to geomorphic and pedological conditions. The distribution was strongly influenced by traditional practices, characterised by a very limited intensive agricultural activity, concentrated in the small coastal plains and near towns and villages, and by a large extension of grassland/maquis used by grazing animals (Le Lannou 1979).

The analysis of land cover change during the period 1955-1996 highlights the following dynamics:

- a large increase in pasture lands (+25.8%);
- a considerable reduction in forested lands (-35.6%);
- a marked expansion of agricultural areas (+19.5%);
- a remarkable increase in urban areas (+611%), but from a very small base.

These dynamics show an intense anthropic influence on the land, mainly due to the agropastoral activities.

The increase in pasture areas is mainly due to the increase in artificial pastures (+1677 ha, +98%) and shrublands (+1246 ha, +29.5%), whereas a decrease in natural pastures (-531 ha, -15.9%) was observed. The marked reduction in forested lands can be totally ascribed to maquis (-3831 ha, -61.4%) and was not compensated for by reforestation (+895 ha, +130%), whereas the relative increase in agricultural areas occurred notwithstanding the reduction in arable lands (-551 ha, -39.2%).

Figure 3 summarises the evolutionary dynamics of land cover. Artificial pastures mainly originated from maquis (777 ha) and natural pastures (1025 ha); the marked reduction of the latter can also be explained by the fact that they were turned into agricultural areas (320 ha). The increase in shrublands mainly originated from the degradation of maquis (2413 ha). These dynamics show that the increase in areas with semi-extensive systems (artificial pastures) and extensive systems (shrublands) is linked to the progressive intensification of the agropastoral systems.

Dynamics are rather different when single municipalities are taken into account (Table 3). In particular, two opposite evolutionary trends occurred in the areas which are well represented by the municipalities of Orosei, located in the coastal area, and Irgoli, in the hilly area (Fig. 1):

- along the coast, industrial and tourist developments has brought about a diversification of economic activities to the detriment of agropastoralism;
- in the hilly areas, agropastoral activities have further consolidated their leading role, also evolving towards semi-extensive systems, with the expansion of artificial pastures favoured by local policies for the development of marginal areas.

The difference between the two municipalities is the result of different choices made by

Table 3:　Land cover changes (ha) in the period 1955-1996; the unchanged areas are reported along the diagonal. ("-" = values lower than 5 ha)

a. Irgoli (7 577 ha)

Land cover classes ↓ / 1996	1955 → Total	1.1	1.3	1.2	2.1	2.2	2.3	3.1	3.2	3.3	4.1	4.2	4.3	6.1	6.2	7
Total		18	-	-	232	-	185	2 192	203	2 591	1 274	771	-	78	-	-
1.1 Urban areas	66	18			20		27									
1.3 Road networks	26		-								17	8				
1.2 Quarries	-			-												
2.1 Arable land	365				185		43	27	44	22				45		
2.2 Permanent crops	16					-		7	5							
2.3 Other agricultural areas	399				9		100	241		37				10		
3.1 Natural pastures	1 088						14	448	20	513	62	23				
3.2 Artificial pastures	1 448							604	126	544	33	130				
3.3. Shrublands	2 185							586	8	1 210	87	293				
4.1 Woodlands	1 181							58		27	1 005	91				
4.2 Mediterranean maquis	416							138		21	46	210				
4.3 Forest plantation	332							82		217	24	10				
6.1 Wetlands	22													21		
6.2 Water bodies	-														-	
7 Beach, dune sands	-															·

Table 3: (continue)

b. Orosei municipality (9 030 ha)

Land cover classes 1996 \ 1955	Total	1.1	1.3	1.2	2.1	2.2	2.3	3.1	3.2	3.3	4.1	4.2	4.3	6.1	6.2	7
Total		32	-	-	1 103	-	724	354	1 226	1 257	8	3 221	666	245	49	142
1.1 Urban areas	235	32	-	-	21	-	77	31	-	-	-	7	53	-	-	10
1.3 Road networks	-	-	-	-	-	-	-	-	-	-	-	-	-	-	-	-
1.2 Quarries	58	-	-	-	-	-	-	-	-	45	-	14	-	-	-	-
2.1 Arable land	336	-	-	-	215	-	16	-	22	-	-	37	-	33	13	-
2.2 Permanent crops	102	-	-	-	57	-	-	8	-	37	-	-	-	-	-	-
2.3 Other agricultural areas	1 449	-	-	-	467	-	584	11	-	48	-	315	8	10	-	-
3.1 Natural pastures	1 485	-	-	-	179	-	7	201	332	562	-	192	-	13	-	-
3.2 Artificial pastures	990	-	-	-	73	-	8	44	533	12	-	320	-	-	-	-
3.3. Shrublands	1 370	-	-	-	60	-	25	34	223	146	-	873	7	-	-	-
4.1 Woodlands	34	-	-	-	15	-	-	8	-	-	5	6	-	-	-	-
4.2 Mediterranean maquis	1 757	-	-	-	-	-	-	14	111	370	-	1 207	30	-	-	-
4.3 Forest plantation	876	-	-	-	-	-	-	-	-	19	-	258	566	5	-	23
6.1 Wetlands	164	-	-	-	7	-	-	-	-	-	-	-	-	136	-	14
6.2 Water bodies	73	-	-	-	5	-	-	-	-	-	-	-	-	35	28	-
7 Beach, dune sands	95	-	-	-	-	-	-	-	-	-	-	-	-	-	-	89

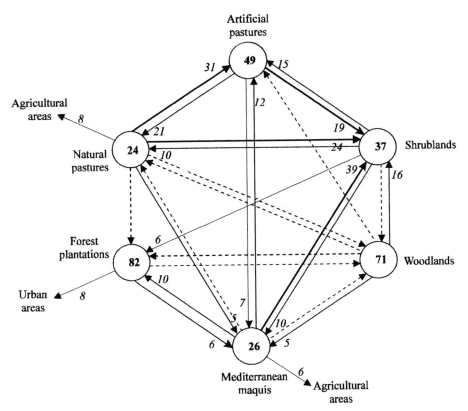

Figure 3. Land cover dynamics, 1955-1996. Values in bold indicate the percentage of unchanged area for each land cover typology; values in italics indicate the percentage of area subjected to changes in the period considered; dotted lines indicate land cover changes involving less than 5% of area.

land owners, as consequences of different development opportunities, supported by local policies.

Along the coastal areas (Orosei municipality), a remarkable tourist development and the related urban expansion (74% of the total increase in the study area) have taken place; the extent of agricultural areas remains almost unchanged (+60 ha, +3.3%) due mainly due to the fact that coastal plains are very fertile and therefore have been cultivated for centuries, but arable lands have been gradually substituted by perennial crops (421 ha).

Pasture lands increased remarkably (+1008 ha, +35.6%), mainly originating from maquis (-1464 ha, -45.3%); this increase can be totally ascribed to natural pastures (+1132 ha, +319%), whereas artificial pastures were subjected to a considerable decrease (-236 ha, -19.2%). The analysis of fluxes shows that pasture lands mainly originate from the abandonment of artificial pastures (net flux 288 ha) and partially from shrublands. In general, production systems seem to evolve towards the utilisation of grasslands in semi-natural conditions. It is worth emphasising that this trend is now reversing, because extensive areas covered with artificial pastures on the basaltic plateau have been created recently.

On hilly areas (Irgoli municipality), pasture areas have undergone a slight decrease (-265 ha, -5.3%), mainly ascribed to the increase in agricultural areas. In these areas the most evident phenomenon is not so much maquis removal but above all the large increase in artificial pastures (+1245 ha, +613.4%).

This analysis showed that many interventions were carried out on unsuitable lands with poor soils and steep slopes by removing the pre-existing natural vegetation (maquis and shrubland). In these areas, field surveys highlighted widespread land degradation processes.

By matching the presence of artificial pastures obtained from the land cover map and suitability classes derived from the model for the evaluation of Land Suitability for the creation of artificial pastures, different degrees of environmental sensitivity to land degradation have been highlighted.

The model was developed according to the methodology proposed by FAO (1976), adapted to the peculiarities of Sardinian agropastoral systems and implemented in the GIS in the course of a three years research program financed by the European Community (Enne *et al.* 1999). The model yields five land suitability classes; suitable (S1), moderately suitable (S2), marginally suitable (S3), currently unsuitable (N1), permanently unsuitable (N2). Only a small part of the current artificial pastures (8%) is classified as unsuitable (N2 + N1) and, as shown in Table 4, can be considered a 'fragile area' (N1) or a 'critical area' (N2). However, about 48% of these areas were classified as S3, and have thus been considered 'potentially sensitive areas' which are severely threatened by land degradation if the current agronomic practices (brush removal and mechanical tillages along the maximum gradient) are repeated over time.

Table 4. Matching of current artificial pastures and suitability classes for the creation of artificial pastures.

Suitability class	Explanation
Highly / Moderately suitable (S1-S2)	*Not threatened by desertification*: current land use can be carried out over time without causing land degradation
Marginally suitable (S3)	*Potentially sensitive areas*: threatened by land degradation if the current agronomic practices are repeated over time
Currently not suitable (N1)	*Fragile areas*: areas undergoing land degradation processes in the short period.
Permanently not suitable (N2)	*Critical areas*: areas quickly undergoing severe land degradation processes (areas already characterised by a low biological potential).

Once more, it is worth emphasising that Irgoli is the municipality with the highest desertification risk: 12% of the total area was classified as unsuitable and 56% as marginally suitable (potentially sensitive areas).

As already highlighted, in this municipality agropastoralism has represented and still represents the most important economic activity. The dynamics of artificial pastures in relation to land suitability are quite interesting (Fig. 4); 54% of the recently created pastures are on marginally suitable lands.

Table 5 also shows the importance of the slope factor (one of the seven factors taken into account by the model; slope, aspect, stoniness, rockiness, vegetation cover, soil depth, soil chemical-physical index): a large part (928 ha, 36.4%) of the interventions are carried out on slopes steeper than 15 %.

The actual presence of land degradation features in sensitive areas was qualitatively assessed during many field excursions. Particular attention was devoted to the evaluation of erosion features (rills and gullies) by taking into account relationships between erosion channels' geometry (length, depth, density) and factors related to morphology (slope, shape and length) and type and date of mechanical interventions. Data on current and past pasture productivity were also considered. Other data referring to soil organic matter content are still being analysed.

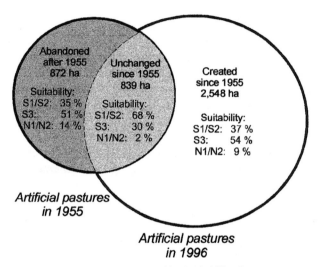

Figure 4. Development of artificial pastures and land suitability classes.

Table 5. Current artificial pastures (ha) by slope class.

Municipality	Slope class		
	<15%	15-25%	>25%
Irgoli	547	360	415
Onifai	638	85	47
Orosei	436	17	4
Total	1 621	462	466

The preliminary results show that in critical and fragile areas, pastures created at least seven to eight years ago always present a high level of degradation and decreasing productivity over recent years. This is particularly evident where slope is the main limiting factor, due to the high erosivity of the climate. In less extreme situations (potentially sensitive areas), the level of degradation is variable. Eighty-five percent of these areas are newly created pastures and most of them are located on Palaeozoic substrata (intrusive and metamorphic; Fig. 1), characterised by hilly morphology and poor soils: here, rill and gully erosion, at various degrees of intensity, can almost always be observed (about 80% of the surface is affected).

5. Conclusions

The research has highlighted the role of agropastoral activities as a major cause of land degradation. The land cover dynamics described are the result of the shift from extensive to semi-extensive livestock production systems, which has altered the landscape of Sardinian rural areas.

This change has been largely favoured by common and regional policies which encouraged agropastoral production by subsidising the extension of the agricultural areas without considering the actual potentialities of the land. The case study presented has highlighted that the lack of technical guidelines supporting the implementation of these policies can result in an intensification of land degradation processes. In particular, the creation of new pastures on unsuitable areas often led to widespread erosive processes, particularly in areas characterised by uneven morphology.

In this context, the analysis of land cover dynamics and land suitability evaluation can allow the identification of areas sensitive to land degradation. This approach can constitute a preliminary tool towards better land management to mitigate land degradation and can be taken into account for the implementation of National Action Plans as provided for by the UN Convention to Combat Desertification.

6. Acknowledgments

The authors have made equal contributions to the present work. Work carried out under EC Medalus III Project, Contract ENV4 – CT95 – 0115 and 40% Fund of the Italian University and Scientific Research Ministry.

7. References

Benedetto, G., Furesi, R. and Nuvoli, F. (1995) La filiera lattiero-casearia, in L. Idda (ed.), *Agroalimentare in Sardegna; struttura, competitività e decisioni imprenditoriali*, PF CNR-RAISA, Sassari, pp.19-124.

CORINE (1992) *Soil erosion risk and important land resources in the southern regions of the European Community*, European Commission, Brussels, EUR 13233 EN.

d'Angelo, M., Enne, G., Madrau, S., Percich, L., Previtali, F. and Zucca, C. (2000) Mitigating land degradation in Mediterranean agro-silvo-pastoral systems: a GIS based approach, *Catena* 40, 37-49.

Enne, G., Pulina, G., d'Angelo, M., Previtali, F., Madrau, S., Caredda, S. and Francesconi, A.H. (in press) Agropastoral activities and land degradation: the case study of Sardinia, in J. Thornes, J. Brandt and N. Geeson (eds.), *Mediterranean desertification - a mosaic of processes and responses*, John Wiley & Sons, Chichester.

Enne, G., d'Angelo, M., Galli, A., Gutierrez, M., Madrau, S., Percich, L., Previtali, F., Pulina, G., Zanolla, C.and Zucca, C. (1999) Land degradation and grazing in Sardinia, in J. Thornes (ed.), *Mediterranean Desertification and Land Use – final report phase III (1996-1999) Contract ENV4 – CT95 – 0115*, Thatcham, UK, pp. 391 – 432.

FAO (1976) *A framework for land evaluation*, Soil Bulletin No. 32, Rome.

IGMI (1993) *Spaziocarta a scala 1:50000; fogli 483 Siniscola, 482 Bitti, 500 Nuoro est, 501 Orosei*, Istituto

Geografico Militare Italiano, Firenze.

ISTAT (1992) *IV Censimento Generale dell'Agricoltura Italiana; Fascicolo provinciale di Nuoro*, Istituto Nazionale di Statistica, Roma.

ISTAT (1996) *Annuario delle Statistiche agricole; Anno 1994*, Istituto Centrale di Statistica, Roma.

Le Lannou, M. (1979) *Pastori e contadini della Sardegna*, Cagliari, Edizione La Torre.

Madrau, S., Loj, G. and Baldaccini, P. (1999) *Modello per la valutazione della attitudine al pascolo dei suoli della Sardegna*, Ente Regionale per lo Sviluppo e l'Assistenza Tecnica in Agricoltura, Cagliari.

Mairota, P., Thornes, J.B. and Geeson, N. (eds.) (1998) *Atlas of Mediterranean environments in Europe: the desertification context*, John Wiley & Sons, Chichester.

Narjisse, H. (1998) Ecological health of Mediterranean rangelands: is grazing livestock the prominent driving force? in V.P. Papanastasis and D. Peter (eds*.), Ecological basis of livestock grazing in Mediterranean Ecosystems*, Proceedings of the International Workshop held in Thessaloniki, 23-25 October 1997, pp. 287-297.

Papanastasis, V.P. (1998) Livestock grazing in Mediterranean ecosystems: an historical and policy perspective, in V.P. Papanastasis and D. Peter (eds*.), Ecological basis of livestock grazing in Mediterranean Ecosystems*, Proccedings of the International Workshop held in Thessaloniki, 23-25 October 1997, pp. 5-10.

Previtali, F. (1996) Soil Degradation Processes and Land Use in North-Western Sardinia (Italy), Paper presented at the International Conference on Land Degradation held in Adana, Turkey, 10-14 June (unpublished).

Seuffert, O. (1992) The project Geodynamik Geoecodynamics in southern Sardinia. The rainfall-runoff-erosion catena: causes, dynamics and effects, *Geokoplus* **III**, 111-128.

Vacca, A., Puddu, R., Tommasi, D. and Usai, D. (1994) Erosion measurement in three areas of Santa Lucia catchment with different land uses, in G. Enne, G. Pulina and A. Aru (eds.), *Land use and soil degradation; MEDALUS in Sardinia*, Università degli Studi di Sassari, pp. 115-127.

AUTHORS

M. D'ANGELO, G. ENNE, S. MADRAU and C. ZUCCA

Nucleo Ricerca Desertificazione – University of Sassari
Via de Nicola, 9
07100 Sassari, Italy

E-mail: nrd@ssmain.uniss.it

Corresponding author: Claudio Zucca

CHAPTER 9

ENVIRONMENTAL PROBLEMS CAUSED BY HEAVY APPLICATION OF NITROGEN FERTILISERS IN JAPANESE TEA FIELDS

TADASHI KATO, XIAOJU WANG and SHINICHI TOKUDA

1. Abstract

Nitrogen fertilisers are widely and heavily used in tea (*Camellia sinensis L.*) fields in Japan, which can result in environmental degradation. The current application level is often over 800 kg N ha^{-1} y^{-1}. Field investigations and plot experiments were carried out in Shizuoka, a major tea growing region in Japan, to assess the environmental problems induced by heavy N fertilisation in tea cultivation. Results showed that long-term tea cultivation caused strong soil acidification. More than 77% of the tea fields investigated had soil pH values lower than 4.0, and more than 60% had pH values below 3.5, with a lowest value of 2.7. Root growth in the top 0-40 cm of inter-row soils was retarded when fertiliser N-input was over 800 kg ha^{-1} y^{-1}. Noticeable water contamination in ponds, rivers and deep wells surrounding the tea fields was observed. The water samples investigated had annual average NO_3-N concentrations of 10-60 mg L^{-1}, and pH values ranging from 3.9 to 7.0 with an average value of 5.6. The water contamination has also caused fish deaths in some ponds. The N_2O production potential (N_2O loss without chemical fertiliser N-input) in the tea fields was as high as 32 kg N ha^{-1} y^{-1}, and the annual N_2O emissions from the tea fields were 32-97 kg N ha^{-1}, increasing with the fertiliser N input rate. It was estimated that the amount of annual N_2O losses from tea fields, which comprise 1.05% of the total area of arable land, was almost equal to that of the total amount lost from all other arable lands in Japan.

2. Introduction

It has been reported that excessive application of nitrogen fertiliser to arable land can widely contaminate water and air, as well as pose a hazard for soils and crops (Bouman *et al.* 1995; Errebhi *et al.* 1998; Mogge *et al.* 1999). The quality of green tea is mainly determined by the content of nitrogenous compounds, especially the free amino acids (Hoshina 1985; Okano and Matsuo 1996). These constituents and the production of new shoots can be increased by the application of chemical fertilisers (Willson and Clifford 1992; Hoshina 1985; Tachibana *et al.* 1995a; Okano and Matsuo 1996). For this reason, nitrogen fertilisers have been increasingly applied in tea (*Camellia sinensis L.*) cultivation in Japan during recent decades. Application rates have been beyond 800 kg N ha^{-1} y^{-1}, and in some cases even above 2500 kg ha^{-1} y^{-1} (Tachibana *et al.* 1995b; Hachinohe 1995; Okano and Matsuo 1996). However, nitrogen absorbed by tea plants in a mature tea field

A.J. Conacher (ed.), Land Degradation, 141–150.

is only about 200 kg ha^{-1} y^{-1}. The large excess of nitrogen can be a serious threat to the environment and even to tea plants.

Furthermore, soil organic matter content tends to increase with the duration of tea growing, because of the large quantity of litterfalls and pruning materials from tea plants as well as the annual application of organic fertilisers by tea growers (Ding and Huang 1992; Wang et al. 1997). Tea fields usually have a thick litter cover and contain large amounts of biomass C (Nioh et al. 1995). Because increasing amounts of nitrogen fertiliser or organic materials cause increasing emissions of greenhouse gases (Bouwman 1990; Mogge et al. 1999), the accumulation of organic matter and heavy fertiliser N-input may also enhance emissions of greenhouse gases.

Environmental problems in tea growing regions have been increasingly noted in Japan in recent years. Kumazawa (1999) observed that groundwater in tea growing regions has a higher nitrate content than in other cropping regions, while Tachibana et al. (1995a) found that tea soils were acidified and root growth was retarded in some tea fields. However, these problems have not been given the attention they deserve. The influence on the environment of the heavy applications of N fertiliser needed to be clarified, in order to improve the current tea growing practices. The objective of this study is to assess some environmental problems induced in tea cultivation in Japan through excessive application of N fertilisers.

3. Materials and Methods

The study was conducted at Makinohara, Shizuoka Prefecture, which is 150 km southwest of Tokyo. Makinohara, covering an area of about 369 km^2, is well known for tea growing. Tea has been grown in this region for more than 100 years, and tea fields cover more than half of the arable land. The climate of the area is humid with an annual precipitation of around 2200 mm and a mean temperature of about 15°C. The soil is mainly a Humic Andosol in the Japanese Soil Classification, with an approximate texture composition of 50% sand, 27% silt and 23% clay.

Seventy tea fields were identified to assess soil acidification. Twenty to 25 samples were collected from each field at a depth of 0-20 cm and mixed together to form a composite sample. Measurements were taken from the composite samples. In order to investigate the influence of nitrogen fertilisation on tea root growth, we selected three adjacent 25-year tea fields with fertilisation rates of 400, 800 and 1100 kg N ha^{-1}y^{-1} respectively. Soil samples were taken in March and December 1998 from 0-40 cm depths in the inter-rows, using a 14 cm diameter corer. White and brown roots in each soil sample were collected as completely as possible and weighed after being carefully washed by water and air dried.

Fourteen drainage canals, ponds and wells adjacent to the tea fields were selected as the water monitoring points. Water samples from these places were collected in the middle of every month from May 1995 to June 1996. In addition, fish populations and water samples were investigated in another 23 permanent ponds, which receive drainage from surrounding tea fields.

NO$_2$ emissions were measured in an experimental tea field plot of the National Institute of Vegetables, Ornamental Plants and Tea from January to December 1998. The experimental plot was set up in a mature tea field in 1996, including five treatments with two replications. The five treatments were: application of N fertiliser at rates of 0, 300, 600, 900 and 1200 kg N ha^{-1} y^{-1}.

NO$_3$-N in water samples was measured using the colorimetric method described by ECAMPN (1990). The pH of soil (1:2.5 soil/water) and water samples was measured using a digital pH meter. Al^{3+} was determined by ICP-AES. Soil particle sizes were determined by the pipette method through dispersion of the sample with NaPO$_3$ after removing organic matter by H$_2$O$_2$ (Black et al. 1965). NO$_2$ emission rates were measured in the duplicate at the surface of inter-row soils once a month, using the closed chamber method (Hutchinson and Livingston 1993). Nitrous oxide in the air phase of the chamber was quantified using gas chromatography.

4. Results and Discussion

4.1. SOIL ACIDIFICATION AND ROOT RETARDATION CAUSED BY HEAVY N APPLICATION

Soil acidification is a major aspect of worldwide soil degradation. Excess soil acidification can cause significantly negative effects on environmental ecosystems including soil, biodiversity, plants and water. Excess use of N fertiliser acidifies soil, and the acidification rate usually increases with increasing N application (Malhi et al. 1991; Wallance 1994; Bouman et al. 1995). Tea soils in Japan generally receive high amounts of N fertiliser. Moreover, the fallen leaves usually accumulate large amounts of aluminium, ranging from 5000 to 30 000 mg kg^{-1} (Philip and Cheruiyot 1989; Ding and Huang 1991). The decomposition of the fallen leaves induces the increase of Al^{3+} in soil, thus leading to soil acidification (Ding and Huang 1991; Wang and Chen 1992).

Our investigation showed that 77% of the 70 tea fields investigated had soil pH values below 4.0, and 60% below 3.5, with a lowest value of 2.7 (Fig. 1). It is noteworthy that most of the soils from orchards, vegetable fields and rice fields in this area have pH values above 6.0. This indicates that long-term tea growing in this region has caused extremely strong soil acidification. Strong acidification of tea soils with heavy use of N fertiliser greatly increases leaching of soluble Ca, Mg, K, Al, Si, NO$_3$$^-$ and SO$_4$$^{2-}$ and some trace elements (Tachibana et al. 1995a; Wang and Chen 1992), which is a serious threat to soil health and local water quality. A large amount of light gray aluminosilicate deposition in the drainage pipes or canals near the tea fields was also observed. The reason for this phenomenon is that the strong soil acidification produces acid drainage with much soluble Al and Si following rainfall or irrigation. When the drainage water runs into rivers and ponds, its pH increases, resulting in aluminosilicate deposition.

Retardation of root growth by heavy N fertilisation was recorded in the three investigated tea fields. White roots in the 0-40 cm layer of inter-row soil were numerous when fertiliser N-input was 400 kg ha^{-1} y^{-1}, but were considerably less when N-input was 800 kg ha^{-1} y^{-1}, and nearly disappeared when N-input was 1100 kg ha^{-1} y^{-1} (Fig. 2). The data support the report of Tachibana et al. (1995a), who found that the high rates of N fertiliser application as currently practiced in some tea fields caused root damage or root death.

4.2 WATER POLLUTION

Water samples collected from the 14 locations in 1995-96 had annual average NO_3-N concentrations ranging from 10 to 60 mg L^{-1} (Fig. 3), all of which exceeded the 10 mg L^{-1} drinking water standard. The pH of the samples ranged between 3.9 and 7.0 with an average value of 5.6, lower than the 6.0 water standard for agricultural use. This indicated that almost all the surface water and ground water in this area had been contaminated by nitrate.

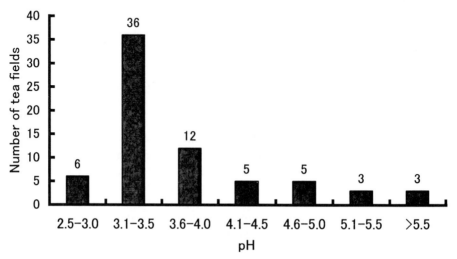

Figure 1. Frequency of soil pH at 0-20 cm in the 70 investigated tea fields

Water contamination has caused the death of some fish in several ponds. Further investigation showed that fish deaths are closely related to high acidity and high concentrations of nitrate and aluminium in the ponds. Figure 4 shows that water with a higher nitrate concentration tends to have a lower pH value. Al^{3+} concentration increased with the decrease of pH. Fish cannot tolerate low pH and high concentrations of Al. Fish deaths occurred with pH below 6.0, and almost no fish were present when pH was below 5.0 (Fig. 4).

    Our findings on water pollution by nitrate concur with the findings of other studies in Japan. In fact, Toda et al. (1997) reported that the nitrate content of a river in the same tea-growing area was 19-33 mg L^{-1} without significant variation during the year. Working in the Iwatahara tea-growing region about 40 km southwest of Makinohara, Kumazawa (1999) reported that the nitrate concentration in a deep well increased from 1 mg L^{-1} in 1971 through 10 mg L^{-1} in 1983 to 25 mg L^{-1} in 1988.

    The serious water contamination in the tea-growing regions is due to the high nitrate leaching in tea fields, which is induced by heavy inputs of nitrogen fertilisers. For a specific soil and land use, the amount of NO_3-N leached usually increases linearly as the proportion of applied N increases. Watanabe (1986) studied nitrate leaching in five types of tea soils under different N fertilisation rates (300-1000 kg N ha^{-1} y^{-1}) for three years, and found that

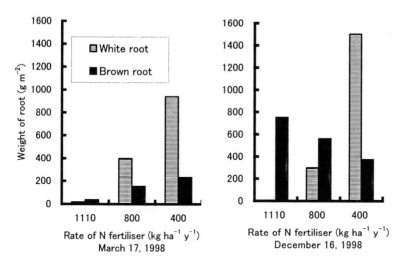

Figure 2. The amounts of white (alive) and brown (dead) roots in the three tea fields with different N fertilisation rates

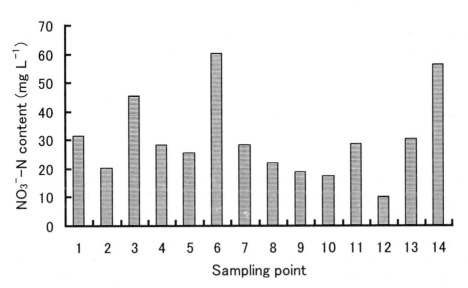

Figure 3. Annual average concentration of NO₃-N in water samples from ponds, wells or drainage pipes in the tea growing region

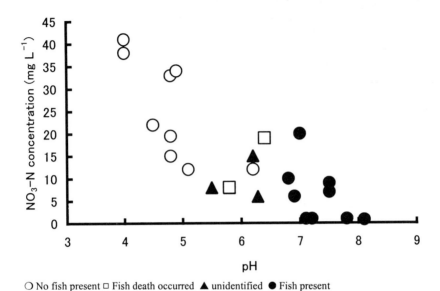

○ No fish present □ Fish death occurred ▲ unidentified ● Fish present

Figure 4. Influence of NO₃-N concentration and pH of pond water on fish growth in the ponds near tea fields

the average amount of NO_3-N leached at 1 m was 421 kg ha^{-1} y^{-1}. Toda *et al.* (1997) reported an average nitrate leaching rate of 244 kg N ha^{-1} y^{-1} from tea fields in a watershed in Shizuoka, and a rate of 547 kg N ha^{-1} y^{-1} in a tea field receiving a N-input of 1080 kg ha^{-1} y^{-1}. Tokunaga *et al.* (1996) estimated a nitrate leaching rate of 450 kg ha^{-1} y^{-1} in a tea field with N-input of 900 kg ha^{-1} y^{-1}. The above reported values of nitrate leaching were much higher than those reported for other crop fields. Indeed, Matushita *et al.* (1969) reported that nitrate leaching in soils with a rotation of potato and wheat averaged 40.3 kg ha^{-1} y^{-1}, while Kikou and Yuita (1991) reported that soil water in tea fields contained a NO_3-N volume 6-200 times higher than in the adjacent forest fields.

4.3. NITROUS OXIDE EMISSIONS

N_2O contributes to global warming and stratospheric ozone layer destruction (Corre *et al.* 1996). It has been estimated that agriculture contributes 65-80% of the total anthropogenic N_2O (Iserman 1994; Thornton *et al.* 1996). It is, therefore, interesting to assess N_2O production originating from agricultural practices.

   The present study reveals that N_2O loss increased with increased fertiliser N-input (Table 1). N_2O loss from inter-row soil in the plots with N-inputs of 1200 kg ha^{-1} y^{-1} was about six times larger than that in the plots with N-inputs of 300 kg ha^{-1} y^{-1}, and about two times that in the plots with N-inputs of 600-900 kg N ha^{-1} y^{-1} (Table 1). However, a loss of 32.28 kg N ha^{-1} y^{-1} was detected in the plot without fertiliser N-input.

In tea fields, inter-row soil accounts for approximately one third of the total field area, while the row soil accounts for two thirds. Fertilisers are usually not applied in the row soil. If the row soil had the same N_2O fluxes as the plot receiving no fertiliser N, then it may be estimated that the annual N_2O emissions from the tea fields would be 32-97 kg N ha[-1], increasing with fertiliser N-input (Table 1).

Table 1. N_2O emissions from tea soils with different chemical fertiliser inputs

Fertiliser N total fertiliser N (kg N ha[-1] y[-1])	Annual average N_2O emission rate in inter-row soil (mg N m[-2] h[-1])	Annual N_2O loss from tea fields (kg N ha[-1])	Emission against input (%)
0	0.37	32.28	0
300	0.42	33.70	0.47
600	1.42	62.90	5.10
900	1.55	66.78	3.83
1200	2.57	96.64	5.36
LSD (0.05)*	0.69	19.63	2.67

* Critical value for comparison in the least significant difference (LSD) test at 95% significance level

Although N_2O emissions in agricultural soils were found to increase with the amounts of fertiliser nitrogen applied, the emission rate in crop fields usually ranges from 0.1 to 9.0 kg N ha[-1] y[-1] (Eichner 1990; Thornton et al. 1996; JSSS 1996; Kaiser et al. 1998; Mogge et al. 1999). Our study showed that tea fields had a much greater N_2O loss. This finding supports the data of Tokunaga (1996) and collective work of JSSS (1996). Tokunaga (1996) reported N_2O emissions of 116-134 kg N ha[-1] y[-1] from the inter-row soil of the tea fields and 39-45 kg N ha[-1] y[-1] from the whole fields. JSSS (1996) reported that annual N_2O emissions from fields of wheat, corn, vegetables and fruit averaged 0.05-0.4 mg N m[-2] h[-1], while those from tea fields were 2.0-8.5 mg N m[-2] h[-1]. According to our data and that from JSSS (1996), it was estimated that the amount of annual N_2O loss from Japanese tea fields, covering 1.05% of the total area of arable land, was almost equal to the total amount lost from all other arable land (Fig. 5).

The large amount of N_2O loss in tea fields was mainly due to heavy nitrogen fertiliser applications and high N_2O production potential. The 0.5-5.4% loss of applied fertiliser N in the present study (Table 1) was similar to the 4.5-5.0% loss of applied N as reported by Tokunaga (1996), which showed that the N_2O loss rate of applied fertiliser N in tea soil is not high compared with that in other arable soils as reported in other countries (Thornton 1996; Kaiser et al. 1998). It was the much heavier fertiliser N input which resulted in a larger N_2O loss than that from other crop soils.

The noticeable N_2O losses in the plot without fertiliser N-input reflect a high N_2O production potential in the tea soils investigated, which is another reason for the big N_2O

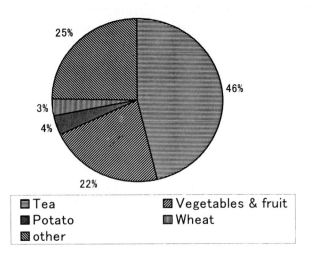

Figure 5. Comparative N_2O losses from different land uses in Japan.

loss in tea soil. Kaiser *et al*. (1998) also reported that the linear reduction of N-fertiliser in arable soil did not result in a linear decrease in N_2O losses because of the high N-mineralisation potential of the soil. The high N_2O production potential might be caused by the large organic matter accumulation. Tea soil usually receives large amounts of organic materials from litterfalls, pruning materials and organic fertilisation (Karasuyama *et al*. 1983; Wang *et al*. 1997). Tea pruning can yield about 30 000 kg N ha^{-1} y^{-1} of organic residues or 500 kg N ha^{-1} y^{-1} of N in mature tea fields. This may explain the high N_2O loss potential occurring in tea fields.

5. Conclusions

The heavy application of nitrogen fertilisers in tea fields in Japan has seriously threatened environmental quality. The soil in more than 77% of the investigated tea fields (n = 70) was acidified to pH values below 4.0. The strong soil acidification induced high losses of soluble nutrients. Heavy nitrogen fertilisation rates also retarded root growth of tea plants. Waters in ponds, rivers and deep wells were contaminated by the nitrate from tea fields, resulting in fish deaths in some ponds. N_2O losses from tea fields were much higher than those from other arable fields. The high N_2O losses were due to the high N_2O production potential and heavy N fertiliser input in tea fields. The study primarily reveals serious problems in the current management of nitrogen fertiliser in tea cultivation in Japan. There is an urgent need to encourage further research into these problems, especially on greenhouse gas emissions and water pollution induced by tea cultivation, and to take appropriate corresponding measures to get this problem under control.

6. Acknowledgments

We thank Ikuo Watanabe for his contribution in this study, and Chikayo Sakurai and Mutsuko Ikeya for their assistance in field investigation and laboratory analysis. We are also grateful for the suggestions and comments from the reviewers and the editor, Dr. Arthur Conacher, which improved the manuscript.

7. References

Black, C.A., Evans, D.D., Ensminger, L.E., White, J.L. and Clark, F.E. (1965) *Methods of Soil Analysis Part 1: Physical and Mineralogical Properties, Including Statistics of Measurement and Sampling*, American Society of Agronomy, Wisconsin, USA.

Bouwman, A.F. (1990) Exchange of greenhouse gases between terrestrial ecosystem and the atmosphere, in A.F. Bouwman (ed.), *Soil and the Greenhouse Effects*, John Wiley and Sons Ltd., Chichester, pp. 61-127.

Bouman, O.T., Curtin, D., Campbell, C.A., Biederbeck, V.O. and Ukrainetz, H. (1995) Soil acidification from long-term use of anhydrous ammonia and urea, *Soil Science Society of American Journal* **59**, 1488-1494.

Corre, M. D., Van Kessel, C. and Pennock, D.J. (1996) Landscape and seasonal patterns of nitrous oxide emission in a semiarid region, *Soil Science Society of America Journal* **60**, 1806-1815.

Ding, R. and Huang, X. (1991) Biogeochemical cycle of aluminum and fluorine in tea garden soil systems and its relationship to soil acidification, *Acta Pedology Sinica* **28**, 229-236.

Eichner, M.J. (1990) Nitrous oxide emissions from fertilized soils: summary of available data, *Journal of Environmental Quality* **19**, 272-280.

Errebhi, M., Rosen, C.J., Gupta, S.C. and Birong D.E. (1998) Potato yield response and nitrate leaching as influenced by nitrogen management, *Agronomy Journal* **90**, 10-15.

Hachinohe, M. (1995) Tea cultivation techniques based on low input of farm chemicals for conservation of environment and sustainable tea production, *Research Journal of Tea* **18**, 29-35 (in Japanese with English summary).

Hoshina, T. (1985) Studies on absorption and utilization of fertiliser nitrogen by tea plants, *Bulletin of the National Research Institute of Vegetables, Ornamental Plants and Tea* **20**, 1-89 (in Japanese with English summary).

Hutchinson,G.L. and Livingston, G.P. (1993) Use of chamber systems to measure trace gas fluxes, in ASA, CSSA and SSSA (eds.), *Agricultural Ecosystem Effects on Trace Gases and Global Climate Change*, Madison, Wisconsin, pp. 63-78.

Iserman, K. (1994) Agriculture's share in the emission of trace gases affecting the climate and some cause oriented proposals for sufficiently reducing this share, *Environmental Pollution* **83**, 1-21.

JSSS (Japan Soil Science Society). (1996) *Promotion Program for Environmental Protecting Soil Management: Report of the Dynamics of Greenhouse Gas Emissions from Soil*, Sato Press, Mito, Japan (in Japanese).

Kaiser, E.-A., Kohrs, K., Kocke, M., Schnug, E., Heinemeyer, O. and Munch, J.C. (1998) Nitrous oxide release from arable soil: importance of N-fertilization, crops and temporal variation, *Soil Biology and Biochemistry* **30**, 1553-1563.

Karasuyama, M., Matsumoto, J. and Fujishima, T. (1983) Correlation between the yields, quality of tea leaves and the amount of available nitrogen in tea field soils supplied with farmyard manure, *Tea Research Journal* **58**, 20-27 (in Japanese with English Summary).

Kikou, N. and Yuita, K. (1991) Vertical distribution of nitrate nitrogen in soil water under tea gardens and the adjacent forests, *Japanese Journal of Soil Science and Plant Nutrition* **62**, 156-164 (in Japanese with English summary).

Kosuge, N. (1992) Some aspects of pH in tea soil, *Study of Tea* **62**, 1-7 (in Japanese with English summary).

Kumazawa, K. (1999) Present state of nitrate pollution in groundwater, *Japanese Journal of Soil Science and Plant Nutrition* **70**, 207-213 (in Japanese with English Summary).

Malhi, S.S., Myborg, M., Harapiak, J.T. and Flore, N.A. (1991) Acidification of soil in Alberta by nitrogen fertilisers applied to bromegrass, in Wright, R.J., Baliga, V.C. and Murrmann, R.P. (eds.) *Plant-Soil Interactions at Low pH, Proceedings of the Second International Symposium on Plant-Soil Interactions at Low pH, 24-29 June 1990, USA*, Kluwer Academic Publishers, Dordrecht, pp. 547-553.

Matushita. K., Fujishima T. and Udagawa, Y. (1969) Productivity and material leaching of upland volcanosol - lysimeter experiment (1): the amount of drainage and material leached, *Japanese Journal of Soil Science and Plant Nutrition* **40**, 337-343 (in Japanese with English Summary).

Mogge, B., Kaiser, E.A. and Munch, J.C. (1999) Nitrous oxide emissions and denitrification N-losses from agricultural soils in the Bornthoved Lake region: influence of organic fertilisers and land-use, *Soil Biology Biochemistry* **31**, 1245-1252.

Nioh, I., Isobe, T. and Osada, M. (1993) Microbial biomass and some biochemical characteristics of a strongly acid tea field soil, *Soil Science and Plant Nutrition* **39**, 617-626.

Okano, K. and Matsuo, K. (1996) Seasonal changes in uptake, distribution and redistribution of ^{15}N-nitrogen in young tea (*Camellia sinensis* L.) plants, *Japanese Journal of Crop Science* **65**, 707-713.

Philip, O.O. and Cheruiyot, D.K.A. (1989) Effects of nitrogen fertilisers on the aluminum contents of mature soils and extractable aluminum in the soil, *Plant and Soil* **119**, 342-345.

ECAMPN (Editorial Committee for Analysis Methods of Plant Nutrition) (1990) *Analysis Methods of Plant Nutrition*, Hakatomo Pressing House, Tokyo, Japan (in Japanese).

Tachibana, N., S. Yoshikawa, and K. Ikeda. (1995a) Behavior of inorganic nitrogen in tea field soils heavily applied with nitrogen and changes in amino acid content of the first crop accompanying the reduction of nitrogen, *Japanese Journal of Crop Science* **64**, 523-528 (in Japanese with English summary).

Tachibana, N., S. Yoshikawa, and K. Ikeda. (1995b) Influences of heavy application of nitrogen on soil acidification and root growth in tea fields, *Japanese Journal of Crop Science* **64**, 516-522 (in Japanese with English summary).

Thornton, F.C., Bock, B.R. and Tyler, D.D. (1996) Soil emissions of nitric oxide and nitrous oxide from injected anhydrous ammonium and urea, *Journal of Environmental Quality* **25**, 1378-1384.

Toda, H., Mochizuki, Y., Kawanushi, T. and Kawashima, H. (1997) Estimattion of reduction in nitrogen load by tea and paddy field land system in makinohara area of Shizuoka, *Japanese Journal of Soil Science and Plant Nutrition* **68**, 369-375 (in Japanese with English summary).

Tokunaga, T., Tanizaki T., Kimura Y., and Fukuda, K. (1996) Emission of nitrous oxide from tea field soil, *Bulletin of Yamaguchi Agricultural Experiment Station* **47**, 59-66 (in Japanese).

Wallance, A. (1994) Soil acidification from use of too much fertiliser, *Communication of Soil Science and Plant Analysis* **25**, 87-92.

Wang, X.J. and Chen, H.Z. (1992) An important problem of soil degradation in tea fields: soil acidification, in Gong, Z.T. (ed.), *Environment Change of Soil*, Science and Technology Publishing House of China, Beijing, pp. 214-216.

Wang, X.J., Hu, X.F. and Chen, H. (1997) Some biogeochemical characteristics of tea soil, *Pedosphere* **7**, 275-280.

Watanabe, N. (1986) The leaching of fertiliser elements from the soil of tea fields and the growth of tea plants: 1. On the leaching of inorganic nitrogen, *Bulletin of Kanagawa Horticultural Experiment Station*, **33**, 54-64 (in Japanese with English summary).

Willson, K.C. and Clifford, M.N. (1992) *Tea*, Chapman & Hall, London.

AUTHORS

TADASHI KATO, XIAOJU WANG* and SHINICHI TOKUDA

*Corresponding author: Center for Environmental Science in Saitama, 914 Kamitanadare, Kisaimachi, Saitama 347-0115, Japan.

Tel: 81-480-73-8374; Fax: 81-480-70-2031

k-oh@kankyou.pref.saitama.jp

Tadashi Kato: National Research Institute of Vegetables, Ornamental Plants and Tea, Shizuoka 428-8501, Japan. Tel: 81-547-45-4924; Fax: 81-547-46-2169; Email: k-oh@kankyou.pref.saitama.jp

Shinichi Tokuda: National Research Institute of Vegetables, Ornamental Plants and Tea, Shizuoka 428-8501, Japan. Tel: 81-547-45-4924. Email: sytoku@tea.affrc.go.jp

Part Three

Society and Land Degradation

Part three extends the discussion to look at some of the social aspects of land degradation. Clearly only a few aspects can be considered by the five papers in this Part. But they serve to emphasise the point that land degradation is not solely or even primarily a technical, biophysical problem; it has important social attributes. Monitoring - how it should be done and by whom - is also a common theme amongst four of the five papers.

The papers fall into three groups. The first two, by Helen Watson, and Fox and Rowntree, examine the relationships between attributes of the biophysical environment and land reform in South Africa. It is encouraging to see this acute problem engaging the attention of researchers, although somewhat depressing when the scale of the issue is considered. As Helen Watson points out, when the Mandela government came into power, some 80% of the population had access to only 13% of the land mass - and an even smaller proportion of productive agricultural land. Her research concerns the sustainability of the land being transferred from white to African ownership, and the means whereby environmental audits can be carried out quickly, effectively and inexpensively. The beneficiaries' perceptions of the land are particularly important in undertaking this assessment. Fox and Rowntree consider the same issue, assessing the constraints (primarily the limited availability of suitable land) on the land reform program in the eastern Cape Province. Most of the land with the highest agricultural potential is already heavily settled. They also comment on the potential for further land degradation if population densities increase in some of these areas.

The next two papers, both from Australia, also discuss farmers' perceptions. Lisa Lobry de Bruyn argues convincingly that for land monitoring to succeed, the farmers who have to implement modified management practices must understand the process and, more importantly, the process must incorporate the ways in which farmers understand and evaluate the health of their land, and not simply attempt to impose the scientists' criteria. Sally Marsh, Michael Burton and David Pannell also evaluate farmer monitoring, this time in relation to one indicator (depth to groundwater) of a particular problem (secondary salinity). They show that farmer monitoring of bores tends to decrease over time. Although this decline in involvement is related to a number of factors, it *could* be suggested that their findings tend to support de Bruyn's argument.

The final paper moves to South East Asia. Based on his extensive experience in the region over many years, Ian Douglas discusses the numerous deficiencies in monitoring by the authorities of various aspects of land degradation, and suggests some of the reasons for this. As he points out, effective management and policy making must rely on good data. Again it might be suggested that de Bruyn's argument is supported; but in this instance the emphasis is placed on improving the science rather than handing over the responsibility for data gathering to the layperson. Serious land degradation is continuing throughout SE Asia in the absence of good data and effective preventative action.

CHAPTER 10

SOIL SUSTAINABILITY AND LAND REFORM IN SOUTH AFRICA

HELEN KERR WATSON

1. Abstract

When South Africa's present government came to power, Africans who comprised about 80% of the population had access to 13% of the country's total land area and to an even smaller proportion of land with high agricultural potential. To bring about a more equitable distribution of land the government embarked on a Land Reform Programme. A number of constraints prevent 'good agricultural potential' from being employed as a prerequisite in this Programme when designating white-owned farms for transfer to peasant farmers, and there is therefore concern that the transferred farms will not be sustainable. The government commissioned the development of methodologies to monitor various aspects of the transferred land.

The first stage of this development involved carrying out full, detailed environmental audits of four land transfers in KwaZulu Natal. The four areas differed from one another in respect of biophysiographic characteristics, spatial extent and land-use histories. The influence of land-use practices over the two decades prior to transfer, the areas' inherent susceptibility to erosion and agricultural potential, the post-transfer status, and the beneficiaries' perceptions of the areas' water, soil, vegetation and animal resources, were assessed.

This paper describes the procedure used to assess sustainability of the soil resource in the four case studies. A comparative analysis of the four audits enabled 14 key indicators of (i) the soil's potential susceptibility to erosion, (ii) contemporary erosion status, (iii) agricultural potential, and (iv) impact of land-use practices, to be identified. The indicators were rated in terms of their potential influence on sustainability, and a composite measure representing the overall sustainability of the soil was derived. In addition to providing a reliable assessment of soil sustainability, the procedure proved to be simple, quick and cheap to use and is intended to provide the basis for a nationally applicable methodology for monitoring soil sustainability in land transfers.

2. Introduction

In 1994 South Africa's present government inherited a system characterised by gross inequalities in the distribution of resources according to race. The progressive loss of land during the country's colonial and apartheid past entrenched poverty and insecurity in its majority African population. In order to redress this situation the government has embarked on a Land Reform Programme (LRP) which has three components: (i) restitution

A.J. Conacher (ed.), Land Degradation, 153–166.

- compensating for or returning land to those dispossessed of it; (ii) redistribution - providing financial assistance to the poor to enable them to purchase land from 'willing sellers', and (iii) facilitating security of tenure. Most of the beneficiaries of the LRP are rural peasant communities who will primarily use the land for subsistence cultivation and pastoralism. Small-scale commercial enterprises are being actively encouraged. As well as bringing about a more equitable distribution of land and alleviating poverty, the LRP aims to ensure that the transferred land is environmentally sustainable and that the transfer process does not intensify or extend environmental degradation (Department of Land Affairs (DLA) 1997). In addition to the urgent need to deliver on the government's election promises and to curb land invasions, a number of other constraints prevent 'good agricultural potential' from being employed as a prerequisite when designating land for redistribution (Turner 1997; Watson 1997, 1998a). In view of this and the fact that mechanisms to regulate land use practices after transfer still have to be formulated, there is concern that the Programme will contribute to scarce, high agricultural potential land being used unsustainably, increasing the severity and extent of land degradation and consequently failing to enable its beneficiaries to improve the quality of their lives (Watson and Ramagopa 1997).

The apartheid government, motivated by the need to curb excessive soil loss and enhance rural livelihoods, implemented the 'Betterment Scheme' which involved resettling peasant farmers. McAllister (1988, 1989), Watson (1996) and others have shown that the scheme caused detrimental environmental and socio-economic consequences. Mindful of the concern noted above and the previous government's failure, and recognising that it would be too expensive and time consuming to conduct environmental impact assessments of each land reform transfer in the country, the DLA commissioned the development of methodologies to monitor various aspects of these areas such as quality of life, sustainability of environmental resources and compliance with development plans. This paper describes a methodological approach developed to monitor the sustainability of the soil resource of land transfer areas, using four case studies.

3. Description of Study Areas

Figure 1 shows the location in the province of KwaZulu Natal of the four previously white-owned farms used in this study; Tembitshe Buthelezi, Misgunst, Labuskagneskraal and Nomoya. Tembitshe Buthelezi is located within a deeply incised river valley in the Mfolozi catchment. Its 900 ha area is dominated by a scarp. It consequently has a large range in altitude (450 m) and steep slopes (81% of the area have slope gradients >10°). It is predominantly underlain by shale and sandstone and covered by well drained sandy clay loam soils which, with a mean erodibility index value (MK) of 0.38, are moderately erodible (Department of Agricultural Technical Services 1976). Its mean annual rainfall is 1136 mm. The rainfall's mean annual erosivity (MAEI$_{30}$) of 400, is moderate (Smithen and Schulze 1982): refer Section 4.1. The vegetation of Tembitshe Buthelezi is predominantly 'sourveld' - a savanna grassland that is unpalatable during the winter dry season as a consequence of the high degree of leaching of the soil in this 'Moist Upland' bioclimatic region (Phillips 1973). Most of the beneficiaries previously lived on land neighbouring Tembitshe Buthelezi. About half of the adults are employed on white

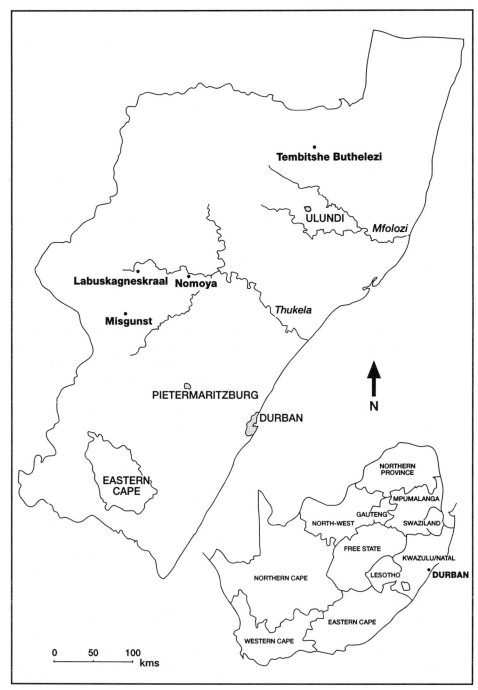

Figure 1. Locations of the four land reform transfers employed in this study: Tembitshe Buthelezi, Misgunst, Labuskagneskraal and Nomoya.

commercial farms, timber and tea estates in the vicinity, or on mines out of the district.

Misgunst is situated along a small tributary in the Thukela catchment. Its topography comprises a dolerite outcrop and an extensive predominantly shale pediment. Although its altitudinal range is only 100 m, 94% of its 148 ha area is steeper than 10°. Its soil is mainly of fine sand to clay texture, poorly drained and highly erodible (MK >6.0). Misgunst has a mean annual rainfall of 778 mm, the average erosivity of which is high (MAEI$_{30}$ = 550). Savanna grassland with scattered shrubs and trees cover the area. As it is located in the 'Dry Upland' bioclimatic region, less bases are leached from its soil and the grass is palatable virtually throughout the year. Prior to transfer the beneficiaries lived on the farm as labourers. Although they still live there, two thirds of the adults are now employed in a nearby town.

Labuskagneskraal is situated on an extensive gentle slope beside the Thukela River. Its altitudinal range is only 60 m and 67% of its 927 ha area has a slope of less than five degrees. Its predominant lithology, soil, climate and vegetation characteristics are virtually the same as those described above for Misgunst. The beneficiaries generally lack farming skills because they came from communities where their subsistence was not dependent on crops and/or livestock. About 80% of the adult males are employed in nearby towns.

Nomoya is also situated beside the Thukela River but in the drier 'Riverine and Interior Lowland' bioclimatic region. Its topography is dominated by a dolerite ridge and consequently has a large altitudinal range (360 m). The fall face of the ridge accounts for 59% of the 908 ha area and is characterised by slopes steeper than 10°, with abundant stone and very shallow, intermittent soil cover. The mid slope occupies the area's central portion. It is underlain by shale and inclined between 5°and 10°. It wanes into a gentle colluvium-covered pediment down to the river. The soil, where present, is predominantly sandy clay loam with a moderate drainage and low erodibility (MK = 0.25). Nomoya receives 700 mm of rainfall per annum on average, the erosivity of which is usually very high (MAEI$_{30}$ = 600). The area is covered with dense shrub thicket under which only sparse pioneer grasses are found. Before 1969, when they were evicted, the beneficiaries grazed livestock and grew crops on Nomoya. Since returning, their subsistence is dependent on these activities as most are unemployed.

4. Methodology

Key indicators of the soil's: (i) potential susceptibility to erosion; (ii) contemporary erosion status; (iii) agricultural potential, and (iv) potential response to land use practices, were identified. A value of 5, 4, 3, 2 or 1 was assigned to each of these indicators based on whether they potentially have a very good, good, moderate, bad or very bad influence on the soil's sustainability. These values were then integrated to provide a composite measure of the soil's sustainability. The rationale for this approach is that changes in this composite measure over time can provide an early warning of deterioration enabling remedial measures to be implemented. Ultimately, it can be ascertained whether the change in land use associated with land reform has had a significant positive or negative effect on the soil's sustainability.

4.1. POTENTIAL SUSCEPTIBILITY TO EROSION

The potential susceptibility of the soils in the study areas to erosion was evaluated by integrating the influence of slope, erodibility, erosivity and cover. Topocadastral maps at a scale of 1:50 000 were used to ascertain the percentage of the study area falling within the five slope angle classes shown in Table 1. This measure for each class was multiplied by the sustainability value assigned to the class (refer Table 1). The resultant scores were summed and divided by 100 to yield the composite measure of the influence of slope shown in Table 4. Watson (1991) provides the basis for the slope angle classes employed and for the sustainability values assigned to them. Her work in African land use areas revealed a strong positive correlation between slope angle and the intensity and spatial extent of erosion for slopes steeper than 5°. However, erosion was better represented on very gentle slopes of less than 2° than on gentle slopes in the 2° to 5° range.

Information on the spatial extent of the different soil types present in the study areas was gleaned from the Land Type Survey Staff (1986) map and associated memoir and verified in the field. In an intensive countrywide program, the Department of Agriculture correlated the Universal Soil Loss Equation's (USLE) erodibility or K value calculated from measured soil losses from standard, simulated rainfall runoff plots, with equivalent values obtained using a wide range of erodibility indices. They concluded that Wischmeier et al. 's (1971) erodibility nomograph was the most reliable, and prescribed the erodibility

Table 1: System used to classify the influence on soil sustainability of the erosion potential indicators.

INDICATOR	UNIT	VERY HIGH 5	HIGH 4	MOD. 3	LOW 2	VERY LOW 1
Slope Angle	Degrees	2 - 5	< 2	5,1 - 10	10,1 - 20	> 20
Erodibility	MK	< 0,13	0,13 - 0,25	0,26 - 0,50	0,51 - 0,70	> 0,71
Erosivity	$MAEI_{30}$	< 251	251 - 350	351 - 450	451 - 550	> 550
Cover	Percentage	< 10	11 - 25	26 - 50	51 - 75	> 75

classes shown in Table 1 developed by Crosby et al. (1983) for general use in South Africa. The average K values of the three most prevalent soils present in the four study areas obtained from Department of Agricultural Technical Services (1976) were derived using the nomograph. Their K values were assigned a sustainability value (Table 1). The sustainability values of the most and second most common soils were multiplied by a weighting of 3 and 2, respectively. The resultant values were summed and divided by six to yield the composite measure of the influence of erodibility shown in Table 4.

The study areas' overall annual rainfall erosivity values were extrapolated from Smithen and Schulze's (1982) iso-erodent map which is officially accepted as the country's most reliable source of information on rainfall erosivity (Crosby et al. 1983). The map is based on the USLE's index of a storm's total kinetic energy and its maximum 30 minute intensity or EI_{30} and is calibrated for South African conditions. The sustainability values assigned to the rainfall erosivities shown in Table 1 were carried forward as the composite measure of the influence of erosivity shown in Table 4.

Combining information gleaned from 1:10 000 orthophoto maps and from interviewing the beneficiaries, representative parts of the rangelands, arable lands and degraded/sheet eroded lands were identified. Three transects were positioned through each. At ten equidistant intervals along each transect a 1 m² grid frame was placed on the ground and the proportion within it covered by rooted basal components of plants, litter and stones was estimated. The length of the transects, the distance between them, and interval between the ten sampling points were all dependent on the total area of these three types of surfaces within the study areas. Watson (1991) reviewed natural and simulated rainfall runoff plot data from a large number of studies conducted throughout the eastern summer rainfall region of the country. These data were used to calibrate the relationship between percentage cover and soil loss described by Stocking (1994), producing the cover classes shown in Table 1. The class represented at each of the 30 sample points in each of the three land surface categories was assigned a sustainability value. The sum of the average value for each category divided by three, yielded the composite measure of the influence of cover shown in Table 4.

4.2. CONTEMPORARY EROSION STATUS

The study areas' contemporary erosion status were established by integrating information on the areal extent and intensity of erosional activity.

4.2.1. Areal Extent

Gullies and degraded/sheet eroded surfaces within the study areas were identified and delineated on the most recent 1:10 000 orthophoto maps. The percentage of each study area affected by these two types of erosion was measured and assigned a sustainability value based on the system developed by Whitlow (1988) in Zimbabwe shown in Table 2. As gullies are likely to be more significant sediment sources over the long term (Watson 1991) and pose the bigger problem to peasant farmers (Garland *et al.* 1994; Brinkcate and Hanvey 1996), their value was weighted by a factor of two to yield the measure shown in Table 4.

Table 2: System used to classify the influence on soil sustainability of the areal extent of erosion.

DESCRIPTION	% OF AREA	RATING	VALUE
Very localised	< 4	very good	5
Localised	4,1 - 8	good	4
Moderate	8,1 - 12	moderate	3
Extensive	12,1 - 16	low	2
Very extensive	> 16	very low	1

4.2.2. Intensity of erosional activity

Within each of the thirty 1 m² sample areas in the rangelands, arable lands and degraded/sheet eroded lands described above, evidence of the indicators of erosional activity listed in Table 3 was recorded. This evidence was also recorded at ten sample areas along transects through three gullies chosen with the aid of orthophoto maps as being representative of those in the study area. The sustainability values assigned to the erosion activity indicators (Table 3) are based on the stages in the development of erosion features described by Morgan (1986) and Dardis *et al.* (1988). Where evidence of an indicator was obvious and dominating, its sustainability value was weighted by a factor of two. Where more than one indicator was present in the sample area, the sustainability values assigned

Table 3: Sustainability values assigned to indicators of the intensity of erosional activity on the surface categories.

INDICATORS	RADSE	GULLIES	ROADS
Muddied grass blades/leaves bent downslope	5	5	5
Splays of coarse material and litter orientated downslope	5	5	5
A slightly higher soil surface on the upslope sides of stones and roots	4	-	4
Sealing, compacting and/or crusting of the soil surface	4	4	4
Soil pedestals capped by stones and/or protected by plants	3	4	-
Exposed plant roots	3	4	-
Deposits of dust	-	-	3
Waterlogging, skid marks and/or alternative tyre tracks	-	-	3
Rills (within or parallel to road)	2	-	2
Gullies (within or parallel to road)	1	-	1
Undercutting of head and slumping	-	3	-
Channel bank scouring and slumping	-	3	-
Subsurface piping	-	2	-
Lack of vegetation in gully channel and interchannel ridges	-	2	-
Cutting into and exposed bedrock	-	1	1

- represents evidence that is so seldom applicable that it is disregarded.

RADSE is abbreviation for Range, Arable and Degraded/Sheet Eroded.

to each were summed. The average sustainability value for each of these four surface categories was carried forward as the composite measure of the intensity of erosional activity on them shown in Table 4.

The lengths of the roads from the farm entrances to the locations of most of the dwellings were measured and allocated to a class of <1, 1 - 2, >2 - 4, >4 - 6 or >6 km and assigned a value respectively of 5, 4, 3, 2 or 1. At ten equidistant transects across the roads their positions relative to the slope profile, its shape and evidence of erosional activity as listed in Table 3 were recorded. The topographic position was categorised as hilltop, middle slope, river valley, upper slope and lower slope and assigned a value of 5, 4, 3, 2 and 1 respectively. The shape was categorised as flat, convex, middle ridge, slope across road and concave, and assigned a value of 5, 4, 3, 2 and 1 respectively. Table 3 shows the

Table 4: Values assigned to the study areas' key indicators of soil sustainability.

INDICATORS	Tembitshe Buthelezi	Misgunst	Labuskagneskraal	Nomoya
SUSCEPTIBILITY				
Slope Angle	2	2	4	3
Erodibility	3	2	3	4
Erosivity	3	2	2	1
Cover	1	5	4	3
STATUS				
Areal Extent				
Gully	4	5	5	4
Sheet	4	5	5	4
Intensity				
Degraded/sheet eroded	2	3	3	2
areas	3	5	5	3
Rangelands	3	3	4	2
Arable lands	1	4	2	1
Gullies	1	4	3	1
Roads				
AGRIC. POTENTIAL				
Scientific	3	2	4	3
Perceived	2	4	3	1
LANDUSE				
Arable lands	2	2	4	3
Rangelands	2	4	3	1
Overall	**2.4**	**3.5**	**3.6**	**2.4**

values assigned to the indicators of erosion intensity. The values assigned to the four data inputs are based on Moodley's (1977) findings regarding sediment production from non-macadamised roads. The composite measure of the intensity of erosional activity shown in Table 4 is the average obtained from these four data inputs.

4.3. AGRICULTURAL POTENTIAL

Information on the agricultural potential of the types of soil in the study area demarcated for cultivation was obtained from Experimental Station of the South African Sugar Association (1984). Agricultural potential classes of very good, good, moderate, poor and very poor were assigned values of 5, 4, 3, 2 and 1, respectively. The composite scientific measure of the study area's agricultural potential shown in Table 4 is the average of the values assigned to the soils represented.

A male with a Masters degree in Environmental Science who is fluent in both English and Zulu was employed to interview the 'head' of 40% of the households in each study area. As the primary responsibility for daily management of natural resources is more often a female responsibility, where the male 'head' was present, every attempt was made to elicit and incorporate the views of the household's women. As quantifiable information was desired, structured interviews comprising predetermined questions asked in the same order, and the same manner to each respondent, were employed. These questions focused on finding out whether, when they settled on the farm, they had received expert advice on where to locate and how to prepare their vegetable gardens and crop fields, whether the arable land available to them was large enough to allow for fallow periods, and whether soil loss, fertility maintenance and tillage posed problems. If there were two possible responses to a question, yes or no, and 'yes' was the more favourable in terms of soil sustainability, it scored 2, while 'no' scored 1. The sum of the scores for the questions was divided by the number of questions. The resultant value dependent on its position in the possible range of such values was adjusted to the scale of 5, 4, 3, 2 and 1, representing very good, good, moderate, poor and very poor, respectively. This value was carried forward as the composite measure of the community's perception of agricultural potential shown in Table 4.

4.4. LAND USE PRACTICES

Responses to three additional questions relating to the fate of the crop residue, were integrated into the evaluation of land use practices on arable lands. The arable land worked by the households interviewed was assessed and the presence or absence of indicators of good soil management practices such as: contour bunds, terraces and drainage channels; contour ploughing; minimum tillage; mulching; multiple, cover and strip cropping, and of bad practices such as: ploughing too deep; planting too far apart, were recorded. These indicators were scored, weighted and analysed as described above to yield the composite measure of the influence of cultivation practices shown in Table 4.

Responses to nine additional questions relating to livestock kept, stocking rates and burning practices were quantified following the same procedure. While obtaining the percentage cover estimates described earlier, the predominant grass species present was recorded and allocated to one of the following categories, based on the work of Camp

(1995):

Decreasers: generally palatable and decrease when overgrazed or under-utilised.
Increaser 1: increase with moderate overgrazing.
Increaser 2: increase when overgrazed.
Increaser 3: pioneer species which dominate when excessively overgrazed.

Values of 5, 4, 2 and 1 were assigned to the above categories, respectively. The average value for the thirty 1 m² sample areas and the average of the values assigned to the nine questions noted above were summed and divided by two to yield the composite measure of the influence of range management practices shown in Table 4.

Values ranging from 5 to 1 were thus obtained for a total of 15 indicators of soil sustainability. The average of these 15 values yielded the overall measure of the soil's sustainability shown in Table 4. Morgan (1986) has discussed a number of weaknesses inherent in the simple scoring technique opted for in this study. Information on the key indicators of soil sustainability is derived from various sources, obtained from various agencies and available at various scales, and is therefore more accurate for some indicators than others. The accuracy of the scores allocated to them and of the composite measure of their influence on soil sustainability is jeopardised by: (i) rating the indicators independently; (ii) weighting the indicators equally; (iii) differences in the threshold values delimiting indicator classes and the arbitrary nature of some of their classifications, and (iv) the fact that the choice to add or multiply individual scores is arbitrary. Despite these weaknesses, Morgan (1986) concluded that this scoring technique is easy to use and has the advantage that indicators which cannot be quantified in any other way, can be readily included. The manual prepared by Watson (1998b) for those employed to monitor the land reform transfers, contains a more detailed description of the procedure used to assess soil sustainability as well as the full questionnaire and observations checklist.

5. Discussion

Application of the methodology described in section 4 above yielded the results shown in Table 4. In addition to values ranging from 5 (very good) to 1 (very poor) for 15 indicators of soil sustainability, an overall measure of this sustainability is provided. Each area's results were presented to gatherings at their respective communities. Without exception they were readily understood and accepted. The results pertaining to all four areas were presented at six workshops comprising audiences with a range of interests in the LRP, including the consultants who computed the sale value of the farms, academics, government (particularly DLA officials), private development planners, non-government organisations and foreign aid agencies. Many of the strengths and weaknesses of the methodology, other concerns and modifications relating to its future wider application as discussed below, came to light in these presentations.

5.1. POTENTIAL SUSCEPTIBILITY TO EROSION

The averages of the four areas' slope angle, erodibility, erosivity and cover indicator value, i.e., 2.3; 2.7; 3.2 and 2.8 for Tembitshe Buthelezi, Misgunst, Labuskagneskraal and Nomoya respectively, provide a reliable estimate of their erosion hazard potential. Although Labuskagneskraal has highly erodible soils, it is the least susceptible to erosion due to the ameliorating influence of its gentle slopes and good cover. While very steep slopes and highly erodible soils are better represented in Misgunst than in Tembitshe Buthelezi, the former's erosion potential is substantially reduced by its very good cover. Nomoya's vegetation cover affords the soil little protection against its highly erosive rainfall. Its cover value is an overestimate, reflecting the abundance of stones. Future applications of the methodology will therefore only include stones actually covering soil in this estimate. As the four study areas were assessed in succession in one month, the inclusion of litter in the cover estimate would not have distorted their comparative composite cover measures. However, in future applications of the methodology litter will be excluded from the cover estimate because of wide seasonal variations in its production and decomposition. It will be impossible to monitor all land reform transfers in the country at the same time of year. Logistics in the monitoring process may dictate that any particular area is reassessed at a different time of the year.

5.2. CONTEMPORARY EROSIONAL STATUS

5.2.1. Areal Extent
Using the most recently available orthophoto maps, both gully and sheet erosion were found to be very localised in all four areas. However, this may no longer be the case as these maps were produced from aerial photographs taken 7 to 12 years prior to this study's assessment of these areas. Other potential sources of these data were investigated. The 1:50 000 multispectral, digital, Quick Maps produced from SPOT imagery with a 20 m spatial resolution and enhanced to improve the discrimination of different surface covers as described by Slabbert (1999), appear to be the best option and are recommended for use in future applications of this methodology. Although the use of Quick Maps will be very much more expensive than using orthophoto maps, this is justified because in addition to providing information on the contemporary erosion status, they are the most reliable source of data on vegetation, land use and settlement patterns.

5.2.2. Intensity of Erosional Activity
The higher levels of erosional activity on the degraded/sheet eroded areas and rangelands of Tembitshe Buthelezi and Nomoya, in comparison with those of Misgunst and Labuskagneskraal (Table 4), are attributed to overstocking and poor burning practices. Both communities should be encouraged to reduce their stocking rates, cover degraded areas with branches from thorn trees to keep animals out, and obtain advice on how to determine when and how often to burn the grass. The intensity of erosional activity was found to be the highest on Nomoya's arable lands (Table 4) and is attributed to the scant plant residue on the soil and excessive spacing between plant rows and between plants within them. It is difficult to maintain mulch cover in Nomoya's semi-arid climate because harvester termites rapidly remove organic matter. The excessive spacing, however, can be readily addressed.

Erosional activity was found to be very intense in Tembitshe Buthelezi's and Nomoya's gullies and intense in Labuskagneskraal's gullies (Table 4). The abundance of subsurface piping and slumping of both head and side walls in Tembitshe Buthelezi's gullies, together with the fact that most of them are located on very steep slopes, suggests that they are primarily regulated by natural factors. The gullies in Labuskagneskraal also appear to be regulated by natural factors. They are all continuous with the Thukela river. The main erosional activities in them of undercutting of the channel walls and downcutting into the soft shale bedrock, appear to be related to fluctuations in the base level of the river. There is little merit in encouraging these two communities to introduce measures to arrest the erosional activity in their gullies. In contrast, the Nomoya community should be encouraged to fence off and place barriers across the channels of their gullies, as the gentle slopes on which most of them are located and the abundant evidence of surface wash activity around and within them, suggest that these gullies are related to land use practices in the area.

The roads in both Tembitshe Buthelezi and Nomoya are poorly sited, poorly constructed and eroding actively. Even if these communities were to receive advice on where and how to construct the roads and how to maintain them, they simply do not have the means to implement it. This provides an example of how this methodology could be used to alert the relevant authorities to aspects requiring their intervention.

5.2.3. Agricultural potential
Using scientific criteria to evaluate the agricultural potential of the four land transfers, Misgunst has the worst and Labuskagneskraal the best. The beneficiaries' perceptions of the agricultural potential of their land did not correspond with the scientific evaluation and appear to be more a function of their level of satisfaction with the land reform process than their knowledge of the soil resource. As noted earlier, prior to transfer Misgunst's beneficiaries lived on the farm as labourers. They carry out the same cultivation practices as had been employed previously and are confident of getting the same good yields as those obtained by the previous landowner. There was general consensus that the better life promised by the majority government had largely been met. In contrast, the beneficiaries of the other three land transfers were all unhappy about aspects of the redistribution process which, although unrelated to agricultural potential, adversely affected their perception of it.

5.2.4. Landuse Practices
The composite measure of the sustainability of rangeland management practices in the four study areas shown in Table 4 revealed that they were managed best in Misgunst. Labuskagneskraal's rangelands were suffering from lack of defoliation. The community should be encouraged to apply a hot burn in order to arrest bush encroachment and to restore the palatability of the grass to livestock. As noted earlier, rangelands in both Tembitshe Buthelezi and Nomoya were overgrazed. The prevalence of unpalatable pioneer grasses in Nomoya's range revealed that its deterioration was more advanced. Both areas would benefit from a substantial reduction in livestock numbers and a burn application once the grass sward has recovered sufficiently to provide adequate fuel for a fire hot enough to kill the woody invaders.

The lower composite measures of the sustainability of cultivation practices obtained for Tembitshe Buthelezi and Misgunst shown in Table 4 are principally a function of their greater vegetable garden and crop production activity in comparison with

Labuskagneskraal and Nomoya. Physical soil conservation measures such as contour ploughing, ridging and terracing were well represented in all four transfers. Additionally, biological soil conservation measures such as inter- and strip-cropping and crop rotation were well represented at Misgunst. In common with Tembitshe Buthelezi and Nomoya, Misgunst's composite measure was depressed by poor practices such as burning or allowing stock to feed on the crop residue, failing to provide a mulch cover, planting too far apart and ploughing too deep.

6. Conclusion

Despite the high degree of subjectivity involved in the selection, quantification, weighting and integration of the soil sustainability indicators, and concerns that they were too focused on erosion, the averages of the 15 indicators employed in this study (Table 4) provide reliable measures of the overall sustainability of the soil resource. This resource was rated moderate in Misgunst and Labuskagneskraal and poor in Tembitshe Buthelezi and Nomoya. While it is useful to be able to compare the soil sustainability status of different land transfers in terms of prioritising service provision, the approach used in this study was adopted primarily to enable the environmental status of a sample of transfers representing different biophysiographic, land use and/or historic scenarios to be monitored. A decrease in the soil's overall sustainability rating the next time it is assessed, should serve as a warning of deterioration and the need to implement appropriate remedial measures. A progressive decrease or increase in this rating over several such assessments will clearly show that the change in land ownership has had a significant negative or positive effect on the sustainability of the resource.

   In addition to being simple and reliable, the procedure outlined above for assessing the sustainability of soils in land reform transfers is quick and inexpensive to carry out. Assessments of 20 other transfers took four days and cost US$2200, on average. Given that the procedure was developed in KwaZulu Natal it will inevitably have to be modified for application in other provinces. This study should be viewed as having provided the basis for a methodological approach, rather than a methodology *per se*.

7. Acknowledgments

I wish to thank the Directorate of Monitoring and Evaluation of the Department of Land Affairs, Pretoria, for funding the research; the University of Durban-Westville for administering the funds and providing other logistical support; Mr A. Somers, the Zulu interpreter; the communities of Tembitshe Muthelezi, Misgunst, Labuskagneskraal and Nomoya for willingly participating in the study; Mrs A Pullan for typing this paper, and the Lewsons for assistance with every stage its production.

8. References

Brincate, T.A & Hanvey, P.M. (1996) Perceptions and attitudes towards soil erosion in the Madebe community, Northwest Province, *South African Geographical Journal* **78**, *75-82.*
Camp, K. (1995) *Valley Bushveld of KwaZulu Natal - natural resources and management,* KwaZulu Natal Department of Agriculture Report N/A95/2, Cedara.
Crosby, C.T. , McPhee, P.J. and Smithen, A.A. (1983) Introduction of the Universal Soil Loss Equation in the

Republic of South Africa, *American Society of Agricultural Engineering,* Paper No. 832072, 1-15.

Dardis, G.F., Beckedahl, H.R., Bowyer-Bower, T.A.S. and Hanvey, P.M. (1988) Soil erosion forms in southern Africa, in G.F. Dardis and B.P. Moon (eds.), *Geomorphological studies in southern Africa,* Balkema, Rotterdam, pp. 187-213.

Department of Agricultural Technical Services (DATS) (1976) *Soil loss estimator for southern Africa,* Bulletin 7, Cedara.

Department of Land Affairs (DLA) (1997) *White Paper on South African Land Policy,* Pretoria.

Experiment Station of the South African Sugar Association (ESSASA) (1984) *Identification of the soils of the sugar industry,* Bulletin 19, Mount Edgecombe.

Garland G.G., Robinson J.R. & Pile K.G. (1994) Policy, perception and soil conservation - a case study from Cornfields, Natal, *Human Needs, Resources and Environment Report,* HSRC, Pretoria, pp. 1-41.

Land Type Survey Staff (LTSS) (1986) *Land types of the Map 2730 Vryheid,* Memoirs on the Agricultural Natural Resources of South Africa, 7, Government Printer, Pretoria.

McAllister, P.A. (1988) The impact of relocation in a Transkei 'betterment' area, in C. Cross and R. Haines (eds.), *Towards freehold options for land and development in South Africa's black rural areas,* Juta, Cape Town, pp. 112-121.

McAllister, P.A. (1989) Resistance to 'beterment' in the Transkei - a case study from Willowvale district, *Journal of Southern African Studies* **15**, 346-368.

Morgan, R.P.C. (1986) *Soil Erosion and Conservation,* Longman, London.

Moodley, M. (1997) Off-road vehicle perturbation effects on the geomorphic environment of Golden Gate Highlands National Park, South Africa, Unpubl. MSc thesis, University of Natal, Pietermaritzburg.

Phillips, J. (1973) *The agricultural and related development of the Tugela Basin and its influent surrounds,* Natal Town and Regional Planning Commission, 19, Pietermaritzburg.

Slabbert, F. (1999) *SPOT based image maps for land reform applications,* Website: dla.pwv.gov.za

Smithen, A.A. and Schulze, R.E. (1982) The spatial distribution in southern Africa of rainfall erosivity for use in the USLE, *Water SA* **8**, 74-78.

Stocking, M.A. (1994) Assessing vegetation cover and management effects, in R. Lal (ed.), *Soil erosion research methods,* 2nd ed., Soil and Water Conservation Society, Ankeny, pp. 211-232.

Turner, S.D. (1997) *Environment and land reform in South Africa,* Land and Agriculture Policy Centre Policy Paper, 33, Johannesburg.

Watson, H.K. (1991) A comparative study of soil erosion in the Umfolozi Game Reserve and adjacent KwaZulu area from 1937 to 1983, Unpubl. PhD thesis, University of Durban-Westville.

Watson, H.K. (1996) Short and long term influence on soil erosion of settlement by peasant farmers in KwaZulu Natal, *South African Geographical Journal* **78**, 1-6.

Watson, H.K. (1997) Geology as an indicator of land capability in the Mfolozi area, KwaZulu Natal, *South African Journal of Science* **93**, 39-44.

Watson, H.K. (1998a) Land reform implications of the distribution of badlands in the Mfolozi catchment, KwaZulu Natal, *Proceedings of the Southern African Association of Geomorphologists Biennial Conference,* June/July, Grahamstown, pp. 201-216.

Watson, H.K. (1998b) *A methodology for assessing the environmental impact of the Land Reform Programme,* Department of Land Affairs Report, Pretoria.

Watson, H.K. and Ramagopa, P. (1997) Factors influencing the distribution of gully erosion in KwaZulu Natal's Mfolozi catchment - Land reform implications, *South African Geographical Journal* **79**, 27-34.

Whitlow, J.R. (1988) Potential versus actual erosion in Zimbabwe, *Applied Geography* **8**, 87-100.

Wischmeier, W.H., Johnson, C.B. and Cross, B.V. (1971) A soil erodibility nomograph for farmland and construction sites, *Journal of Soil and Water Conservation* **5**, 189-193.

AUTHOR

HELEN KERR WATSON

Department of Geography and Environmental Studies
University of Durban-Westville
P/Bag X54001, Durban 4000, South Africa. hwatson@pixie.udw.ac.za

CHAPTER 11

REDISTRIBUTION, RESTITUTION AND REFORM: PROSPECTS FOR THE LAND IN THE EASTERN CAPE PROVINCE, SOUTH AFRICA

R C FOX and K M ROWNTREE

1. Abstract

South Africa's history has left the country with a complex, racially skewed distribution of land resources. The aim of current government land policy is to bring about a more equitable distribution of land among the different population groups and to improve rural livelihoods through three initiatives; land restitution, land redistribution and land tenure reform. This paper explores the constraints on the Land Reform Programme in the Eastern Cape Province with reference to the current relationship between population density and potential biological productivity and considers the implications for land degradation. An examination of these issues at the provincial and district level, backed up by a local case study, points to important implications with regard to both historical patterns of land use and land degradation and the potential for land reform. It is concluded that the land reform program is constrained by the limited areas of low density, high potential land into which the population can move.

2. Introduction

"Land ownership in South Africa has long been a source of conflict. Our history of conquest and dispossession, of forced removals and a racially-skewed distribution of land resources, has left us with a complex and difficult legacy" (Department of Land Affairs 1997, Foreword). Current government policy intends to facilitate a more equitable distribution of land amongst the different population groups in the country and to improve rural livelihoods through three initiatives: land restitution, land redistribution and land tenure reform.

South Africa's land has been subject to degradation processes which are often the result of inappropriate land management systems in marginal areas. If the reform process is to be successful in creating sustainable rural livelihoods, it will be necessary to take due consideration of the potential of the land to support the population. This paper examines the triangle of population, biological productivity and land degradation in the Eastern Cape in order to assess the potential success of these land reform initiatives in the province.

Given the broad nature of this paper it will be impossible to deal in-depth with all population and environmental factors. Accordingly, the following sections contextualise, rather than give detailed examinations of, two aspects of our research. The first deals with an analysis of the relationship between people and the environment in the Eastern Cape from 1936 to 1991; data from the controversial 1996 census are not used as they are spatially incompatible with those from earlier censuses. This analysis is aimed particularly at showing how, at the regional scale,

167

A.J. Conacher (ed.), Land Degradation, 167–186.

increasing population densities occurred throughout the two homelands of Ciskei and Transkei (Fig. 1) regardless of their agro-ecological potential. This provides a situation where, without appropriate agricultural innovations and sustainable land use practices, degradation is extremely likely. The second examines more detailed case study material relating to the Peddie District over roughly the same time span. Peddie District will have been contextualised from the first section as one district where we could expect, *a priori*, degradation to have occurred. The district also presents what is almost a microcosm of the whole province as it possesses a range of tenure and management systems found throughout the province. In the final discussion we relate our findings about the district and province back to the current policy arena.

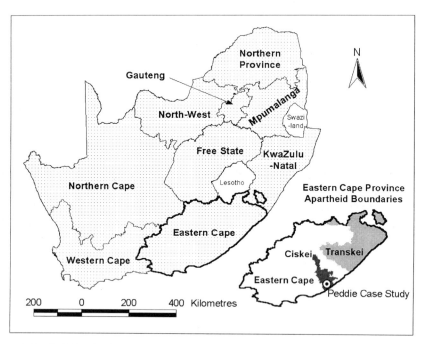

Figure 1. South Africa showing post-*apartheid* provincial boundaries and the Eastern Cape Province's *apartheid* divisions.

Land degradation in South Africa has recently received exhaustive attention by Hoffman *et al.* (1999). It is interesting to note that through using different methods we arrive at a different identification of districts most at risk from degradation. In particular, our study was able to isolate longer population growth trends and also focus more closely on rural population densities. The two studies, however, follow a comparable broad approach to the examination of the relationships between people and land and identify similar causes of degradation.

3. The Rural Context

3.1 AGRICULTURE IN THE 1990s

South African agriculture has the appearance of success. For example, the country was self-sufficient in virtually all staple crops between 1985 and 1993, although there were exceptions in some meat commodities and sunflower oil. This is a particularly striking performance given the country's inadequate agricultural resource base. Arable land, for example, represents only 13.5% of the surface area of the country (van Zyl *et al.* 1996) but only 3% is considered as high potential (approximately four million hectares). The problem, however, is that self-sufficiency has been achieved through extremely dualistic policies which have been at the expense of social equity and environmental sustainability. In 1990, 68% of rural African households were living below the minimum subsistence level (Republic of South Africa 1995).

A set of five policy instruments was responsible for this agricultural performance (van Zyl *et al.* 1996). Price support above market levels, particularly for maize and wheat, coupled with subsidised storing facilities, enabled crop production in very marginal areas such as in the drier parts of the North West Province and western Free State (Fig. 1). Subsidised investments in irrigation facilities led to expansion into areas of marginal potential and depletion of groundwater resources. Agro-chemical inputs were subsidised, such that fertiliser use grew at the rate of 7.7% per annum from 1947 to 1979 and dipping and spraying at the rate of 20.6% per annum from 1973 to 1980. Subsidised irrigation water led to the perception that it was a plentiful resource which could be used inefficiently. Favourable taxation and interest rates encouraged the over-mechanisation of the production process and extended production down the ecological gradient into areas with lower moisture availability.

At the same time as these policies were favouring (and often degrading) the white sector, the black homelands, such as Ciskei and Transkei in the Eastern Cape, were being systematically underdeveloped in a variety of ways. Enforced population movements both into homelands and within them through villagisation programs - known as betterment planning - created much dislocation. The increased population was in part responsible for the breakdown of the traditional rotational systems of mixed farming in Transkeian districts (Andrew 1991), especially since black agricultural innovation was stifled. Black small-holder agriculture suffered from inadequate market access, poor levels of farmer outreach and insufficient infrastructural services. Not surprisingly, agricultural production in homeland areas was very low in comparison to the white farming sector.

The challenges for South African agriculture today are complex as the country struggles to redress the inequities of the past. It has a fragile and often degraded natural environment and is experiencing the effects of much more open global competition now that the commercial sector has become effectively deregulated. Furthermore, the whole agricultural sector, black and white, is characterised by very low employment levels which have prevented the development of local regional economies and small towns or service centres, such as are found elsewhere in Africa.

3.2 LAND POLICY SINCE 1994

Policy statements and new laws enacted by the post 1994 governments have understandably focussed on improving rights of access to land by the rural poor (Republic of South Africa 1995, 1997). The three cornerstones of the *White Paper on Land Policy* (Republic of South Africa 1997) are: land restitution through the Land Claims Court for those found to be dispossessed on racial grounds; land redistribution to overcome homelessness and prevent land invasions, and land tenure reform to overcome insecurities such as experienced by people utilising communal land in the former homelands and labour tenants on white farm land.

The restitution program intends to restore land rights to people forcibly removed after 1913 through the implementation of racially based legislation. The extent of the problem is indicated by the 63 455 claims which have been lodged throughout the country in both urban and rural areas. The Eastern Cape has examples of communities moved through Group Areas enactments in urban areas and forced removals of black people from white South Africa into homelands. There is also a limited number of white farmers claiming restitution for their farms which were incorporated into the province's former homelands of Ciskei and Transkei. It is very difficult to calculate just how much land is involved in the Eastern Cape or other provinces until the cases are brought to the Land Claims Court. We can estimate, however, based on the size of the claims so far settled, that in the Eastern Cape and Free State the amount of land may be as high as 10% of the land area. By 1999, only 241 claims had been finalised for the country as a whole (Department of Land Affairs 1999), indicating that this is going to be a long drawn out process. Development of contested land will be blocked until the case has been judged by the Land Claims Court. This is an important point as land restitution cases have the potential to prohibit the redevelopment of areas (both urban and rural) throughout the Eastern Cape wherever people were forcibly moved into or out of black or white areas.

More implementation has occurred in the redistribution programme which gives settlement grants of R15 000 (US$2500 approximately) to help poor and marginalised families obtain access to land for settlement and/or farming purposes. Some 427 redistribution projects were being implemented across the country by early 1999, involving almost 300 000 people in the transfer of 480 400 hectares of land (Department of Land Affairs 1999). The average size of holding was just over 10 hectares per family. Most of these projects are found on former white freehold land or state land which has been released for settlement.

Tenure reform in communally managed areas has yet to get beyond the discussion of policy stage, though it appears that statements on land rights and tenure security in these areas are imminent. The rights of labour tenants, at least, have been the subject of various Acts since 1995. Since most of the land in Ciskei and Transkei lies under a form of communal tenure it is clear that in the immediate future there will be little change in these areas.

3.3 AGRO-ECOLOGICAL POTENTIAL

The potential biological productivity of the Eastern Cape can be mapped using Enpat (1998), the digital database produced by the Department of Environmental Affairs and Tourism. In this database productivity is assessed in terms of the tonnes of potential biomass production per hectare and is a reflection of the moisture availability in the soil. Figure 2 shows that in the Eastern Cape there is a marked spatial gradient, with very high values in the range of 9-10 t ha^{-1} found in the wetter east of the province and very low values from 0.5-1.0 t ha^{-1} found in the much drier west of the province. The coastal areas in the south are an exception to this trend as high productivity values are found along the extreme western portion of the province's narrow coastal strip.

Within the limits set by available water, the soils of the Eastern Cape further constrain land use. Soils over large areas of the province are prone to degradation (Laker 1999). Shallow soils dominate the Karoo to the west and the adjacent coastal plateau. In these drier parts of the province overgrazing of rangelands has contributed to surface water erosion. The more humid area to the east is dominated by highly erodible solonetz type soils, soils that are prone to both

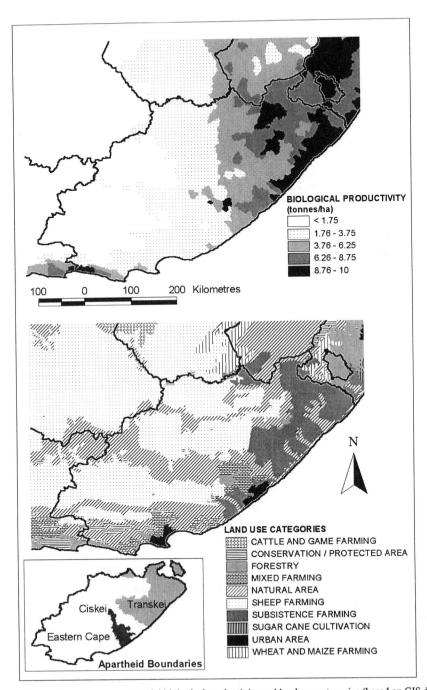

Figure 2. Eastern Cape Province: potential biological productivity and land use categories (based on GIS data base from Enpat 1998).

surface and tunnel erosion. Wind erosion is a problem in some coastal areas, while mechanised agriculture on commercial farms has led to soil compaction. Irrigation has caused salinisation of otherwise productive bottomlands.

3.4 LAND USE

Patterns of actual land use can also be derived from the Enpat data source (Fig. 2) and a certain degree of correspondence may be found between the biological potential described above and the actual use made of the land. Sheep farming is prevalent in the dry central and western parts of the province, while forestry is found in the wet south-western and north-eastern extremities. Mixed farming (including dairy farming) is found in coastal areas, particularly close to the major urban centres. Large areas of land were classed, however, as being either natural areas or under subsistence farming, and it is this latter category which presents the biggest anomaly. Most of the land with the highest biological productivity, in the east of the province, is under subsistence farming. Much of this high potential land lies in what was, under apartheid policies, the Transkei and, to a lesser extent, Ciskei homelands; Ciskei's subsistence areas tend to be drier and have a lower biological potential than Transkei.

It is important to note at this point that the dominant land use was subsistence farming in the Ciskei and Transkei and that the majority of their populations were classed as rural or peri-urban. The homelands' populations, however, were dependent on transfers of capital from the urban sector for survival (Fox and Nel 1998). In the Ciskei and Transkei it was estimated in 1985 that only 6.6% and 31.2%, respectively, of household earnings were derived from agriculture (Mashile 1988). The widespread abandonment of arable agriculture has been described as 'sub-subsistence farming' or 'under-farming' in attempts to capture the nature of these rural areas (Cobbett 1988; Hoffman *et al.* 1999).

3.5 LAND TENURE

There is not the space here to tackle the complex topic of land tenure. What must suffice will be something of a simplification to enable us to get to the prospects for the Land Reform Programme. Cross and Haines (1988) give a good overview of the evolution of different land tenure systems in rural South Africa. The Eastern Cape Province is essentially divided into three major types of land tenure; freehold land under private ownership, state land and communal land. Virtually all of the freehold land lies in former white South Africa and is in white hands. Small pockets of freehold land (known in *apartheid* times as 'black spots') are owned by black communities in the districts that lie between Ciskei and Transkei (known as the Border corridor).The *apartheid* boundaries are shown in Figures 1 and 2.

The creation of consolidated homeland areas in the 1970s and 1980s witnessed much freehold land in areas designated for black consolidation passing from white control to the control of the independent black state of Ciskei. This process was far less important in the former Transkei as it consisted of a much larger and contiguous piece of territory in the first place. State land under the control of various government departments was also found in white South Africa. Finally, most of the land in the former homelands is under communal tenure regimes where land is allocated by the representative of a tribe/ethnic group to individuals for their use as arable land, residential land or grazing land. The areas under different types of tenure have a different potential for the success of South Africa's land policies, as the following sections will show.

4. The Population Context

4.1 SPATIAL DISTRIBUTION

The distribution of people in the present Eastern Cape Province has exhibited a marked east-west trend since 1936 (Fox and Tipler 1996). There are higher populations in the easternmost magisterial districts of the former Transkei and along the southern coast, with a steady reduction in population size westwards (Fig. 3a). A subsidiary trend has been the decline in numbers northwards from the coastal strip in the western half of the province. Thus, on the one axis, the extreme western districts of the province had population densities of only 1 person per square kilometre in 1991, whereas the extreme east had 73 persons. On the other axis Peddie District, which lies on the south-eastern seaboard, had 43 persons per square kilometre, but districts to the north-west declined to 3 persons per square kilometre.

4.2 AGE-SEX DISTRIBUTION

The population of the province increased from some 2.2 million in 1936 to 5.8 million by 1991 (Fox and Tipler 1996). The migrant labour system resulted in a marked gender imbalance, with fewer males to females present in each census year since 1936. In 1991 the province was 44.6% male and 55.4% female. In numerous areas the number of women greatly exceeded that of men, particularly in areas far from the major urban centres which are located along the south-eastern seaboard. The absence of men has been cited as one of the factors leading to a decline in rural productivity in areas of the Transkei (Andrew 1991). Clearly, land policy should impact most on women as they are numerically dominant, particularly in the black rural areas.

4.3 RURAL-URBAN BALANCE

In 1991 the population of the province was 34.2% urban and 65.8% rural, with the two cities of Port Elizabeth and East London accounting for most of the urban population. These, and the very few other predominantly urban magisterial districts, have been excluded from the analysis of the relationship between rural populations and environment undertaken below. The former Transkei and, to a lesser extent Ciskei, have never been strongly urbanised owing to their role as labour reserves for the South African economy. In contrast, the westernmost districts are highly urbanised as their rural populations have diminished with changes in commercial agricultural practices such as mechanisation (Cobbett 1988).

4.4 RURAL POPULATION DENSITIES

The rural population has always exhibited a marked pattern of diminishing numbers per district along a north-westerly axis away from the seaboard. This distribution mirrors, in part, the declining agricultural potential (mentioned above) as one moves away from the more well-watered coastal and eastern parts of the province. *Apartheid* policies reinforced this pattern, sponsoring a decline in numbers of people in commercial farming areas, and Figure 3b shows

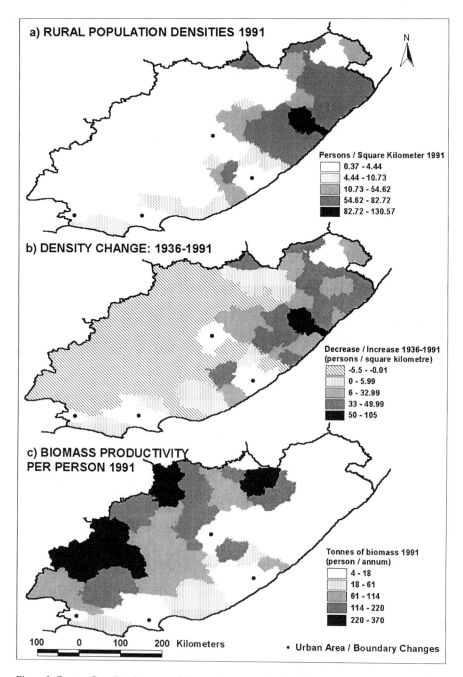

Figure 3. Eastern Cape Province: population-environment relationships.

the marked declines in densities throughout most of the west of the province. Mechanisation was promoted at the expense of labour (van Zyl *et al,* 1996), while forced removals, influx control and a whole battery of other legislation ensured large rural population increases in the eastern part of the province (Fig. 3b).

4.5 RACIAL DISTRIBUTIONS

Apartheid policies were the driving force behind the racial distribution of the province's population (Fox and Tipler 1996). In 1991 the black population formed the majority, some 5 081 159 people in comparison with the 374 608 whites and 287 549 'coloureds' (mixed race). The Asian population, in terms of number, was largely insignificant with a total of 14 657. By 1991, Transkei and Ciskei were the main centres of black population, black people being largely excluded from residing in what was then 'white' South Africa. Districts with particularly high black population concentrations were adjacent to the 'white' Border corridor. The major concentration of 'coloured' people was in the west of the province, since this was one of the areas where coloured people were given employment preference during apartheid (Fox and Tipler 1996). The white population has been located overwhelmingly in the former Eastern Cape and Border areas since the 1930s, with very few people located in former Ciskei and Transkei districts. By 1991 whites were mostly prevalent in districts with large urban centres (reflecting a trend of long term rural depopulation).

Later we will see that the relationship between people and the land has been in favour of the white population, distributed as above. The areas of very dense black population are those where the potential for land degradation due to natural factors may well have been highest. Adding to the complexity of this issue is the fact that the three aspects of land policy will have different impacts in the racial zones.

5. Population-Environment Relationships 1936 to 1991

The relationship between people and their environment in Africa has received noteworthy attention recently (Turner *et al.* 1993; Tiffen *et al.* 1994). The following analysis broadly supports the view of the regional political ecologists that the relationship between people and their environment has been disrupted since access to resources has been highly racially discriminatory and politically driven (see Jolly (1994) for a summary of other approaches to this relationship). Consequently black rural populations have increased whilst agricultural innovations have been stultified, producing land fragmentation, environmental deterioration, poverty and famine. This presents South Africa's policy makers with many problems.

In an attempt to ascertain the areas of the province which have experienced the most stress on the environment, an examination of rural population growth from 1936 to 1991 was undertaken and then related to biological potential. Enpat (1998) was the source for the biological data since it presents a measure of productivity, the tonnes of potential biomass production per hectare, through modelling the moisture availability in the soil. Figure 3c shows the results of this analysis, which was undertaken at the level of the magisterial district; urban districts and those whose boundaries changed a great deal over this time period were excluded from the analysis. Other studies have also examined the correspondence between population density and land degradation, but arrived at varying conclusions as to whether there is an association or not (Hoffmann *et al.* 1999; Stocking and Elwell 1973; Tiffen *et al.* 1994; Marker 1988; Weaver 1988). One of the major reasons for this may be the different measures of population and environment used, as well as the scales of examination and time frames used.

Figure 3a indicates the vast disparities in rural population densities found across the province in 1991. The westernmost districts had less than five persons per square kilometre while the easternmost had well over 100. To a certain extent this reflects the biological potential (Fig. 2), but it is a much clearer spatial analogy to the boundaries of the homeland areas shown in Figures 1 and 2. The process of rural depopulation in the former white areas is clearly shown in Figure 3b, where decreases in population density from four to two persons per square kilometre were common in the north-west of the province between 1936 and 1991. The more fertile coastal areas, however, recorded modest increases. The black homelands exhibited spectacular increases in rural densities, with densities commonly doubling over this period.

Dividing the potential biomass productivity by the population density gives a measure of the land available to support that population and therefore can be used as one indicator of the potential for land degradation at the district level (Fig. 3c). The authors acknowledge that this is a simplistic measure which takes no account of the different land use systems, social structures and so on present in the province, but nonetheless it is thought to provide a useful regional-scale analysis.

Figure 3c shows that in the former white areas there are high values recorded despite the very low biological potential, reflecting the extremely sparse rural populations. It is worth noting that over the 60-year time span of this study, the potential biomass productivity per person increased in these districts due to the decline in population numbers. It is very important to note that in the black areas biological productivity varied considerably from east to west. Thus the westernmost black districts (including Peddie District in Ciskei) experienced very high increases in population densities whilst their biological productivity was much lower than the districts in eastern Transkei. The black districts in most danger of serious degradation are therefore those which lie towards the centre of the province.

The study by Hoffman *et al.* (1999) reveals a somewhat different set of Eastern Cape districts defined as priority areas in terms of their current and potential land degradation status; the chief difference with our assessment is his identification of a few districts in north-eastern Transkei. Hoffman *et al.*'s (1999) national study, however, singles out eight Eastern Cape districts in their list of the 20 districts most at risk in the whole country (which consists of 348 districts). KwaZulu-Natal had seven districts and Northern Province five to make up the balance. The commonality between these three provinces is that they each have very large areas of former homeland territory within their borders juxtaposed with much lower density white areas.

6. People and Land in Peddie District

Peddie District (Fig. 4) lies in the former homeland of Ciskei and is one of the districts which our analysis above suggests as being highly likely to experience problems of degradation. In a number of respects it is transitional and represents in microcosm the problems relating to people and the land in the Eastern Cape. Perhaps for this reason Peddie has attracted the attention of a number of geographical and anthropological researchers (Branch 1994; Kakembo 1997; Tanser 1997). Because of its political history it is also representative of areas in the province where former white land, taken over by the state in 1981, is now available to the land reform program.

Peddie lies on the transition between the more humid areas to the east and the semi-arid areas to the west. Rainfall, averaging 489 mm a year, is highly variable and droughts are common. Natural vegetation mirrors this moisture gradient. Peddie lies on the transition between the savanna grassland biome and the karoo biome. Its vegetation is composed of a topographically related mix of grassland and thicket, the latter with a high predominance of succulent shrubs (Tanser, 1997). Karoo species (low woody or succulent shrubs) are common, especially in the drier or more degraded areas where they achieve invader status (Tanser 1997).

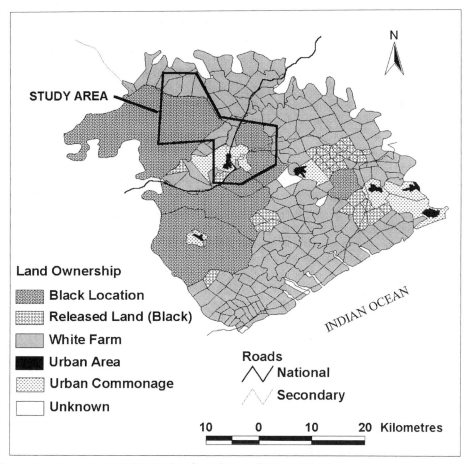

Figure 4. Land ownership in Peddie District prior to the consolidation of Ciskei.

Peddie District also lies in the transition zone between the historically white areas to the west and the historically black areas to the east (Fig. 1). The district formed part of the buffer zone created between the Great Fish River and the Kei River in the early to mid eighteenth century – British Kaffraria. By the 1850s this area consisted of a mosaic of white commercial farms and tribal locations (Branch 1994). This pattern persisted with some modification until 1974 when, following the National States Act of 1971, the consolidation process started which resulted in the formation of the 'independent' state of Ciskei in 1981. Extensive areas of white farmland plus a number of towns such as Peddie were incorporated into Ciskei (sowing the seeds for future land restitution claims). Of significance is the fact that most of the former white land was not opened up for black settlement at 'independence', but was retained as state land, either under white management as a commercial enterprise or essentially taken out of production. Thus in 1994, when Ciskei was reincorporated back into South Africa, Peddie District consisted of a mix of densely populated communal areas (tribal and released lands) and sparsely populated, often under-utilised, former white areas (Fig. 4).

Significant changes in the redistribution of the population occurred after the passing of the National States Act in 1971. While white farmers were bought out and left the area, the resettlement (often forced) of black communities from the Republic of South Africa took place in the opposite direction (sowing the seeds for land restitution claims in the opposite direction). The effect of this influx of people into the communal areas can be seen from the graph in Figure 5, which shows a steep rise in the number of dwellings between 1965 and 1975 continuing into the late 1980s.

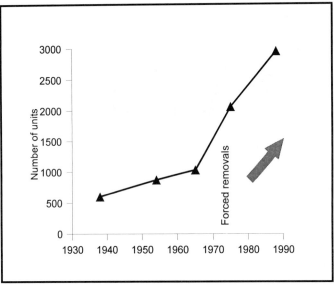

Figure 5. Increase in population in the study area based on number of dwelling units (based on data from Kakembo 1997).

Kakembo (1997) has studied the effect of land tenure and population growth on the historical pattern of land degradation in Peddie District. Taking a relatively small area which was considered to be homogeneous in terms of its natural environment, Kakembo assessed the distribution and intensity of degradation in relation to land use in the two land tenure types – communal land (indicated as black locations in Figure 4) and white commercial farmland/state land. Land use, vegetation cover and soil erosion were mapped from aerial photographs taken in 1938, 1954, 1965, 1975 and 1988. Kakembo's results are summarised in Figures 6, 7 and 8.

6.1 LAND USE CHANGE

Three land use classes were identified; grazing land, cultivated land and abandoned land. From Figure 8 it can be seen that in the study area the white commercial farms were largely used for grazing, with small areas of cultivation. After independence virtually all of the cultivated land was abandoned. In 1938 the black communal areas were evenly divided between grazing and cultivation. From 1938 to 1988 progressively more of the cultivated land was abandoned and left as fallow land. The grazing land was reduced in area as some was taken up for cultivation. It is not known why so much of the cultivated land was abandoned, but it could have been due to a number of factors. These include exhaustion of fertility, lack of manpower to plough land due to the migrant labour system, and loss of oxen power following drought and the cattle disease rinderpest (Branch 1994). Another important factor, as noted by Kakembo (1997) and Fox and Nel (1998), has been the reduced dependency on local food production due to the availability of external income from remissions and pensions. Together, these factors have led to the phenomenon of 'under-farming' of land as described by Hoffman *et al.* (1999).

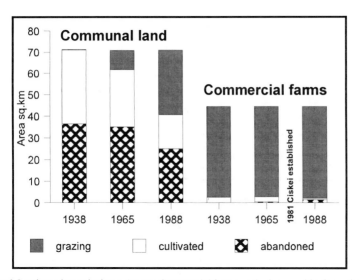

Figure 6. Land use change in the study area. Commercial farms are the former white farms that became state land after Ciskei became independent (based on data from Kakembo 1997).

6.2 VEGETATION COVER

A marked difference in vegetation cover can be discerned between the communal areas and the commercial farms (Fig. 7). The commercial farms comprised a mixture of open woodland and wooded grassland, presumably reflecting a sustainable grazing pressure. After independence access to this land from the communal areas remained closed and, with the removal of the white owned stock, bush encroachment resulted in an increase in open woodland.

The communal areas present a very different picture. As early as 1938 the grazing lands were in poor condition, with sparse grass cover being the most widespread category. A century of confined livestock grazing had already taken its toll. By 1988, small areas of wooded grassland present in 1938 had become almost non-existent, probably as a result of wood harvesting for fuel, fencing animal enclosures and building dwellings. The extraction of wood resources persists to the present day. Hoffman *et al.* (1999) note that while agricultural use of the communal land may be at a low level, the exploitation of natural resources, especially

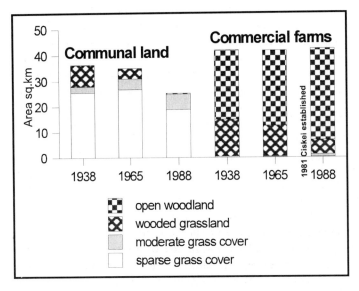

Figure 7. Vegetation change on grazing land (based on data from Kakembo 1997).

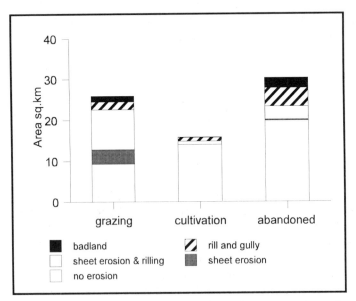

Figure 8. The relationship of erosion class to land use in the communal areas (based on data from Kakembo 1997).

wood, continues. With the loss of woodland in the communal areas of Peddie District, there is a strong likelihood that wood gathering will shift to the former commercial areas.

6.3 SOIL EROSION

Figure 8 shows the relationship between erosion intensity and land use on communal land in 1988 as mapped from the aerial photographs by Kakembo (1997). It should be noted that throughout the period of record, 1938 to 1988, no erosion was discernible on the commercial lands. Sheet erosion was widespread in communal areas in 1938, but rilling and gully erosion were insignificant. By 1988 these more intense forms of erosion were widespread. Gully erosion took the form of badlands, with complete loss of topsoil over extensive areas. From Figure 8 it can be seen that all classes of erosion were concentrated in the grazing land and abandoned cultivated land. Grazing land suffered from widespread sheet erosion and rilling, whereas abandoned land was the locus of much of the gullying in the area, often in the form of highly dissected badlands. These gullied areas represent a serious degradation of resources, well beyond the threshold for rehabilitation back to cultivable land. This association between badland gully erosion and abandoned land has been shown to be a widespread phenomenon throughout the Eastern Cape (Andrew 1991; Whisken 1991). Analysis of the aerial photographs by Kakembo indicated that gully erosion took place after the fields were abandoned. Gully erosion, therefore, was not the reason why cultivation ceased.

Kakembo's study of land degradation in Peddie District demonstrated that the areas under communal land tenure have become severely degraded while those under commercial livestock farming showed no photographic evidence of erosion in any of the years examined. He attributed degradation to heavy grazing pressure, unsustainable cultivation methods which led to the land being abandoned, and collection of wood resources. This conclusion was also reached by Marker (1988), who studied the impact of increasing population densities on soil erosion in another area of the former Ciskei. The erodible nature of the soils adds to the problem, while the transitional nature of the vegetation means that invasion by unpalatable woody shrub species is a further consequence of degradation (Tanser 1997).

6.4 REDISTRIBUTION, RESTITUTION AND LAND REFORM IN PEDDIE DISTRICT

High population densities can be attributed to high natural growth rates, apartheid resettlement policies and the confinement of this growing population into a restricted area. Incorporation of white farms into Ciskei did little to relieve this pressure, as the land was not opened up for settlement. Policies presented in the 1997 White Paper (Republic of South Africa 1997) now create the possibility for resettlement through redistribution into land such as the former white farms. The question remains, however, as to what population densities are sustainable in this environment so as to be able to assess how many people could be redistributed.

According to Branch (1994), by the end of the 19th century about 14 000 people were living at a density of approximately 25 persons per km² in areas which were later to become scheduled land through the 1913 and 1936 Land Acts. This was perhaps already exceeding sustainable levels, although the agricultural practices in the area had previously resulted in the export of agricultural surpluses to adjacent white districts (Bundy 1979). In the mid 1990s the population densities in the communal areas were nearer 75 people per km² and for the whole of Peddie District (including the urban area of Peddie town) the population density was 43 persons per km².

It is highly unlikely that the vacant, former white areas in Peddie can absorb the population from adjacent communal areas at a sustainable level without very major innovations in smallholder agricultural systems. In the 1950s the Tomlinson Commission Report (South Africa 1955) estimated that a black smallholder family practising mixed farming needed approximately 89 morgen (76 hectares) to have an economically viable unit. This works out (for a family of six) as a density of around 8 persons per km². The former white farms of Peddie District covered 870 km², which means that some 6960 people could be settled on them to practise the kind of farming the Tomlinson Commission envisaged. Given that the population of the district in 1996 was estimated at 60 288 (Statistics South Africa 1998), there is only limited potential for expansion in the former white areas and there is certainly no scope for redistribution into Peddie from other parts of the province. Reinvigorating the collapsed irrigation schemes in the western part of the district would be another option to pursue although, again, the numbers of people who can benefit from them will be small relative to the size of the population in the district. Even if much of the population were resettled in residential rather than agricultural schemes, degradation of natural resources, particularly of fuelwood, will continue.

Restitution could follow two directions. The biggest impact would be if displaced white farmers were to put in successful land claims and return to their farms, thus decreasing the amount of land available for redistribution. A number of Peddie's communities have also put in claims to land from which they were forcibly removed in 'white' South Africa. This would, however, only have a small impact in reducing local populations. It is also likely to be some time before all land claims are settled.

Tenure reform, with a move towards individual land ownership, has been proposed as a way to promote improved land husbandry. This is possible if an environment for investment is created and if farm units are large enough to be economically viable. The real danger of granting land tenure within the communal areas is that it creates a division between those with land and the landless, a factor which has been recognised since the days of the Tomlinson Commission Report. Care must also be taken to ensure that tenure reform includes women since they are the majority in these areas. The landless group will still continue to use natural resources unless alternative fuel and building materials are made available (Hoffman et al. 1999).

7. Redistribution, Restitution and Land Reform in the Eastern Cape

Having considered Peddie District, what of the Eastern Cape as a whole? It is clear from Figure 3c that the areas of high biological productivity to the east already have high population densities and, in comparison to Ciskei, there was much less apartheid-driven population displacement in that area. Redistribution or restitution are therefore unlikely options in these areas, although there are cases of communities reclaiming nature reserves. Tenure reform may be a more likely option for bringing about a sustainable, productive agriculture in these high potential communal areas but, as noted above, it brings with it the real danger of creating a landless class of people who still need to harvest environmental resources.

Some redistribution could take place into the high potential former 'white' districts in the centre and north-east of the province and also into the coastal strip in the extreme west. Care, however, must be taken to ensure that increased population densities are not accompanied by degradation of the erodible soils found in these areas. As the case study of Peddie illustrates, it is highly unlikely that there will be enough suitable land to settle all of the people from the black rural areas.

Elsewhere in the province the land does not have a high biological productivity and the scope for redistribution is extremely limited. White livestock farms in the semi-arid north west have themselves suffered widespread degradation under population densities of only two to six people per km^2 (Roux and Opperman, 1986). The economic and security climate has led to wide-scale redundancy of farm workers and a move towards game ranching and tourism as alternative livelihood strategies. Thus there has been a decline in population in these areas. At present the biological productivity per person in this area is amongst the highest for the province, but it must be realised that this is the result of a low biological productivity and an even lower population density. It might be possible to increase the population density from around 0.5 to 2.5 people per km^2, but this only amounts to an increase of 2 people per km^2. The shrubby vegetation can only support herbivores; the land cannot be used for sustainable crop production without irrigation, and perennial rivers are limited to the Great Fish River - Sundays River interbasin transfer schemes which import water from Lesotho.

Land reform policies are concerned with opening up access to land and therefore with a redistribution of population. They reflect a concern with the imbalance in population densities relative to the potential of the land to sustain agriculture. Figure 9 presents a model of land degradation for the Eastern Cape which focuses on problems resulting from high population densities. Studies such as Kakembo's for Peddie District (Kakembo 1997) have shown that in some areas of South Africa's former homelands more people can be equated to more erosion.

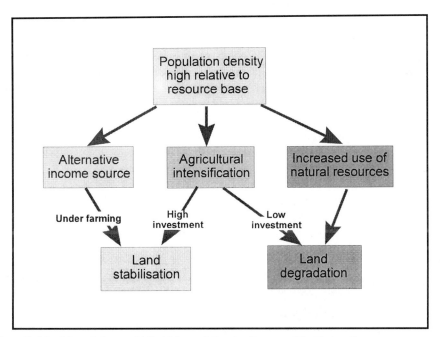

Figure 9. A land degradation model for high population density areas of the Eastern Cape.

In contrast to the findings for example of Tiffen *et al.* (1994), access to outside remittances andpensions have, in part, negated the need to adapt agricultural practices to more sustainable systems as population increases. The adaptation has been rather to abandon cultivation, thus creating a system of 'under-farming' (Hoffman *et al.* 1999). It is likely that the present level of 'under-farming' is in a reasonable balance with present day land potential so that we must now be very careful not to restart another cycle of degradation through poorly informed land policies.

Land degradation under the present system can be linked to two causes. One is the use of unsustainable farming systems which depleted fertility and led to soil degradation. The second is the use of natural resources, especially wood, for fuel and building. High population densities can only be sustained from the land if they are accompanied by an investment of time, labour and resources such as fertilisers. This almost certainly would require a change in the tenure system towards individual ownership, an issue to be addressed by the land tenure reform program. For those with no access to land a sustainable source of fuel and building materials will need to be provided. It is important to acknowledge the full scope of resources provided by the land, not simply to look at its agricultural productivity.

8. Conclusions

This paper has presented a regional scale analysis of prospects for land reform in the Eastern Cape. The implications for degradation have been illustrated by one case study from Peddie District in the former Ciskei. Within this province, white farmland and state land within and contiguous to the former homelands offer some prospect for redistribution or restitution, though the actual areas of suitable land are limited relative to the available excess of population. Moreover it has been shown that most of the highest potential land is already heavily settled. The historical association between increasing population densities and land degradation evident from Kakembo's study (Kakembo 1997) points to the need for a sensitive approach to resettlement, especially in areas of marginal biological productivity. Land tenure reform may promote the adoption of more sustainable farming methods, but does little to address the issue of high population numbers which continue to put pressure on the land's various resources. Restitution is unlikely to have a major effect in the short term, and may well block land development for some time to come.

9. Acknowledgments

The authors wish to acknowledge the helpful comments of the two referees who reviewed the first draft of this paper.

10. References

Andrew, M. (1991) A geographical study of agricultural change since the 1930s in Shixini Location, Gatyana District, Transkei, Unpublished MA thesis, Department of Geography, Rhodes University.
Branch, B. (1994) Land management and control in the rural areas of South Africa: a case study of the Peddie District, Unpublished Honours dissertation, Department of Geography, Rhodes University.
Bundy, C. (1979) *The rise and fall of the South African peasantry*, Heinemann, London.
Cobbett, M. (1988) The land question in a post-apartheid South Africa: a preliminary assessment, in C.R. Cross and

R. Haines (eds.), *Towards Freehold: Options for Land and Development in South Africa's Black Rural Areas*, Juta, Cape Town, pp. 60-72.

Cross, C.R. and Haines, R. (eds.) (1988) *Towards Freehold: Options for Land and Development in South Africa's Black Rural Areas*, Juta, Cape Town.

Department of Land Affairs (1999) Land Reform in South Africa [Online]. Available: http://land.pwv.gov.za/briefin2.htm [1999, September 7].

Enpat. Environmental Potential Atlas [CD-ROM]: (1998). Available: Department of Environmental Affairs and Tourism [1998, November].

Fox, R.C. and Tipler, D.J. (1996) Rural and urban population trends in the Eastern Cape Province 1936-1991, Institute of Social and Economic Research, Rhodes University, Development Studies Working Paper 68.

Fox, R.C. and Nel, E.L. (1998) Pension payouts, periodic marketing and the continuance of urban dependence in rural South Africa, *South African Geographical Journal* **80**, 108-110.

Hoffman, T., Todd, Ntshona S. Z. and Turner, S. (1999) *Land degradation in South Africa*, Department of Environmental Affairs and Tourism, Pretoria.

Jolly, C.L. (1994) Four theories of population change and the environment, *Population and Environment* **16**, 61-90.

Kakembo, V. (1997) A reconstruction of the history of land degradation in relation to land use change and land tenure in Peddie District, former Ciskei, Unpublished MSc thesis, Department of Geography, Rhodes University.

Laker, M.C. (1999) Soil resources: distribution, utilization and degradation, in R.C. Fox and K.M. Rowntree (eds.), *The Geography of South Africa in a Changing World*, Oxford University Press, Cape Town, pp. 326-360.

Marker, M. (1988) Soil erosion in a catchment near Aalice, Ciskei, southern Africa, in G.F. Dardis and B.P. Moon (eds.), *Geomorphological Studies in Southern Africa*, Balkema, Rotterdam, pp. 267-276.

Mashile, G.G. (1988) The economic environment of rural land policy and planning, in C.R. Cross and R. Haines (eds.), *Towards Freehold: Options for Land and Development in South Africa's Black Rural Areas*, Juta, Cape Town, pp. 50-59.

Republic of South Africa (1995) *Rural Development Strategy of the Government of National Unity*, Government Gazette 16679, Pretoria.

Republic of South Africa (1997) *White Paper on South African Land Policy*, Department of Land Affairs, Pretoria.

Roux, P.W. and Opperman, D.P.H. (1986) Soil erosion, in R.M. Cowling, P.W. Roux and A.J.H. Pieterse (eds.), *The Karoo Biome: a Preliminary Synthesis. Part 1 - Physical Environment*, South African National Scientific Programmes Report No. 124, Foundation for Research and Development, Council for Scientific and Industrial Research, Pretoria, pp. 92-111.

South Africa (1955) *Summary of the Report of the Commission for the Socio-Economic Development of the Bantu Areas within the Union of South Africa*, Government Printer, Pretoria, UG 61/1955.

Statistics South Africa (1998) *The People of South Africa. Population Census 1996* [On Disk] Available: Statistics South Africa [1999, March].

Stocking, M.A. and Elwell, H.A. (1973) Soil erosion hazard in Rhodesia, *Rhodesia Agricultural Journal* **70**, 93-96.

Tanser, F.C. (1997) The application of a landscape diversity index using remote sensing and Geographical Information Systems to identify degradation patterns in the great Fish River Valley, Eastern Cape Province, South Africa, Unpublished MSc thesis, Department of Geography, Rhodes University.

Tiffen, M., Mortimore, M., and Gichuki, F. (1994) *More People, Less Erosion: Environmental Recovery in Kenya*, John Wiley, Chichester.

Turner, B.L., Hyden, G. and Kates, R.W. (1993) *Population Growth and Agricultural Change in Africa*, University Press of Florida, Gainesville.

Van Zyl, J.M., Kirsten, J.M. and Binswanger, H.P. (eds.) (1996) *Agricultural Land Reform in South Africa: Policies, Markets and Mechanisms*, Oxford University Press, Cape Town.

Weaver, A van B. (1988) Factors affecting the spatial variation in soil erosion in Ciskei - an initial assessment at the macroscale, in G.F. Dardis and B.P. Moon (eds.) (1988) *Geomorphological Studies in Southern Africa*, Balkema, Rotterdam, pp. 215-227.

Whisken, J.B. (1991) An assessment of the effectiveness of betterment planning in combating soil erosion, Unpublished Honours dissertation, Department of Geography, Rhodes University, Grahamstown.

AUTHORS

R.C. FOX[1] AND K.M. ROWNTREE[2]

Department of Geography
Rhodes University
PO Box 94, Grahamstown
South Africa

[1] R.Fox@ru.ac.za

[2] K.Rowntree@ru.ac.za

CHAPTER 12

ESTABLISHING FARMERS' UNDERSTANDING OF SOIL HEALTH FOR
THE FUTURE DEVELOPMENT OF 'USER-FRIENDLY' SOIL MONITORING
PACKAGES

LISA ALEXANDRA LOBRY DE BRUYN

1. Abstract

The main purpose of examining farmers' understanding of soil health is to incorporate this understanding into monitoring packages on land condition. The findings of this project will allow farmers to develop a soil health checklist which is commensurate with their understanding of soil health - its definition, recognition and measurement. By developing a soil health checklist with farmers the research process acknowledged the importance of local conditions, including farmers' existing knowledge about soils. The examination of farmers' understanding of soil health was conducted in the north west cropping region of New South Wales, Australia. The project took a qualitative research approach and employed several techniques (interviews, soil testing and focus group discussions) to acquire and validate farmers' understanding of soil health.

This paper reports on some preliminary results, focussing on farmers' soil health checklists - the features they use, how they recognise those features, especially the language they use to describe a healthy and unhealthy soil, and finally the techniques they use to measure those features. The most spoken about features farmers used to identify soil health were plant growth and soil feel. Other properties which were consistently mentioned by farmers in relation to identifying soil health were organic matter, plant roots and soil life. Interestingly, farmers in workshops then dismissed some features, such as weeds, which had been mentioned frequently by farmers in interviews as being more related to identifying a soil type than the health of a soil. Ways of measuring or determining the health of a soil were informal and related closely to observation skills, but are usually carried out while conducting other farming operations. Nearly all farmers were unaware of any formal soil monitoring packages. To encourage and motivate farmers to monitor soil health will require the development of a package or soil health checklist which allows for adaptation to local conditions and soil types, provides a model or protocol for interpretation, and is disseminated to farmers by people from whom they regularly seek advice.

2. Introduction

Over the years the accepted approach to solving environmental problems has been to derive a technical solution and to transfer the solution to the end users. In a commentary

187

A.J. Conacher (ed.), Land Degradation, 187–206.
© 2001 *Kluwer Academic Publishers. Printed in the Netherlands.*

by Molnar *et al.* (1991:87) they, as scientists, claim that: "for the most part, we already know how to protect the environment for the long term ... . What is lacking is widespread adoption of what is already known". This type of attitude has disenfranchised farmers from the knowledge construction process and reduced their involvement in land management and problem solving to being passive recipients of expertise from outside (Röling 1994). As Cornwall *et al.* (1994:101) stated: "conventional research and extension aims to produce and convey recommendations to remedy the absence of knowledge about certain processes ... the process assumes that farmers are ignorant about certain elements of their practice and, therefore, renders their knowledge invisible". However, the transfer of technological solutions to farmers has failed to solve their environmental problems, and its failure can be attributed largely to ill-defined problem identification, disregard of local knowledge and design failure rather than adoption failure by 'recalcitrant' end users.

The 'blame the client' syndrome has been repeated in the development of sustainability indicators and protocols for monitoring land condition, as shown by the poor acceptance by practitioners of science-derived indicators or protocols for environmental monitoring (Bosch *et al.* 1996b). Most of the information and procedures being generated on environmental condition of soils are by policy makers and scientists, and are perceived to be for scientific use and government reporting, and not for use by practitioners or decision makers. Bosch *et al.* (1996a:12) emphasise that many "environmental programmes fail to become an integral part of management because they are not designed to help decision-makers". If farmers are to determine whether their farm goals are being met, and/or whether changes in farm management are leading towards a more sustainable farming system, then they need to be able to monitor these areas. Yet their needs are not being addressed in the appropriate manner (Webster 1999). An easy to use and reliable checklist to monitor trends in soil health would be a demonstrable aid to achieve this understanding, especially if it was commensurate with farmers' understanding. However, few practical guides exist in Australia, although attempts have been made by farmer-led organisations such as the Land Management Society (LMS) by developing a Farm Monitoring Kit (Hunt and Gilkes 1992).

One reason for the slow take-up rate of sustainability indicators at the farm level is that the soil health measures derived by scientists are not 'locally relevant' to farmers. Scientists have ignored the distinctiveness of place (including the cultural, social and ecological spaces) by creating generic monitoring packages, and they have not acknowledged that there are 'multiple ways of knowing' (Kloppenburg 1991, 1992). Farmer groups, such as the Liverpool Plains Land Management Committee (located in the study region), have also been critical of research and extension efforts into sustainable land management. They feel that, in the past, scientists have treated "all farmers as a homogenous group", and they have recommended 'one size fits all' solutions to environmental problems (Industry Commission 1998:207). This lack of understanding and acknowledgment of the individuality of farmers and their problems has led to questioning of "the appropriateness of extension methodologies" (Industry Commission 1998:207). Extension as defined by Röling (1988:49) is: "the professional communication intervention deployed by an institution to induce change in voluntary behaviours with a presumed public or collective utility". Other reasons for farmers not to embrace soil sustainability indicators developed by scientists are that the indicators: require too much technical expertise; are not cost-effective; are time-consuming to conduct, and the results are not easily interpretable by farmers. An example of a complex, 'user unfriendly' framework is outlined in Dalal *et al.* (1999).

Additional explanations for lack of interest in sustainability indicators are detailed by Pannell (2000:135), but in summary he believes that their derivation has been *ad hoc* and not based on any consistent, underlying conceptual framework and with "no close link to management decisions". Another example of a framework for monitoring land resources, but which has limited acceptance, in Australia, is Environmental Management Systems (EMS) (Carruthers and Tinning 1999). In *Australian Farm Journal,* the acceptance of such a scheme was queried unless incentives were provided to induce landholders to participate. Voluntary investment in environmental management was unlikely to occur when the majority of broadacre agriculture businesses are unprofitable (Francis 1999). However, the ability of farmers to monitor land condition could act as a strong incentive to change management practices or confirm to the farmer that their practices are maintaining or enhancing the soil resource. In addition, preventative management of soil problems is preferable to curative management, especially before the problem becomes intractable. This is another important reason to assess farmers' understanding of soil health.

For this research project and its outcomes to be relevant to farmers it needed to be inclusive and engage participants in a two-way interaction as demonstrated by other areas of land management, such as Landcare (Campbell 1994; Chamala and Keith 1995), adult education, extension and development programs (Röling 1994). The marginalisation of farmers' knowledge is apparent in the literature (Kloppenburg 1991; Murdoch and Clark 1994), and the research goal was to engage in pedagogy "with, not for" farmers (Freire 1970:33), or to try to learn from farmers (Chambers 1983). Analogies can be drawn between a traditional education and research process which features "a one way flow of facts to fill empty vessels" [farmers] and which persists in much of the world, in what Friere has called the "banking system of education" (Chambers 1997:60). The intention of this study was to engage in a two-way process with farmers and to make their understanding of soil health 'visible'. The use of 'patronising' language in relation to farmers' understanding of land management practices and their influences on soil health has also been identified by some researchers and practitioners. In a recent forum on soil health (Daniells 1997:3), the analogy was made, by a scientist, between soil health and human health, and that a farmer would seek 'expert' advice on how to solve a soil problem. The process being advocated here is more aligned to self diagnosis and empowerment, thereby enabling farmers to identify and solve their own soil problems, and relying less on 'specialists' to solve their problems for them. However, soil science as a discipline finds the popularisation of soil health demeaning, and actively discourages "premature institutionalizing [of] a concept for which there remains significant questions and criticisms" (Sojka and Upchurch 1999:1050).

By developing a soil health checklist with farmers, the research process acknowledged the importance of local conditions and included farmers' knowledge and interpretation of the soil health concept. The examination of farmers' understanding of soil health was conducted in the north west cropping region of New South Wales,

Australia. The project used a qualitative research approach and employed several techniques (interviews, soil testing and focus group discussions) to acquire and validate farmers' understanding of soil health. This paper reports on some preliminary results, focussing on farmers' soil health checklists - the features they use, how they recognise those features, especially the language they use to describe healthy and unhealthy soil, and finally what techniques they use to measure those features. In conclusion the paper examines which format and mechanisms are best used to disseminate a farmer-derived soil health checklist.

3. Methodological Approach and Research Phases

A decision was made that information would be most useful if collected and understood from an insider's point of view. Hence the participants - farmers - were interviewed on their land and the conversation recorded and transcribed so as to faithfully reproduce their words. Qualitative research allows the researcher to investigate sub-cultures, to seek out links with the larger social and political environment and to tease out from the tangles of customs, values and concepts some alternative solutions. Freire's concept of praxis (1970) was used as a conceptual framework for this study. The intellectual history of the term praxis, "provides grounding for activist, collaborative, constructive science" (Seng 1998:37).

Chambers' (1997) criteria for rigour - trustworthiness and relevance - were also applied throughout the research process. Trustworthiness is the quality of being "believable as a representation of reality" (Chambers 1997:158), and relevance refers to the practical utility for learning and action. He states that "the purpose of rigour is to assure quality" (Chambers 1997:158). In science, under the positivist paradigm, quantification, statistical validity, repeatability and objectivity define rigour. These requirements in positivist rigour result in simplification of the 'objects' being studied and "miss or misrepresent much of the complexity, diversity and dynamism of system interrelationships... In consequence, they are often not useful" (Chambers 1997:158).

The research program was organised in four phases:

- Interviews.   75 farmers and 26 service-providers were interviewed.
- Soil Testing.   A healthy soil and a control soil were sampled on each farm.
- Data Analysis.   A qualitative software package called QSR NUD*IST4 (details to follow in section 3.4) was used to analyse the interviews, and soil analysis was conducted on the soil samples.
- Workshops.   Eight workshops were conducted and farmers' advice sought on interpretation of interviews and relevance of existing soil monitoring packages.

3.1. SELECTION OF PARTICIPANTS IN THE INTERVIEWS AND WORKSHOPS

Seventy five interviews were conducted over a large region of north-west New South Wales, Australia, covering an area 400 by 500 kilometres (Fig. 1). Every effort was made to ensure that participants were representative of the broader farming community. Initial contacts with 50 farmers (mostly from the eastern portion of north-west NSW)

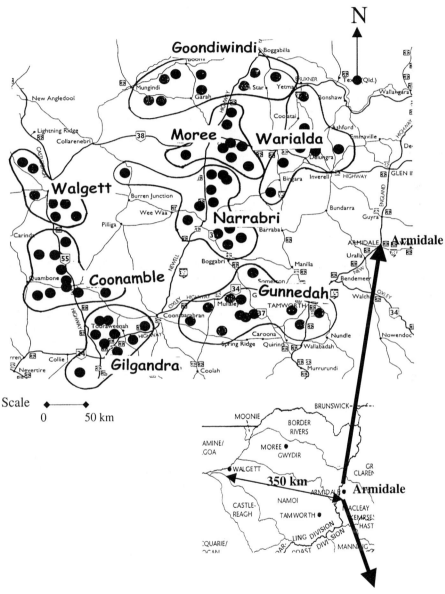

Sydney, Australia (505 km south east)

Figure 1: Locations of farmers who were interviewed in the north-west cropping region of New South Wales (not to scale). The dots represent farmers and the lines define the borders of farmers involved in workshops centred around the rural town (place name shown).

were obtained through NSW Agriculture, which first conducted a survey in the late 1980s (reported in Daniells *et al.* 1996). When these farmers were interviewed it was found that, in the main, there was an air of prosperity and success amongst them. It seems that although these farmers had originally been selected randomly by picking farms from a grid, a number of farmers (40%) had left the district, retired or dropped out of farming. In order to obtain a representative sample, within the north-west region, various other means were tried. The first method was to select names randomly from an electoral roll. This, however, proved to be very time consuming and not very fruitful. When making numerous phone calls to people to ask if they were cropping farmers, nevertheless, people were very willing to talk and made comments such as "You might like to try Mr X up the road". This selection method was taken further as it accorded with the qualitative philosophy and gave the project access to farmers who were not necessarily members of local agricultural organisations (the traditional first contact for a list of addresses for survey purposes). Ultimately, local organisations and members of the community became contact sources for wheat farmers in the district. After visiting the selected properties and interviewing the farmers it became apparent that the selected cohort was a mixed group.

In the final analysis the types of farmers interviewed were quite diverse. They are very similar in demographic profile to that found in State or Australia wide data presented in the Australian Bureau of Agriculture and Resource Economics (ABARE) *Farm Surveys*. The greatest proportion of farmers owned the farm, a quarter were part-owners, 5% (or four in number) were managers and only one person was a tenant. Over the eight regions the majority of farmers were between 40 and 55 years old, which explains why their farming experience was, on average, 28 years (range 1-62 years). Most of the farmers had also been farming in their district for at least 24 years, which means they had experience with and knowledge of the soil types in their district. Farm sizes varied from 66 to 30 000 ha (average 4173 ha). Most of the participants (63%) considered that they made all management decisions on the farm, 20% of the participants considered that they made 75% of the decisions, and the other 17% were minor parties in decision making. No participant indicated that they had no part in any decision making.

3.2. INTERVIEW PROTOCOL

Apart from the eleven interviews which were recorded over the phone, most farmers were interviewed in their homes over a two hour period. The dialogue between the farmer and interviewer was recorded and transcribed verbatim. Field notes were made later to reflect on how the interview had proceeded and the quality of the information given by the farmer. Holding on to the emotional energy of what transpired from the interviews was important because it retains the nature of the interview which would otherwise be little more than cold and stark words on paper.

During in-depth interviews the interviewer had a guide consisting of core questions and a series of follow-up prompts, but the interview was essentially a conversation as recommended by experienced researchers in the field (Spradley 1979; Minichiello *et al.* 1990; Kleinman 1991). The intent was to draw out the interviewee's thoughts and information which they use in recognising soil health and how they apply their understanding. The interviews were intended to gain 'thick descriptions' (meaning reporting details in context in sufficient detail to allow others to determine when another situation is similar enough to transfer findings to another setting (Lincoln and Guba 1985)) of farming systems, present and future problems and methods of problem identification. Farmers were also asked to describe healthy and unhealthy soils and were encouraged to use the words they were comfortable with or the kind they would use when talking to other farmers. The use of the pre-determined agenda of many surveys which claim to be measuring opinions or attitudes (e.g. Jenkins 1998) is rarely made explicit when results are presented. In the present study it was also found that farmers have a negative attitude to the presumption that they would freely give their time to fill out an anonymous survey - for little thanks or brief explanation of purpose, as was also reported in the work for an RIRDC (1997) report:

> ... the process of consulting with farmers about sustainability indicators was just as informative as the outcomes themselves. Early on, facilitators found it necessary to 'bring the indicators to life' by explaining their background and justifying the need for them. Without such justification, say the project coordinators, farmers were profoundly uninterested! (RIRDC 1997:5).

A decision was made to break-up the interview as it proceeded from the more familiar subject areas to issues which were discussed less frequently, by using a value setting exercise. The value setting activity provided variety and a 'fun' element, allowing the second half of the interview to be more relaxed and informal when more complex and reflective questions were asked. The 'exercise' consisted of values which were considered important for a sustainable farming system, and farmers were asked to rank them from most important to least important. Each major value had subsidiary values. These values were mounted on coloured card, laminated, and cut out. Farmers ranked the major values first, followed by the subsidiary values (Table 1).

Table 1: Hierarchy of objectives for achieving a sustainable farming system

Major Values	Subsidiary Values
PHYSICAL ENVIRONMENT	Water, Soil, Air
FOOD PRODUCTION	Quantity, Sustainability, Quality
NATURAL ENVIRONMENT	Flora, Fauna, Landscape
BASIC INCOME / PROFIT	Farm level, Regional level, National level
HEALTH /WELL BEING	Farm animals, Urban people, Rural people
EMPLOYMENT	Farm level, Regional level, National level

Source: Vereijken (1992).

Farmers enjoyed the activity, were usually very thoughtful about the process, and invariably explained in great detail their reasoning to the interviewer as they went along. It also provided lively debate if other members of the household were around. Where farmers had been hesitant to share information, this activity provided a 'no knowledge, only feelings' break, an opportunity for general conversation and proved to be an effective communication bonding tool.

3.3. SOIL TESTING - IDENTIFYING CHANGES IN SOIL HEALTH

The purpose of the soil survey was to compare the intuitive understanding of each farmer's assessment of soil health against standard soil science procedures. This is not to say that soil science can judge the health of a soil more accurately than farmers, but to give credence to the features/characteristics that farmers use to identify a good soil. By validating their appraisal techniques it is hoped that these features can be used by the farmer to monitor soil condition more consistently and formally.

The following procedure was used for soil testing at each farm. The farmer would be asked to locate what he or she believed to be a healthy soil, based on the features they had spoken about in their interview. The rationale for phrasing the question this way was to avoid leading the farmer, and to prevent them choosing a paddock to satisfy 'the researcher's requirements'. Often they would respond by saying "What do you want?", and it would be reiterated that the selection of the 'good soil' paddock was their choice, based on their understanding of soil health. Some farmers had reflected on their selection, and their decision making process was discussed as well as the land management history of the paddock. The farmers also located a undisturbed site on the same soil type where the soil had not been cultivated, and this was referred to as the 'control soil'. The purpose of the control soil was to act as a reference point for the 'good' soil, and to determine whether soil condition had declined, remained stable or improved since the land has been farmed. In places this meant choosing an area which may have been grazed. The intention of the soil sampling was not to tell the farmer what levels or amounts of nutrients or percentage soil carbon are required to maintain a healthy soil. Rather the objective was to empower farmers to collect their own data and build their own soil health trend lines, leading to better informed decisions on soil condition and a greater understanding of how their management practices were impinging on soil health. Field notes were taken of the farmer's comments, after the farmer had left, and these were added to their interview data. There were also occasions when the farmer assisted in the soil profile sampling and description. This was also noted.

3.4. QSR NUD*IST4 ANALYSIS OF INTERVIEWS AND WORKSHOPS WITH FARMERS TO VALIDATE SOIL HEALTH CHECKLIST

The interview data were coded - building themes and interpreting farmers' responses to the semi-structured questions (14 core questions) using a qualitative software program called QSR NUD*IST4 (Qualitative Solutions and Research, Non-numerical, Unstructured, Data, Indexing, Searching and Theory building). The software program allowed for thematic analysis and could accommodate the building of evidence, reiteration and an organic construction and reconstruction of ideas. This process required expertise in soil science to categorise the farmer responses, and to understand that the answers to the questions may be distributed throughout the interview transcript and not just be located in one place. The decision trail was documented whereby each theme could be verified by tracing it back to the farmer's original language and be "explained by the researcher's interpretive scheme" (Koch 1994:985). QSR NUD*IST4 allows data to be coded as text units (which are farmers' responses to a question) from any part of the interview transcript and located in one or a number of nodes (themes). Each text unit is defined by a header which clearly shows each farmer's code and location, and hence ensures that the integrity of each farmer's voice is retained.

For the data analysis and workshops the region was divided into eight districts which have at their centre a rural town, with eight to ten interviewed farmers residing in a 50 km radius from the town centre (Fig. 1). The workshop discussions with farmers were designed to assist in the development of a soil health checklist which can be validated at the local level rather than deriving a generic package for the whole region. Initially, 67 of the 75 farmers indicated they would attend their local workshop, but due to conflicting interests some were unable to participate. However, the deliberations and amendments to each district soil health checklist were sent to each farmer who had indicated they would attend, for checking and further validation. In the end, 45 farmers and three associates attended, on average six farmers per workshop. The purposes of the workshops with farmers were to:

- verify with farmers that the interpretation of their interview transcripts accurately reflects their current understanding of soil health;
- give farmers the opportunity to add to, amend and validate the soil health identification and measures checklists, especially on the terms they use to distinguish between a healthy and unhealthy soil;
- present farmers with soil/agronomic information and ask them to critique these brochures, manuals, soil test kits and decision support systems in terms of their presentation, usefulness and relevance; in addition, to relate the outcomes of this session to how best the soil health checklist could be packaged and disseminated to farmers;
- view the soil testing findings and a district profile on soil health, and examine the role of soil testing in monitoring land condition, and
- obtain farmers' opinions on the need for a soil health checklist and whether they think it would help them to monitor land management practices and their influence on soil health.

4. Preliminary Results of Soil Testing, QSR NUD*IST4 Analysis of Interviews and Workshops

4.1. SOIL TESTING

Soil testing revealed several issues associated with the frequency, use and effectiveness of farmers' own soil testing procedures and findings. One aspect was that few farmers conducted their own soil tests, and many had little idea where the soil sample/s had been taken. The usual procedure was to have the soil tested three to four months before planting, mainly for phosphorus, nitrogen and sometimes micro-nutrients such as zinc and sulphur requirements for the new season's crop. Soil testing was conducted every three to five years, and while records were kept, very few farmers compared the results from one soil test to the next. Hence any comparisons of soil tests to form a trendline of soil conditions would be meaningless, largely as a consequence of soil variability and not knowing where the previous soil test had been conducted. To make soil testing meaningful for monitoring soil health, soil samples would need to be taken in a consistent and repeatable manner. District agronomists often take soil samples, arrange for their analysis and interpret the results, usually for fertiliser requirements. Many farmers in this study have not considered the concept of comparing a control soil with a paddock soil to examine the impact of their farm management. When soil test results were presented and explained at workshops, it became clear that many farmers had difficulty interpreting the soil tests without translation into layman's terms.

The soil testing revealed the following characteristics about the control and 'good' soil. The most obvious differences between control and 'good' soil (as located by farmers, see section 3.3) were in soil pH, percentage carbon and available phosphorus (Fig. 2). Usually the good soils were slightly more alkaline than the control soils, which is a consequence of cultivation which brings more alkaline subsoil to the surface. One expected finding was the decline in percentage carbon in the 'good' soil, which was typically 50% lower than in the control soil. What surprised most farmers was the loss of available phosphorus, with control soils usually having two to three times more available phosphorus than the 'good' soil. This reflects the lack of fertiliser use by farmers and their lack of appreciation of the amount of phosphorus which had been removed in cropping.

4.2. QSR NUD*IST4 ANALYSIS OF INTERVIEWS AND WORKSHOPS

The following section summarises from farmers' interviews the features they use to identify a healthy soil, and how they measure soil health. Table 2 shows a range of features used by farmers to identify whether a soil is healthy or unhealthy. The features are derived and ranked on the basis of a qualitative software analysis package - QSR NUD*IST4 - which allows responses to 11 core questions from 75 interviews to be placed under identified themes. In addition to interview data, Tables 2 and 3 include farmers' priorities as ascertained from workshop discussions with 45 of the 75 farmers interviewed. Twenty two features which farmers used to recognise soil health were identified through analysis of interview transcripts. There were clear district variations in features used but also some strong commonalities. The most frequently and consistently mentioned features amongst farmers were plant growth, soil feel, and organic matter (Table 2). Table 2 shows that nearly 100% of the farmers mentioned plant growth and soil feel with at least five to six times the number of text units coded than any other feature. Plant growth featured frequently in farmers' conversations about soil health, with a clear link between a healthy plant and a healthy soil. Within 'plant growth', the form and colour (dark, deep green) of the plant was the best indicator of its vigour. In conjunction with plant growth farmers often examined the plant roots (development and biomass) to identify hardpans and soil structure decline. By far the most animated terms for a healthy and an unhealthy soil were recorded under the theme 'soil feel'. For instance, a healthy soil would be described as friable, loose and easy to work, seeds germinate freely in it, soft and crumbly, fluffy, going to fall apart, and nice and soft. In contrast, an unhealthy soil would be described as big chunky pieces, shiny to pull apart, glazed, powdering and running together, it's not flowing, powdered

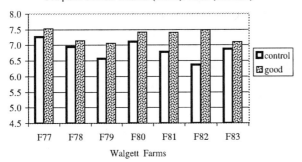

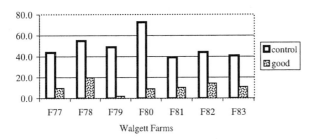

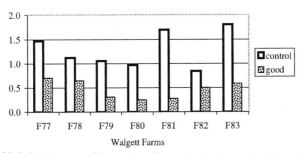

Figure 2: Variation between soil properties of control soil and 'good' soil samples (0-5 cm) as identified by farmers located near Walgett. F (77 to 83) identifies individual farms.

concrete, ground is packed hard, just massive, ground tighter, bit of concrete, setting down, setting like concrete. Farmers were aware that the presence of organic matter was very important to soil health, but usually associated organic matter with the biomass (live or dead) above the soil rather than root biomass or humus. They were also unclear as to how much organic matter was required in a good soil. It was also only through workshop meetings with farmers that it became apparent that several features widely mentioned by farmers in response to questions about soil health were later disregarded as soil health parameters and used more in the recognition of soil type. This applied particularly to the presence or absence of certain weeds.

Table 3 outlines the types of measures farmers use to evaluate soil health. Farmers'

Table 2: Ranking of soil health features used to identify soil health, as mentioned by farmers in interview, and in addition their evaluation in workshops to ease of recognition and importance in identifying soil health of each feature.

% of farmers who mentioned the feature	Number of times a feature was mentioned	Features used to identify soil health	Easy to recognise?		Important to use in identifying soil health?	
No of Farmers interviewed n= 75			No of farmers in workshops n = 45		No of farmers in workshops n = 45	
			YES	NO	YES	NO
99	610	Plant Growth*	22	23	38	7
97	527	Soil Feel ✓	40	5	36	9
80	215	Soil Erosion #	41	4	33	12
79	241	Organic Matter *	18	27	40	5
72	216	Weeds #	34	5	20	19
63	185	Plant Roots ✓	37	8	39	6
61	145	Soil Life *	25	20	40	5
59	113	Soil Texture ?	35	10	33	12
59	124	Hardpans ✓	39	6	43	2
49	85	Plant colour ?	44	1	34	11
49	77	Water Absorption *	35	10	42	3
44	151	Plant disease?	32	13	31	14
39	62	Soil Colour #	20	3	10	13
31	56	Sodicity *	13	27	27	13
28	43	Native Vegetation #	28	7	20	15
25	40	Salinity ?	1	7	7	1
13	14	Scalding ?	13	5	17	1

These annotations are based on feedback from 45 farmers at workshops.
* Soil health features which farmers found difficult to recognise, but appreciated their importance to deciding if a soil was healthy.
# Soil health features which farmers felt were relatively easy to identify but were not considered critical in identifying a healthy soil.
✓ Soil health features which farmers found easy to recognise and felt they were also important in deciding if a soil was healthy.
? Soil health features which farmers found were relatively difficult to identify and which they considered relatively unimportant for identifying a healthy soil.

assessment of soil health was often incidental and done while completing other tasks such as cultivating, driving around the paddock, testing for soil moisture, checking for weeds, putting in a dam, taking a soil test or building a fence. A common phrase used by farmers was 'scratching around'. They would often pick up the soil and see how it broke up, or use the 'boot test' where they would kick the soil and see how much stuck on the end of the boot. However, both these measures were used more to assess soil moisture levels than soil health.

In direct response to the question: "How are you monitoring the condition of the soil?", the 75 farmers gave a variety of responses ranging from soil testing (42 farmers), observations (22 farmers), yield (22 farmers), protein of crop (18 farmers), plant growth (19 farmers), paddock history (15 farmers), to soil moisture probe (13 farmers). Even though soil testing was mentioned the most frequently, little detail on how the results were used was given, and usually soil testing was combined with other indicators such as crop growth, observation and yield and protein levels of crop. For instance F40: "Well, our records give us our yields and protein. We'll probably do soil tests every two or three years or something. But our yields and protein, that's the main one". It is also clear that most farmers use a combination of techniques (see quote below from F35), and only a few farmers said they were not monitoring their soil condition. Other techniques or measures used infrequently by farmers (only one or two farmers mentioned when interviewed) to monitor soil health were signs of erosion, earthworms, smell, organic matter levels and using a computer.

Q12: How are you monitoring the condition of the soil?
F35: I think we sort of covered it. We scratch, we work it, we soil test it, watch for worms, just the feel of it you know the way it looks, texture, fluffy or not, that its easily worked, whether it would grow lots of wheat and have a high protein, what sort of weeds that grows, whether yours is better than the neighbours, or as good as your neighbours you know.
I: Just from those techniques you've mentioned, which one would you use most often in monitoring the condition of the soil?
F35: Whether the texture and the fluffiness were right.
I: What is your opinion of using yields as an indicator of soil condition?
F35: Good, but its after the event isn't it?

However, there was, once again, a hands-on knowledge of monitoring techniques. One of the most commonly mentioned ways of monitoring the soil relied on observation skills with little formal recording structure, with records kept in their pocket diary or head. Some of the farmers' language supports this as shown by the following excerpts: "you see it with your eyes", " I guess look", "with my eyes", "basically visually …", "just basically by observing …", "looking at it" , "mainly once again visually. I keep a careful eye on the places where I see water running …", " I would rely mainly on my visual appraisa l…", "just general observations ...", "that's basically it. Generally looking at it …". Observations ranged from just "looking at the crops", to "pulling up plants, examining roots, probing the soil …". Farmers were often looking at the crop, and its growth through the season, as well as the way the soil feels and its workability when cultivating or seeding.

Yield and protein levels of the crop seemed to go hand in hand, and often they were talked about in unison. Farmers' opinion on yield as an indicator of soil condition was asked as a prompt after the questions on how they monitor soil condition and what they used most. Opinion on yields as an indicator of soil condition was equally divided between those who thought it was "the best" and the remainder who were unsure about its reliability to monitor soil condition, mainly due to seasonal variations. F2: "... you don't put too much emphasis on yield because different varieties suit different climates in different areas but its a good indication and the quality is another one…and of course

Table 3: Ranking of measures used to monitor soil health, as mentioned by farmers in interviews, and in addition their responses in workshops to questions on use, usefulness and relevance to the development of a soil health checklist of the measures they identified for monitoring soil health.

% of farmers who mentioned the feature No of farmers n= 75	Number of times a feature was mentioned	Technique used to measure soil health	Do you use? Number of farmers in workshops n = 45		Easy to use? Number of farmers in workshops n = 45		Need to include in soil health checklist? Number of farmers in workshops n = 45	
			YES	NO	YES	NO	YES	NO
99	292 or 263	Yield or Protein Level of Crop *	40	1	34	7	34	7
97	394	Soil Testing *	44	1	36	9	45	0
92	451	Observation ✓	45	0	41	4	41	4
87	167	Soil Moisture *	38	4	38	4	34	8
73	180	Digging ✓	42	3	43	2	41	4
68	125	Paddock History ✓	43	2	40	5	43	2
67	102	Soil testing – for deficiencies *	43	2	34	11	43	2
64	126	Cultivating #	34	11	38	5	33	11
58	150	soil pH #	43	2	35	10	39	6
49	83	Driving across paddock*	38	7	42	3	36	9
47	47	Soil Depth #	28	14	27	15	28	14
40	90	Soil testing - organic carbon*	36	9	35	10	41	4
31	39	Walking across paddock #	40	5	39	5	35	10
23	52	Aerial Photographs ✗	12	32	14	30	19	25

These annotations are based on feedback from 45 farmers at workshops.

✓ Soil health measures which farmers use, find easy to use, and think should be included in the development of a soil health checklist

* Soil health measures which farmers use, but find difficult to use yet value their importance to the development of a soil health checklist.

Soil health measures which are used by no more than 75% of farmers, but which they either find difficult to use (at least 22%), or easy to use but farmers are diffident about their contribution to the development of a soil health checklist.

✗ Soil health measures, which most farmers do not use, which they find difficult to use (at least 50%), and which they think are not required for the development of a soil health checklist.

if there is some sort of problem in that crop, you investigate through soil tests, knowledge and try and eliminate". Those farmers who thought yield was an "excellent monitor" usually felt that the "higher the yield the better the soil".

Another frequently mentioned mode of monitoring soil condition is exemplified by F27: "probably by visual assessment of exactly how my crops are travelling. I use the crops as an indicator of whether the soils are doing their job or not". In addition to crop growth, farmers use paddock (field) history such as the type of crops grown, and rotating crops to build up nitrogen, reduce weed and disease problems to monitor soil condition. The following quote is an example of how paddock history can be used to monitor the soil condition. F69: " ... and the history of what we are doing from that point on is obviously going to be very relevant so we will be forming an historical thing of what paddocks have produced and what they've done and try and take it from there. The yield and quality of the grain obviously and protein is a reasonably good indicator of nitrogen". One of the difficulties in using paddock history is the lack of "paper records". "Paper records" could include information on stocking rates, fertiliser and herbicide applications, tillage type, yield records, soil test results and monthly rainfall information. If a farmer retires or sells the farm the paddock history information is lost (see quote below). In discourse with farmers who had taken over the property in the last 20 years, they were asked what records had been left behind. Usually the answer was "None".

> F26: I haven't done it the last couple of years, I've just been too busy .... so I am not doing enough. I've got a computer ... and I've got a program called Wheatman ... . My daughters say to me that I don't put enough down on paper, but basically is the history of what I've done with the paddocks. Now I've got country, I have been here 25 years, some of the country of where we live here, I broke up, so I know the history of the this land. The point is when I move on, if I haven't written it down, nobody else does, they can only guess. The Wheatman addresses a lot of those things.

A final point is that farmers consider monitoring to be an "on-going" or "continual" process, yet they have not embraced a formalised process for monitoring soil health. Indeed interviews confirmed that soil kits which are currently available are not used widely by farmers to monitor their soil conditions. When farmers were asked about soil monitoring and whether they knew of any existing soil kits or packages, 83% said "No but ...". Thirty seven of those farmers who denied knowledge of soil kits and packages had heard of soil testing kits, and seven of the 37 farmers were specifically aware of soil pH testing kits. How can farmers be encouraged to formalise their intuitive collation of soil health status and record their observations, either on paper or on the computer, so that a trendline of soil condition can be formed?

4.3. FARMER WORKSHOPS

What will be the most effective way of packaging and disseminating the soil health checklist? Several themes appeared in the first session of the workshop, where soil/agronomic information was reviewed, and these are relevant to the development of a 'user-friendly' soil health checklist. The most prevalent themes were:

- Farmers are overwhelmed with mail - and so develop a habit of "read it and bin it". Otherwise they said their offices were covered with paper, so much so that they could not get into them.
- Farmers' habits are difficult to alter, unless the alternative is more 'attractive'. For example, one habit which was common amongst farmers was carrying a pocket diary for recording details of farming operations. Some farmers might transfer this information to a computer, but more than likely the diary became the only written record. It would be difficult for a new form of record keeping to usurp old habits, and the habit of keeping all records in a diary does not allow for longitudinal examination of various paddocks.
- It seems that although farmers want to gain knowledge, especially about new research ideas, they are reluctant to spend time reading about them. They often made remarks to the effect that they would get an agronomist to undertake any analysis of soil health, or soil problems.
- Flowing from this it seems that what is missing is a link between the knowledge they have of their land condition, financial situation and goals in farm management, and the technical knowledge they can obtain from an agronomist.
- One of the most favoured publications was *The Field Guide to Faba Bean Disorders* (Moore *et al.* 1995), mainly because it is compact, robust, an excellent guide, with very good pictures, informative and easy to locate the section required. Most farmers said they would use it if they were growing such a crop. The characteristics exhibited by this publication, if replicated in the soil health checklist, would ensure a favourable response by farmers.
- Even though farmers were interested in other publications which they were shown they stated that difficulties in locating or obtaining a copy were another impediment to their use.

Those farmers who mentioned where information on soil packages could be obtained from would go to their district agronomist (68%), and the remainder said they would make inquiries at New South Wales Agriculture. Table 4 shows the range of information sources which all 75 farmers were asked to comment on, ranging from personal contacts to media such as television and radio. Interestingly, more farmers responded to the category 'somewhat useful' (almost 100%), while only 73% placed some information sources in the 'extremely useful' category. Those information sources which were most valued by farmers seemed to be those with which the farmer had one to one contact, such as the district agronomist, neighbours and family members.

The next band of information sources which were valued by farmers was print, radio and television media. Those information sources which farmers appear to view as 'somewhat useful' or even 'not useful' are organisations or committees such as the National Farmers Federation, Total Catchment Management committees, TOPCROP and even Landcare groups. When further probed about the lack of credibility of these groups as sources of farming information, farmers responded by suggesting, that "the big say, the little pay" or they were just not involved in a TOPCROP or Landcare group. Farmers seem wary of or reticent to deal with organisations which represent bureaucratic 'red tape' or regulation. This may be the reason why the Department of Land and Water Conservation (DLWC), which is presently involved in native vegetation and water reform, rated poorly as a source of information for farming. The National Parks and Wildlife Service (NPWS) was also not widely regarded by farmers as a useful source of information, even though NPWS, in conjunction with New South Wales Agriculture and DLWC, conducts an education program for farmers called 'Farming for the Future'. Another source of information which farmers were critical of were company representatives for fertiliser or livestock, because they considered company representatives were not always objective and had 'something to sell'. Farmers felt that company representatives acted in a way which would promote the company's product, which was not necessarily in the farmers' best interests. It appears the best strategy to promote a soil health checklist would be through a farmer network or trained district agronomists.

Table 4: Information sources about farming which farmers view as credible and useful (n=75).

Information Source	Extremely Useful		Somewhat Useful		Not Useful		Not Used	
	n	%	n	%	n	%	n	%
District Agronomist	36	68	10	19	7	13	0	0
NSW Agriculture	29	54	20	37	5	9	0	0
Family members	28	53	18	34	7	13	0	0
Ag Consultant	28	54	9	17	7	13	8	15
Farm neighbours	27	49	23	42	4	7	1	2
Farmer as Expert	27	54	15	30	7	14	1	2
Reference Books	24	44	25	45	4	7	2	4
Magazine eg Australian Farm Journal	20	34	31	53	5	9	2	3
Radio eg ABC Country Hour	19	35	32	58	4	7	0	0
TV eg Landline, News	17	33	31	60	4	8	0	0
Rural Newspaper eg The Land	16	30	31	58	5	9	1	2
University researcher	15	27	20	36	9	16	11	20
Stock agent	14	26	26	49	13	25	0	0
Dept of Land & Water Conservation	11	21	30	58	10	19	1	2
Landcare Group	10	20	24	47	8	16	9	18
Scientific Journals	9	16	29	53	12	22	5	9
Company Rep	7	13	34	64	9	17	3	6
Fertiliser Company	7	14	30	60	9	18	4	8
NSW Farmers Federation	6	11	26	49	20	38	1	2
TopCrop group	6	11	19	36	16	30	12	23
Greening Australia	4	7	15	27	28	51	8	15
Total Catchment Management Committees	2	4	19	36	19	36	13	25
National Parks & Wildlife Service	2	3	8	14	44	76	4	7

5. Conclusions

The most spoken about features farmers used to identify soil health were plant growth and soil feel. These two features were mentioned nearly six times more frequently than other features, and by 97% of farmers interviewed. In comparison, soil salinity was mentioned by 25 percent of farmers and on only 40 occasions (Table 1). Other properties which were mentioned consistently by farmers in relation to identifying soil health were organic matter (78%), plant roots (63%) and soil life (61%), but they were mentioned three times less frequently than plant growth or soil feel (Table 1). Interestingly, some frequently mentioned features used to identify soil health, such as weeds, were dismissed later on as 'red herrings' by farmers in workshops; they said these features were used to identify a soil type rather than recognise soil health status.

Ways of measuring or determining the health of a soil lacked any formal process, records were not kept, and monitoring was incidental (usually carried out while conducting other farming operations) and related closely to observation skills (92% of farmers). Soil testing was primarily for fertiliser requirements, and conducted, analysed and interpreted by a service provider. Farmers need to capture, formalise and rely on their own observational skills to: diagnose their soil's health; interpret their soil test results, and use monitoring techniques they feel comfortable with. Whilst some farmers use testing kits to obtain soil samples or to assess soil pH, nearly all farmers (83%) were unaware of any formal soil monitoring packages which would assist them to monitor soil health or condition in an holistic farm context.

To encourage and motivate farmers to monitor soil health will require the development of a package or soil health checklist which allows for adaptation to local conditions and soil types, provides a model or protocol for interpretation, and is disseminated to farmers by people from whom they regularly seek advice. Also fundamental in developing a useful and viable soil health checklist is a participative research process which provides farmers with several avenues and forums to become involved. Once the final product is available - the soil health checklist - it will only be used by farmers if they perceive a benefit in its application. Through interviews, soil testing and workshops the researcher and farmers involved can resolve ambiguities and clarify how the soil health checklist should be developed and who would use it. Thus, analysis, reflection and critical review by researchers, service providers and farmers of work in progress on the development and application of the soil health checklist will lead to effective action and continued learning.

6. Acknowledgments

I would like to extend my thanks to the participants of the research project and to GRDC for funding this work, as well as to Graham Marshall and Rebecca Spence for commenting on an earlier draft, to Arthur Conacher for his tireless editing and advice, and to the two anonymous referees.

7. References

Bosch, O.J.H., Allen, W.J. and Gibson, R.S. (1996a) Monitoring as an integral part of management and policy making, Proceedings of Symposium *Resource Management Issues, Visions, Practice*, Lincoln University, New Zealand, 5-8th July, pp 12-21.

Bosch, O.J.H. Allen, W.J. William, W.J. and Ensor, A.H. (1996b) Monitoring and adaptive management paper, *The Rangeland Journal* **18**, 23-32.

Campbell, A. (1994) Community first: Landcare in Australia, in I. Scoones and J. Thompson (eds.), *Beyond Farmer First - rural people's knowledge, agricultural research and extension practice*, Intermediate Technology Publications, London, pp 252-257.

Carruthers, G. and Tinning, G. (1999) *Environmental Management Systems in Agriculture*, Proceedings of a National Workshop, May 26-28, 1999, RIRDC, Canberra, Publication No 99/94.

Chamala, S. and Keith, K. (1995) *Participative Approaches for Landcare - perspectives, policies and programs*, Australian Academic Press, Brisbane.

Chambers, R. (1997) *Whose Reality Counts? - putting the first last*, Intermediate Technology Publications, London.

Cornwall, A., Guijt, I. and Welbourn, A. (1994) Acknowledging process: challenges for agricultural research and extension methodology, in I. Scoones and J. Thompson (eds.), *Beyond Farmer First - rural people's knowledge, agricultural research and extension practice*, Intermediate Technology Publications, London, pp 98-116.

Dalal, R., Lawrence, P., Walker, J., Shaw, R.J., Lawrence, G., Yule, D., Doughton, J.A., Bourne, A., Duivenvoorden, L., Choy, S., Moloney, D., Turner, L., King, C. and Dale, A. (1999) A framework to monitor sustainability in the grains industry, *Australian Journal of Experimental Agriculture* **39**, 605-620.

Daniells, I. (1997) *Soil Structure Assessment Workshop - getting the Message Right for Farmers*, Proceedings of a workshop at Condobolin, NSW 2-3 rd December 1997.

Daniells, I., Brown, R. and Hayman, P. (1996) Benchmark values for soil nitrogen and carbon in soils of the northern wheat-belt of NSW, NSW Agriculture, Orange.

Francis, P. (1999) Special Report: Environmental Management Systems, *Australian Farm Journal* **9**, 5-13.

Freire, P. (1970) *Pedagogy of the Oppressed*, Continuum Publishing, New York.

Hunt, N. and Gilkes, B. (1992) *Farm Monitoring Handbook - A practical down-to-earth manual for farmers and land users*, University of Western Australia Press, Perth.

Industry Commission (IC) (1998) *A Full Repairing Lease - Inquiry into Ecologically Sustainable Land Management*, Report No. 60, Commonwealth Government, Canberra.

Jenkins, S. (1998) *Native Vegetation on Farms Survey 1996 - a survey of farmers attitudes to native vegetation and landcare in the wheatbelt of Western Australia*, National Development Program on Rehabilitation, Management and Conservation of Remnant Vegetation. Research Report 3/98.

Kleinman, S. (1991) Field-Workers Feelings, what we feel, who we are, how we analyze, in W. Shaffir and R. Stebbins (eds.), *Experiencing Fieldwork: An Inside View of Qualitative Research*, Sage Publications, Newbury Park, pp. 184-195.

Kloppenburg, J. (1991) Social theory and de/reconstruction of agricultural science, local knowledge for an alternative agriculture, *Rural Sociology* **56**, 519-548.

Kloppenburg, J. (1992) Science in Agriculture: a reply to Molnar, Duffy, Cummins, and Van Santen and to Flora, *Rural Sociology* **57**, 98-107.

Koch, T. (1994) Establishing rigour in qualitative research: the decision trail, *Journal of Advanced Nursing* **19**, 976-986.

Lincoln, Y.S. and Guba, E.G. (1985) *Naturalistic Inquiry*, Sage Publications, Newbury Park.

Minichiello, V., Aroni, R., Timewell, E. and Alexander, L. (1990) *In-Depth Interviewing. Researching People*, Longman Cheshire, Melbourne.

Molnar, J.J., Duffy, P., Cummins, K.A. and Van Santen, E. (1991) Agricultural science and agriculture counterculture: paradigms in search of a future, *Rural Sociology* **57**, 83-91.

Moore, K., Nikandrow, A., Carter, J., Ramsey, M. and Mayfield, A. (1995) *Field Guide to Faba Bean Disorders in Australia*, NSW Agriculture, Tamworth.

Murdoch, J. and Clark, J. (1994) Sustainable knowledge, *Geoforum* **25**, 115-132.

Pannell, D. J. (2000) A framework for economic evaluation and selection of sustainability indicators in agriculture, *Ecological Economics* **33**, 135-149.

RIRDC Short Report No.20 (1997) *Developing Indicators for Sustainable Agriculture*, retrieved March 4,

1998 from the World Wide Web, http://www.rirdc.gov.au/pub/shortreps/sr20.html

Röling, N. (1988) *Extension Science - information systems in agricultural development*, Cambridge University Press, Cambridge.

Röling, N. (1994) Facilitating sustainable agriculture: turning policy models upside down, in I. Scoones and J. Thompson (eds.), *Beyond Farmer First - rural people's knowledge, agricultural research and extension practice*, Intermediate Technology Publications, London, pp 245-247.

Seng, J.S. (1998) Praxis as a conceptual framework for participatory research in nursing, *Advances in Nursing Science* **20**, 37-48.

Sojka, R.E. and Upchurch, D.R. (1999) Reservations regarding the soil quality concept, *Soil Science Society of America Journal* **63**, 1039-1054.

Spradley, J. (1979) *The Ethnographic Interview*, Holt, Reinhart and Wilson, Fort Worth.

Vereijken, P. (1992) A methodic way to more sustainable farming systems, *Netherlands Journal of Agricultural Science* **40**, 225-238.

Webster, P. (1999) The challenge of sustainability at the farm level: presidential address, *Journal of Agricultural Economics* **50**, 371-387.

AUTHOR

LISA ALEXANDRA LOBRY DE BRUYN

School of Rural Science and Natural Resources, Ecosystem Management, University of New England, Armidale, NSW, 2351.

llobryde@metz.une.edu.au

CHAPTER 13

UNDERSTANDING FARMER MONITORING OF A 'SUSTAINABILITY INDICATOR': DEPTH TO SALINE GROUNDWATER IN WESTERN AUSTRALIA

SALLY P. MARSH, MICHAEL P. BURTON and DAVID J. PANNELL

1. Abstract

Dryland salinity is a serious resource conservation problem in Western Australia. A number of projects are in progress to provide more comprehensive information about the location and extent of potential saline areas in the landscape. Associated with some of these projects, a large number of bores have been installed or are being installed throughout the agricultural area to provide information on depth to groundwater. This information helps to forecast the rate and extent of salinisation of soils and to assess impacts of current treatments. However, many farmers choose not to install bores, or having installed them, do not continue to monitor them. Using data from the Jerramungup Land Conservation District (LCD) we explore factors influencing the behaviour of farmers in choosing whether or not to monitor their existing bores. In 1989, 110 bores were sunk in seven catchments in the district. Monitoring was initially exceptionally high, with 96% of bores observed in 1990, but then fell steadily to 44% by 1997. Our statistical analysis indicates that the probability that a bore will be monitored decreases with time and is influenced by the current depth to groundwater, the amount of salt stored in the soil and the farm location. As well as these physical factors, we explore some of the sociological and economic factors which influence bore monitoring behaviour. Monitoring is more likely to continue when farmers are clearly able to link the collected information with land management practices, such that the information collected is of potential economic value.

2. Introduction

The clearing of native vegetation in Australia to support an agricultural system based on annual crops and pastures has reduced the amount of water captured and transpired by vegetation, resulting in rising water tables. Because the groundwater in most agricultural regions of Australia is naturally highly saline, rising water tables eventually result in salinisation of surface soils. Forecasts of areas threatened with dryland salinity are large (Anon. 1999), particularly in Western Australia (Anon. 1996). However, the exact degree and distribution of salinity threat across the landscape is still being investigated. In Western Australia a number of projects are in progress to provide more comprehensive knowledge of the current location and potential extent of salinised land (National Land and Water Resources Audit, 1999). Associated with some of these projects, a large number of bores have been installed and more are being installed

A.J. Conacher (ed.), Land Degradation, 207–222.

throughout the agricultural region.

Data from the drilling of bores provides information to hydrologists and farmers on factors such as the total salt storage in the soil, salt concentration of groundwater, depth to water and depth to bedrock. If farmers continue to monitor the depth to groundwater, the bores provide information about whether and when the groundwater will reach the surface, causing losses of agricultural production through salinisation of soils. This information is one of many possible 'sustainability indicators' which have been recommended for farmers to monitor. Sustainability indicators are environmental attributes which measure or reflect environmental status or condition of change (Smyth and Dumanski 1993). Monitoring of sustainability indicators has increasingly been advocated as a means of improving agricultural resource management in Australia and elsewhere (Syers *et al.* 1995; Walker and Reuter 1996). Despite this, farmers have generally not adopted monitoring of indicators, even for very serious problems such as dryland salinity (Kington and Pannell 1999).

In their analysis of sustainability indicators in agriculture, Pannell and Glenn (2000) argue that the value of a sustainability indicator is directly related to its potential to improve decision making. An indicator is valuable if the information it provides leads to a change in management. Additionally, they conclude that many of the sustainability indicators recommended by scientists are strongly technical in focus, with no close link to farm management. Commenting on the lack of monitoring by farmers, they observe that, "given the lack of a management focus of most indicators proposed so far, this is not surprising" (Pannell and Glenn 2000:136).

The aim of this study is to better understand the influences of a range of physical, economic and social factors on the decisions of farmers to continue or discontinue monitoring of groundwater levels. The primary focus is on whether the physical characteristics of bores can adequately explain monitoring behaviour. A range of social and economic influences is also examined less formally. The analysis is a case study, using data from the Jerramungup Land Conservation District (LCD).

3. Background

The Jerramungup region, located near the south coast of Western Australia (Fig. 1), is a comparatively new farming district. Some agricultural development took place in the western part of the district in the 1920s (Twigg 1987) but the great majority of the district has been settled and cleared since the 1950s (Davis 1997).

Dry seasons in the early 1980s resulted in the district experiencing severe wind erosion problems and this, together with an awareness of increasing dryland salinity, provided the impetus for the formation of a Soil Conservation District Advisory Committee in 1983 (Twigg and Lullfitz 1990). The Land Conservation District Committee (LCDC) which formed subsequently was one of the first to establish in Western Australia. The first catchment group within the LCD, Jacup, formed in 1984, and the second, Corackerup, in 1989.

In 1989 the Jerramungup LCDC obtained funding from the National Soil Conservation Program (NSCP) for a network of bores to monitor groundwater levels. The original impetus to set up the monitoring scheme came from individuals within the LCDC who were anxious to raise awareness. They were concerned that "some people thought that they did not have a problem ... they did not believe that they would have a saline watertable" (Jerramungup LCDC members, pers. comm. 1999). The project was

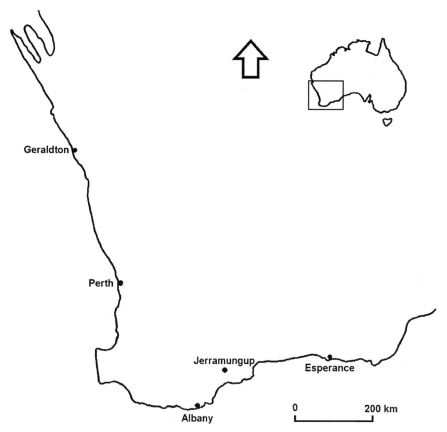

Figure 1. Location of the study region, Jerramungup, Western Australia

supported strongly by Agriculture Western Australia (AgWest), which was responsible for drilling the bores and collecting the initial data, and was committed to providing feedback to individual farmers about their particular bore(s).

In 1989/90, supported by the NSCP funding, 110 bores were sunk on 81 farms in seven catchments in the Jerramungup LCD: Gairdner-Bremer, Carlawillup, Needilup North, Corackerup-Ongerup-Nawainup, Fitzgerald, Jerramungup North, and Jacup. The LCDC was keen to involve as many farmers as possible so most farms received only one bore. AgWest had no fixed method for choosing locations for the bores. Farmers were consulted about location and encouraged to be present when the bore was drilled. "Involvement was the biggest thing we wanted, so bores went where farmers wanted them" (Jerramungup LCDC members, pers. comm. 1999). A consequence of this was that some bores were not ideally sited. Farmers were (and still are) sent quarterly reminders from the local LCDC co-ordinator to observe the depth to groundwater in their bore(s) and to pass on the information to AgWest. The project had an exceptional response, with close to 100% of the bores being monitored in its early years.

The first detailed feedback on the bores was given to farmers individually in 1992. The salt profile associated with each bore was presented in a graphical format. Comment was made on the total soluble salt content, depth to groundwater and any discernible trend.

A plot of individual bore water levels over time has been made available to farmers each year since 1992. By 1993, 82 bores had sufficient data to enable trends in groundwater levels to be estimated and these trends were presented at the 1993 Jerramungup Agricultural Science Exposition (JERAC). The trends which were displayed at JERAC were not encouraging, but not surprising to AgWest hydrologists. Although not intended to be so, the data could have been disturbing to individual landholders whose farms were potentially under threat. For example, analysis of the bore data (McFarlane and Ryder, 1993) indicated that:

- the average rate of rise had been 14 cm yr^{-1};
- water tables in individual bores were rising by up to one metre every year;
- the average depth of the watertable was only 6.5 m;
- on average, there were over 2500 tonnes of salt stored under each hectare in the Jerramungup region. Some areas had over 10 000 t ha^{-1}, and
- groundwater salinity averaged 14 867 mg L^{-1} or 2703 mS m^{-1}, almost half as saline as sea water (35 500 mg L^{-1}).

The data were also presented on a catchment basis, and this clearly illustrated that trends in some catchments were worse than others. These catchment differences are discussed later.

At JERAC in 1994, AgWest presented data from the bores on a landform rather than a catchment basis in the form of salinity hazard maps. The maps illustrated that salinity in some areas would be harder to control than in others. There was some negative reaction by a few farmers to this public disclosure of what was considered to be sensitive information. For example, there was concern about the potential effect of such information on land values. Because of these concerns, a field trip was organised and issues and management options were discussed with a senior AgWest hydrologist.

The Upper Gairdner area of the Jerramungup LCD became a Focus Catchment in 1996. A Focus Catchment is a catchment designated by AgWest to receive extra inputs of money and technical assistance over a limited period (usually three years) to address land management issues. The bore monitoring data had indicated that farms in the Upper Gairdner area had higher rates of rise of groundwater and higher salt storage and salt concentrations (McFarlane and Ryder 1993).

By 1993 the number of bores being monitored had fallen to 74% and by 1995 it was 52% (Fig. 2). This approximate monitoring level has continued until the present. Although this is considerably less than the original monitoring rate, it still represents a high on-going monitoring rate by many standards. Since 1989/90, more bores have been installed in better locations in conjunction with new projects (40 in the Upper Gairdner, 20 in the Fitzgerald), but similarly to the original bore network, not all are monitored regularly (C. Daniel, Jerramungup Landcare District Committee Co-ordinator, pers. comm. 1999).

4. Methodology and Analysis of the Data

This analysis is an attempt to identify factors which explain the change in monitoring percentage over time. Two approaches were used to do this. First, we conducted Probit analyses using the statistical package STATA (StataCorp. 1999) to relate the probability that an individual bore would be monitored to the physical characteristics of the bore,

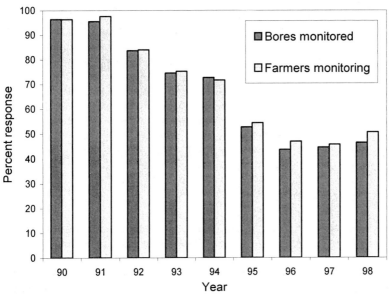

Figure 2. Bore monitoring in the Jerramungup Land Conservation District (110 bores on 81 farms installed in 1989/1990)

such as salt storage and depth to groundwater. Probit analysis is a form of multivariate regression analysis used when the dependent variable is a dichotomous variable with the value of either 1 or 0. In this case we consider an index variable, Y, which takes a value of 1 if the bore is monitored during a specific period and 0 otherwise. A set of technical and socio-economic factors (x), might explain the farmer's decision, so that:

$$\mathrm{Pr\,ob}(Y = 1) = F(\beta x) \tag{1}$$

The function F was assumed to follow the normal distribution, giving the Probit model:

$$\mathrm{Prob}(Y = 1) = \int_{-\infty}^{\beta x} \phi(t)dt \tag{2}$$

$$= \Phi(\beta x)$$

where ϕ and Φ are the standard normal density and distribution functions respectively.

We expected that the monitoring behaviour of farmers would be explained, at least partially, by characteristics of the initial drilling and on-going trend data from the bore.

Hypotheses were that:

- The probability that a bore would be monitored would be increased by a higher groundwater level, a more rapidly rising groundwater level, a higher level of salt storage and higher conductivity readings.
- Differences in the monitoring rates between groups of bores in different catchments would be explained by the physical characteristics of the bore data.

The second element of the study was a series of informal interviews with AgWest personnel, the Jerramungup LCDC co-ordinator and a small number of Jerramungup LCD farmers. The interviews used a semi-structured approach with open-ended questions. Some reasons for monitoring behaviour which were suggested by interviewees were subsequently tested statistically. Similarly, ideas originating from the statistical analysis were discussed in the interviews.

4.1. DESCRIPTION OF THE DATA

AgWest provided both the initial physical data obtained when the bores were drilled and quarterly water level readings for individual bores (if taken) since 1989. Additionally, trend analyses conducted by AgWest in 1992 (Greenham 1992) and 1996 and by CSIRO in 1999 were provided. As some farms had more than one bore, we investigated whether the monitoring percentage was different when expressed in terms of percentage of farmers monitoring. Figure 2 illustrates that it is similar. The percentage of farmers monitoring bores varies by catchment. In 1998 it ranged from 36% in the Gairdner-Bremer-Carlawillup catchment to 70% in the Corackerup-Ongerup-Nawainup catchment (Table 1).

Table 1. Percentage of farmers monitoring bores by catchment and year*

| Year | Catchment | | | | | |
	Gairdner/ Bremer/ Carlawillup (n=14)	Needilup North (n=8)	Corackerup/ Ongerup/ Nawainup (n=10)	Fitzgerald (n=7)	Jacup (n=28)	Jerramungup North (n=14)
1990	100	100	100	71	96	100
1991	100	100	100	71	100	100
1992	93	88	90	29	93	79
1993	71	63	80	43	82	86
1994	71	63	80	29	86	64
1995	43	63	70	29	64	43
1996	36	50	50	29	57	43
1997	43	50	70	29	50	29
1998	36	38	70	57	54	50

* To count as monitoring, farmers must make at least one bore reading in the year.

The physical data associated with the bores varied greatly between the catchments (Table 2). This reflected different land forms, soil types and climate variables (McFarlane and Ryder 1993).

Table 2. District groundwater data in 1993* (Source: McFarlane and Ryder, 1993)

	Catchment					
	Gairdner/ Bremer/ Carlawillup (n=26)	Needilup North (n=8)	Corackerup/ Ongerup/ Nawainup (n=17)	Fitzgerald (n=9)	Jacup (n=34)	Jerramungup North (n=16)
Rate of rise in groundwater levels (cm yr^{-1})	6	16	13	13	28	24
Depth to the water table (m)	5.5	6.8	6.1	21	6.3	5.5
Salt storage (t ha^{-1})	1600	2900	3500	2600	2000	3900
Salt concentr'n (kg m^{-3})	8.5	12	17	11	11	13
Groundwater salt (mg L^{-1})	14 000	22 000	24 000	12 000	17 000	24 000
Depth to bedrock (m)	18	18	19	24	16	25
Average annual rainfall (mm)	470	390	410	400	410	400

* The groundwater trends for the Fitzgerald and Needilup North districts may not be accurate as they are based on only six or seven well-monitored bores.

The trend analysis conducted by AgWest in 1993 showed that only 16% of bores (of those with sufficient water level readings) had falling water levels. On average, water level in the bores was rising at the rate of 14 cm per year, although some were rising at rates of greater than 60 cm per year. Jacup and Needilup North catchments showed the highest rates of rise (Table 2). The trend analysis done by AgWest in 1996 showed that 37% of bores had falling water levels. A preliminary analysis done by CSIRO in 1999, using a different methodology to estimate groundwater trends (Shao *et al.* 1999), estimated that of 68 Jerramungup bores with sufficient readings, only 10% showed a falling trend. Another 18% of the bores, however, measured shallow water tables with strong seasonal fluctuations where the water is within one metre of the surface, and the remainder displayed a rising trend (L. Crossing, Agriculture Western Australia, pers. comm. 1999).

A number of variables were defined for the purpose of the Probit analysis. The dichotomous Yes=1/No=0 dependent variable was defined as whether the bore was monitored in each quarter (February, May, August, November) for the years 1989 to 1998. The first reading for each bore was ignored in the analysis as it represents the 'test' reading done at installation rather than a decision by the farmer to monitor.

There could be a number of practical reasons why a bore might not be monitored. For example, the bore might be dry or have been damaged so that water level could not be read, but the data do not distinguish which bores, if any, are not monitored for such reasons. No socio-economic data are available for the statistical analysis, so only technical data related to bore readings and bore location are included.

The independent variables investigated are listed in Table 3. As a general modelling strategy, quadratic (i.e. squared) terms were included to allow for flexibility in the response function. Furthermore, the coefficients for GWLEVEL and its square GWLEVEL^2 were allowed to vary as a function of (logged) salt storage. Both TIME and TIME^2 were used, and a dummy variable was also included to identify if there was any change in monitoring after the public presentation of results in 1993.

Table 3. Independent variables used in the Probit analyses

Independent variable	Description	Expected sign
CATCHMENT	Dummy variables to specify particular catchments	?
DISTANCE	Distance of the bore from the coast (km)	?
AVGSALT	Salt concentration in the soil (kg m^{-3})	Positive
SALTSTORE	Salt stored in the soil under each hectare of land (tonnes ha^{-1}). ln(SS) is the natural log of this variable	Positive
GWCOND	Groundwater conductivity (mS m^{-1})	Positive
DEPTH	Depth to bedrock (m)	?
TIME *	Time in quarter-years from the first recorded reading	Negative
DUM93	Dummy variable = 1 for dates after 1992, 0 otherwise	Negative
GWLEVEL *	Depth to groundwater at last reading (expressed in positive terms) (m)	Negative
GWCHANGE *	Change in groundwater level between the last two readings (m)	Positive
SEASON*	Dummy variables representing seasons of the year in which the readings occurred	?
MULTI	Number of bores potentially monitored by farmer monitoring this bore	Positive
RAINFALL *	Rainfall for the quarter recorded at the Jerramungup Post Office (mm)	Positive

* Only these variables vary across time for each bore: Other variables were measured only at the time of drilling the bore.

From casual inspection of the data, it was hypothesised that the probability of reading a bore will decline over time. This may be due to failure of the bore, a loss of interest in the project, or a perception that there is no further information of value to be gained from monitoring. Given the different dates at which bores were drilled, the measure of time elapsed is conditional on the date of the first reading, which occurred when the bore was installed. However, it was expected that greater severity of the problem (that is, higher water tables and increased salt) would increase monitoring. The appropriate measurement of these variables was explored within the analysis.

5. Statistical Results

Two results are statistically very robust across all model specifications: (a) monitoring behaviour is related to current depth to groundwater but not to the rate of change in groundwater level, and (b) monitoring behaviour is more closely correlated with total salt storage than with either groundwater conductivity or average salt concentrations. This is surprising since, in theory, total salt storage is less valuable as an indicator of the likely severity of dryland salinity. (All three salt variables were made available to farmers at the start of the program.) Results for the preferred model specification are reported in Table 4.

The value of dF/dx represents the change in probability of monitoring following a unit change in the variable or, in the case of dummy variables, a switch from 0 to 1. It is calculated with all other variables set at mean levels. This value gives an indication of the magnitude of the influence of the variable. It is not reported for variables that have quadratic or interaction terms, as it has no straightforward interpretation in those cases.

Table 4. Results of the Probit analysis

Number of observations = 3615

Variable	Coefficient	Standard error[A]	Z[B]	P[C]	dF/dx[D]
CATCHMENT-NN[E]	0.566	0.432	1.31	0.19	0.21
CATCHMENT-CON	1.06	0.356	2.98	0.00	0.37
CATCHMENT-FITZ	-0.0829	0.451	-0.18	0.85	-0.03
CATCHMENT-JACUP	0.857	0.377	2.27	0.02	0.32
CATCHMENT-CW	0.595	0.209	2.85	0.00	0.22
CATCHMENT-JN	0.457	0.407	1.12	0.26	0.17
SEASON-2[F]	-0.154	0.0495	-3.10	0.00	-0.06
SEASON-3	0.0146	0.0648	0.23	0.82	-0.01
SEASON-4	-0.0964	0.0663	-1.45	0.15	-0.06
RAINFALL(lagged)	-0.00192	0.000542	-3.54	0.00	-0.0005
DISTANCE	-0.00750	0.00739	-1.02	0.31	-0.003
TIME	0.0342	0.0188	1.82	0.07	
TIME^2	-0.00190	0.000405	-4.70	0.00	
DUM93	-0.587	0.118	-4.98	0.00	-0.13
GWLEVEL	-0.634	0.227	-2.79	0.01	
GWLEVEL^2	0.0306	0.0116	2.63	0.01	
$ln(SS)$*GWLEVEL	0.0893	0.0309	2.90	0.00	
$ln(SS)$*GWLEVEL^2	-0.00441	0.00159	-2.78	0.01	
$ln(SS)$	-0.124	0.112	-1.12	0.26	
Constant	1.84	0.800	2.30	0.02	

[A]Standard errors corrected for clustering by bore.
[B]Z is the ratio of coefficient to standard error.
[C]P is significance level for test of H_0 that coefficient equals zero.
[D]dF/dx is the change in probability of monitoring, for a discrete change of dummy variable from 0 to 1, or for a unit change in other variables, all other variables measured at their mean. See text for further explanation.
[E]Baseline catchment is Gairdner-Bremer.
[F]Baseline season is January-March.

The results point to rejection of the hypotheses that

- the probability of monitoring would be increased by a more rapidly rising water table (this variable was not significant in any model specification examined), and
- differences in the monitoring rates between catchments could be explained solely by differences in physical characteristics of the bores. Even after accounting for physical differences, bores in the Corackerup-Ongerup-Nawainup, Jacup and Carlawillup catchments are more likely to be monitored than those in the baseline catchment, Gairdner-Bremer.

The dummy variable for periods after 1992 is statistically highly significant, and indicates that a substantial fall in monitoring occurred after this date. Rainfall in the previous period is negatively correlated with monitoring. Bores are less likely to be monitored in May and November - times which coincide with peak workloads on farms for sowing and harvesting. There is also a very robust relationship between monitoring and distance from the sea. Given the inclusion of catchment dummy variables, this result is not catchment-specific and may reflect some additional hydrological or topographical trend.

Interpretation of the impacts of time, water level and salt storage is complicated by the non-linear and interaction terms included in the model. For bores with moderate to high salt storage, the influence on monitoring of depth to groundwater is similar to that shown in Figure 3. Monitoring remains high for a wide range of depths and then falls for deep water tables. The interaction term also means that for a wide range of depths, higher salt loads are associated with higher rates of monitoring (not illustrated). For bores with very deep groundwater, this relationship breaks down, possibly due to the low numbers of observations in the relevant ranges.

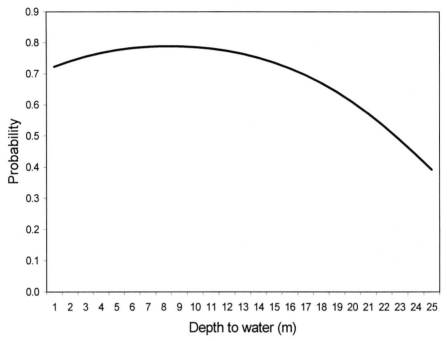

Figure 3. Estimated relationship between probability of monitoring and depth to groundwater (for case of salt storage level = 2500 t ha^{-1})

The standard Probit model is based on the assumption that the error terms are independent, but in this case the data set includes repeated observations on the same bores and so the standard assumption may be violated. In response to this concern, the modelling procedure used defined bores as 'clusters', allowing estimation of robust standard errors assuming independence between clusters without requiring assumptions about within-cluster correlation (StataCorp. 1999). The overall results were remarkably robust to other statistical specifications (such as a Random Effects Probit model, or explicitly modelling the structure of within-bore correlation of errors).

    Overall, the set of physical data relating to the bore is an incomplete predictor of whether a bore will be monitored. The distribution of actual versus predicted monitoring of bores generated by the model (using a 50% cut point) is reported in Table 5. Although illustrative of how the model works, such a table, or estimates of the proportion of correct predictions, should not be used as a measure of the goodness of fit of the model (Veall and Zimmermann, 1994).

Table 5. Actual monitoring versus prediction of preferred model
(0 = not monitored, 1 = monitored)

Predicted value	Actual value	
	0	1
0	1157	462
1	557	1438

Instead we report one of the "R^2" type measures, given by

$$\sigma_n = \frac{p_{11} + p_{22} - p_{\sim1}^2 - p_{\sim2}^2}{1 - p_{\sim1}^2 - p_{\sim2}^2} \tag{3}$$

where p_{ij} is the fraction of times the realisation was outcome i when the model predicted outcome j, and $p_{\sim j}$ is the fraction of times alternative j is predicted. σ_n is positive for a model with any predictive power, and has an upper limit of 1.0. A value of 0.434 is calculated for the model reported in Table 4, indicating a useful degree of predictive power.

The analysis has, somewhat mechanistically, modelled the process of bore monitoring but it does not explain why people do or do not continue to monitor. The following section explores this issue further.

6. Socio-Economic Considerations

In this section we report on insights obtained from semi-structured interviews with a range of stakeholders.

6.1. THE INITIAL HIGH SUCCESS RATE OF THE MONITORING PROJECT

Involvement of farmers in the project was obtained by the efforts of key individual farmers in each area who had been nominated by the LCDC. Reporting results from a survey conducted in two Jerramungup catchments in 1993, Davis (1997:57) says that "most participants cited the same two or three people as the key person who invited them to join the Farm Planning Project". That project was instigated at a similar time and by the same people as the bore drilling. There were some mixed responses about the level of overall participation, with one comment received being that the "initial project was ill-conceived, some people were not asked, the bores were not well-sited".

6.2. THE VALUE OF MONITORING AND REASONS FOR CONTINUING TO MONITOR

Farmers considered that the monitoring project had been useful in contributing to a general awareness of the extent of saline rising groundwater within the region. It was also evident that farmers had learnt from monitoring their bores. Examples of the learning which had occurred include:

- Some hydrological concepts were better understood. One farmer explained that he had previously perceived the water table as flat (like a table). There was a realisation by some that recharge happens throughout the landscape rather than in discrete areas.
- Farmers who had plotted their rainfall against the groundwater level had gained an appreciation of the fluctuations in groundwater level caused by rainfall variation.
- One farmer indicated that he had learnt that surface water management was not going to be enough to control a rising watertable. "I thought that annual pastures and surface water management would be enough but the piezometers have shown otherwise."
- Some farmers had begun to experiment with their use of bores and realised that they did not all have to be drilled to bedrock. More bores are being used. One farmer had installed 20, some shallow, along a transect across a field in which he had established perennial pasture in an effort to reduce recharge.

Some farmers considered that monitoring is useful for management and that data from regular monitoring reveal useful trends. One farmer indicated that he had stopped monitoring but plans to start again in conjunction with installation of salinity treatments. A number of farmers said that if bores form part of a land management plan, it is likely that they will then be read more frequently. However, there were some farmers who could not see any further value for themselves in monitoring. This was evident in comments such as "What are we to do with this information?" and "Measuring is most use for the LCDC, not for us".

Interestingly, AgWest personnel say that farmers should be monitoring for their own benefit, but some farmers say that they are monitoring for the catchment group or AgWest. It seems that some farmers perceived AgWest's considerable initial involvement as the main reason for the project.

6.3. REASONS FOR NOT MONITORING

Reasons given for not continuing to monitor were, in many cases, simple and pragmatic. For example, the bore was in the wrong place, the bore was dry, the water table was very low, or the bore was damaged (for example by stock). As a result of our visit and discussions with Jerramungup farmers, the LCDC suggested that it would be useful for AgWest to revisit and service bores. The LCDC was prepared to co-ordinate such an activity.

It was pointed out that properties get sold. New owners are less likely to continue with monitoring as they were not originally involved in the project.

Other reasons for not monitoring related to the value of the information collected:

- A farmer commented that data from one bore is not representative and that "a lot of piezometers would be needed to tell the truth".
- Both AgWest personnel and farmers suggested that a succession of dry years may have reduced the perceived need to monitor. The early signs of salinity are often linked to waterlogging, and after a wet year in 1993, rainfall had been below average in 1994, average in 1995 and below average in 1996 . To test the impact of this we introduced quarterly rainfall data from the Jerramungup Post Office into the statistical analysis. Rainfall during the immediately preceding quarter did significantly affect the probability of monitoring, but not in the way that had been

suggested. Higher rainfall in the preceding quarter reduced the probability of monitoring.

- A comment was made that, with only one bore, the monitoring job is too small and is easily overlooked. Failure to monitor on this basis suggests that the information is not highly valued. To test the relevance of the comment, we introduced into the statistical model a multiple bore variable (MULTI), but it was not a significant variable in the regression.

- AgWest personnel commented that farmers seem less interested now in feedback (for example, of groundwater trends) than earlier in the project. This seems consistent with the argument of Pannell and Glenn (2000) that the value of monitoring is likely to decline after initial observations have been made. This would indicate that for these farmers the bores have been more valuable for raising initial awareness than for ongoing management decision making.

- For information to have practical value, it must be possible for managers to respond to it productively (Pannell and Glenn 2000). There is evidence that some farmers do not feel empowered to prevent or manage their salinity problem, so that information about rising groundwaters is distressing without being of practical value. Both AgWest staff and the LCDC co-ordinator said that salinity had caused some despondency and despair amongst Jerramungup farmers and their families.

- Finally, scepticism of government information about salinity leads some farmers to discount the value of information from monitoring. There was evidence of some hostility towards professionals working on land conservation, and anger at the official predictions of saline area. An earlier report for a particular site was mentioned as having predicted dire consequences which failed to fully eventuate.

6.4. REASONS FOR DIFFERENCES IN MONITORING BEHAVIOUR BETWEEN CATCHMENTS

There are distinctive soil and water differences between the catchments. Soils in the district vary from north to south, ranging from well-drained silty sediments at Gairdner-Bremer-Carlawillup to poorly drained heavy clays at Corackerup-Ongerup-Nawainup. Near the coast, rainfall is higher and water tables tend to be lower (McFarlane and Ryder 1993). These soil and water differences are reflected in the characteristics of bore data, such as depth to water, salt storage and groundwater conductivity, as reported in Table 2. However, the statistical analysis showed that there are differences in monitoring behaviour which cannot be explained by differences in the technical bore data. Farmers suggested that there are different types of water problems in different areas, such as water erosion, inundation and waterlogging, which might affect the perceived value of information from monitoring groundwater levels. This may provide an explanation of the significance of the DISTANCE variable in the statistical analysis, as these problems are related to land form, soil and climatic factors.

Additionally, there are social differences between the catchments. Areas within the district were settled at different times, resulting in different social characteristics. In contrast to communities around Jerramungup, which were settled since the 1950s, there have been families in the Needilup area since 1912 (Twigg 1987). The recently settled communities often have a strong sense of unity and common purpose (Davis 1997).

Finally, some of the catchments, namely Jerramungup North, Corackerup-Ongerup-Nawainup and Needilup North, formed part of the Upper Gairdner catchment which became a Focus Catchment in 1996. It is possible that this affected farmers' bore monitoring behaviour. Table 1 does indicate that a greater percentage of bores in the Corackerup-Ongerup-Nawainup catchment were monitored than in other catchments. Fitzgerald and Jacup are likely to join the Focus Catchment program in the near future and it is possible that anticipation of this has contributed to the higher monitoring percentage seen in Fitzgerald in 1998 (Table 1). Jacup has maintained a consistently high monitoring percentage.

7. Discussion

The Jerramungup LCDC has been recognised for its Landcare efforts, winning the national award for Landcare groups in 1991. Although this study has examined reasons for farmers failing to monitor, the Jerramungup LCD has achieved a very high level of bore monitoring. A number of factors have contributed to this. A high degree of community ownership of the program was achieved through the efforts of the Jerramungup LCDC, which focused on involving as many farmers as possible in the project. Ideal bore siting was sacrificed for the sake of involvement and ownership. Other key reasons for the success of the Jerramungup program have been the commitment of AgWest to providing support and feedback to the project, and especially the co-ordinating and motivating role played by the LCDC co-ordinator. Her enthusiasm and persistence were recognised and praised by both farmers and agency personnel.

Our analysis shows that the physical characteristics of the bore data have some influence on monitoring. Data related to the bore, such as higher water level and higher salt storage, influence the likelihood of a bore being monitored more frequently. The interaction between water level and salt storage is also significant. This makes intuitive sense as it is rising water levels in soils with high levels of salt that pose the most serious salinity threat.

Groundwater monitoring does appear to be a powerful tool for raising awareness. Some farmers discontinue monitoring after their awareness has been raised, even though they have a rising saline water table. Pannell (1999) argued that the usefulness of information for farm management is related to its ability to reduce uncertainty. The information from initial groundwater monitoring can quickly and dramatically reduce much of the uncertainty about current groundwater levels, and any further monitoring then needs to provide information that reduces other areas of uncertainty, such as the performance of management options (Pannell and Glenn 2000; Kenny 1998). Farmers in the Jerramungup LCD who spoke enthusiastically about continued monitoring were evaluating farming systems options such as lucerne, perennial grasses and surface water management. Promotion of bore installation in Western Australia appears still to be motivated primarily by awareness raising. There is a need for greater involvement of farmers in research and development related to the implementation of high-water-use systems on farms and for linking this work to groundwater monitoring.

Social and psychological factors also appear to be influences on monitoring. Although we have not explored these factors in depth, it appears that social factors are at least partly responsible for differences in monitoring behaviour between catchments, and that feelings of discouragement or despair can dissuade farmers from continuing to monitor. A growing awareness of the problem and an inability to see a solution puts people in a vulnerable position, both to despair and to persuasion to invest in technically unsound measures. Furthermore, the strong social networks present in this agricultural region appear to explain the greater success of efforts to initiate bore monitoring by farmers here than elsewhere.

It may also be possible that farmers are monitoring bores but choosing not to make the information available to AgWest. There are a number of reasons why farmers may perceive that this information should remain confidential. Our analysis shows a significant fall-off in monitoring after 1992. This could be related to the public release of information at that time, or it could be related to a perceived reduction in the value of monitoring after awareness had been achieved.

8. Conclusion

Farmers will only monitor bores in the long term if the information they get from doing so contributes to their ability to make better decisions about their farm business. This means being able to clearly link the collected information to land management practices, such that the information is of potential economic value. In addition, we have identified a range of social and psychological factors which have influenced bore monitoring in this region.

Increasing efforts have been made to encourage farmers to monitor environmental indicators of various types. In general, there has been little attention paid to the issues raised here. It is, therefore, not surprising that most programs have not succeeded in achieving high levels of farmer uptake of the recommended monitoring practices. Greater attention to these issues may either enhance monitoring levels achieved, or reduce to realistic levels expectations about the level and duration of monitoring which are likely in any given situation.

9. Acknowledgments

The authors are grateful for the contributions to this work made by Don McFarlane and Arjen Ryder (Agriculture Western Australia), Carolyn Daniel (Jerramungup LCDC), farmers in the Jerramungup Land Conservation District, and two anonymous reviewers. We acknowledge financial assistance from the Grains Research and Development Corporation.

10. References

Anonymous (1999) *Dryland Salinity and its Impact on Rural Industries and the Landscape*, Prime Minister's Science, Engineering and Innovation Council, Occasional Paper Number 1, Department of Industry, Science and Resources, Canberra.

Anonymous (1996) *Salinity: A Situation Statement for Western Australia*, Government of Western Australia, Perth.

Davis, J. (1997) *The Effectiveness of Public Funding for Landcare: A Case Study From Jerramungup, Western Australia*, Thesis, Master of Science (Natural Resource Management), Faculty of Agriculture, University of Western Australia.

Greenham, K. (1992) *Collation and Interpretation of Jerramungup LCDC bores, 1992*, Unpublished report compiled for the Jerramungup LCDC and Western Australian Department of Agriculture.

Kenny, N. (1998) Avon water quality: Sustainable community water quality monitoring, in *Proceedings of the National Waterwatch Conference: Getting Better at Getting Wet*, University of Adelaide, Roseworthy, 21-24 July 1998, pp. 159-164.

Kington E. and Pannell, D.J. (1999) Dryland salinity in the upper Kent River catchment of Western Australia: Farmer perceptions and practices, Sustainability and Economics in Agriculture Working Paper 99/09, Agricultural and Resource Economics, University of Western Australia, http://www.general.uwa.edu.au/u/dpannell/dpap9909f.htm

McFarlane D. and Ryder, A. (1993) *Groundwater Trends in the Jerramungup Region*, Unpublished report compiled for JERAC EXPO, October 1993.

National Land and Water Resources Audit (1999) http://www.nlwra.gov.au

Pannell, D.J. (1999) Social and economic challenges in the development of complex farming systems, *Agroforestry Systems* **45**, 393-409.

Pannell D.J. and Glenn N.A. (2000) A framework for economic evaluation and selection of sustainability indicators in agriculture, *Ecological Economics* **33**, 135-149.

Shao, Q., Campbell, N.A., Ferdowsian, R. and O'Connell, D. (1999) Analysing Trends, in Groundwater Levels. Report Number CMIS 99/37, CSIRO Mathematical and Information Sciences, Floreat, Western Australia.

Smyth, A.J. and Dumanski, J. (1993) FESLM: *An International Framework for Evaluating Sustainable Land Management*, World Soil Resources Report 73, Food and Agriculture Organisation of the United Nations, Rome, Italy.

StataCorp. (1999) *Stata Statistical Software: Release 6.0*, Stata Corporation, College Station, TX, pp. 65-79.

Syers, J.K., Hamblin, A. and Pushparajah, E. (1995) Indicators and thresholds for the evaluation of sustainable land management, *Canadian Journal of Soil Science* **75**, 423-428.

Twigg, B. (1987) *Settlement and Landuse History Fitzgerald – Western Australia*, Unpublished report, Murdoch University, Perth.

Twigg, B. and Lullfitz, B. (1990). Landcare in W.A. – A look back ... then forward, *Australian Journal of Soil and Water Conservation* **3**, 54-55.

Veall, M.R. and Zimmermann, K.F. (1994) Goodness of fit measures in the Tobit model, *Oxford Bulletin of Economics and Statistics* **56**, 485-499.

Walker, J. and Reuter, D.J., (1996) Key indicators to assess farm and catchment health, in J.W. Walker and D.J. Reuter (eds.), *Indicators of Catchment Health – A Technical Perspective*, CSIRO Publishing, Melbourne, pp 21-33.

AUTHORS

SALLY P. MARSH*, MICHAEL P. BURTON and DAVID J. PANNELL

Agricultural and Resource Economics, University of Western Australia, Nedlands, 6907 Australia

*Current address: Department of Agricultural Economics, University of Sydney, NSW 2006, Australia

s.marsh@agec.usyd.edu.au

CHAPTER 14

NEEDS AND OPPORTUNITIES IN EVALUATING LAND DEGRADATION AND EROSION IN SOUTH-EAST ASIA

IAN DOUGLAS

1. Abstract

Many, diverse government, commercial, academic and aid agencies in south-east Asia have conducted short term, and occasionally long-term, research on natural and accelerated erosion. Seldom do agencies depart from traditional, standard, textbook plot studies or catchment monitoring. The upscaling necessary for working at the management scale seldom occurs. Key issues, such as erosion along unsealed roads or on construction or mining sites, are often neglected because they fall outside the responsibilities of the agricultural, forestry and water resource agencies which normally make measurements of soil loss or sediment transport. Often the data are not collected sufficiently frequently or for a long enough time-span. River sediment sampling is often insufficiently frequent and seldom at high discharges. Many plot studies are short-term and do not analyse the data in relation to storm rainfall events.

Such haphazard monitoring makes it difficult to assemble representative data on potential sediment sources in major river basins and to direct attention the land cover changes and land development practices which create major offsite, down-slope and down-river problems. Hence information is limited and gaps occur in the most basic hydrologic records for many upland and interior areas of South East Asia, as the rainfall records for parts of Sarawak and Sabah indicate. New commercial pressures on land transform erosion conditions rapidly. Mining and the cultivation of vegetables and flowers, for both local and global markets, are transforming the uplands and highlands of most SE Asian countries leading to high sediment discharges and problems of agricultural chemical residues in the headwaters of major rivers. The necessary detailed hydrologic information is usually lacking. However, in northern Thailand, long-running plot studies indicate the severe consequences of deforestation by shifting cultivators on State Forest land. In the Central Highlands of Vietnam, resettlement of people from areas occupied by new water reservoirs and of migrants from the north of the country coming to farm the relatively underused, yet stable, basaltic soils of the plateau is displacing traditional farmers into the hills where the soils are unstable and much more erodible. Already the area is prone to flash floods that wash away bridges and disrupt communications. Further forest clearance and access road construction will aggravate these hazards. Yet, the only detailed soil erosion investigations in the area are standard plot studies examining the effects of different cropping systems on the basalt soils of the plateau. For the more erodible steepland soils there are no suitable data.

In these and other parts of the region, the challenge to all soil erosion and land degradation scientists is to develop more purposeful studies which carry out high quality science, in the places where it matters, for long enough to gain adequate understanding to be able to make reliable erosion forecasts at the management scale.

A.J. Conacher (ed.), Land Degradation, 223–235.

2. Introduction

Over 30 years of field research in Southeast Asia and adjacent regions have suggested to the author that land degradation and soil erosion are only partially understood and that often institutional views and professional narrowness mean that whilst some aspects of the problem are studied in great detail, others are virtually neglected. This paper sets out to examine why that may be so and to suggest some challenges to make the science of land degradation more relevant to the problems that affect people's lives. It aims to answer the question: Why is it that despite all the studies of soil erosion and land degradation, the best information is not being used to make predictions and appropriate studies of all forms of erosion and all sediment sources are not being made?

In SE Asia, as elsewhere, the impact of land degradation is twofold, on-site in terms of loss of the organic soil and plant nutrients, and offsite, in terms of the effects of sediment on water resource systems. The on-site degradation is the prime concern of farmers,, rangeland managers and foresters. The offsite impacts concern water engineers, fishermen, irrigators and all others with a stake in rivers and water supplies. They are different facets of the hydrologic and sedimentary cycles but all too often they are investigated and analysed separately. Even though major international assessments of soil degradation in South-East Asia recognise the roles of industrial activities, urbanisation, road construction, and mining and quarrying as causative factors, their expert assessment teams are often drawn almost exclusively from agricultural and soil conservation agencies (Van Lynden and Oldeman 1997).

Many advances to increase, or restore, the sustainability of agricultural systems are greatly assisted by field research and modelling investigations, at the plot scale, into which crop combinations produce the least erosion (e.g. Rose et al. 1998). Nevertheless, extrapolations from these investigations to the river catchment scale are difficult and sometimes misleading. Yet, in SE Asia consultant reports and environmental impact statements frequently use simple models based on plot data to forecast the erosion which will occur from land development and major construction projects. The Universal Soil Loss Equation (USLE) is applied time and time again to produce erosion risk maps that essentially are maps combining rainfall intensity with soil types (Watthanasuk 1976; Sombatpanit 1992; Fors and Chantaviphone 1994; Eiumnoh et al. 1997). Not only is this making gross assumptions about the form of soil loss that occurs in these environments, but it is ignoring the non-linearity of the rainfall:erosion process. Frequency-magnitude distributions of rainfall are poor predictors of distributions of runoff or soil loss at the plot scale (Boardman and Favis-Mortlock 1999). However, Rose and colleagues have tested an event-based process model widely in SE Asia using plot studies in the Philippines, Thailand and Malaysia (Ciesiolka et al. 1995). The work has demonstrated changing soil properties through a growing season and their impact on the availability of sediment for entrainment by surface runoff as well as the importance of extreme events and variations from year to year in storm magnitude and occurrence.

Frequency-magnitude predictions are even less reliable at the catchment scale, as a simple analysis of the suspended sediment:discharge relationship in the Nam Mae Kok, a northern Thailand tributary of the Mekong, shows (Fig. 1) (Douglas 1999). In this river, the sediment concentrations at given discharges in the early part of the wet season from May to July are generally much higher than later in the year, probably because there is more loose and friable sediment on the land surface and in river channels to be entrained at the end of the dry season. Antecedent and seasonal conditions greatly influence degradation processes in these monsoonal climates.

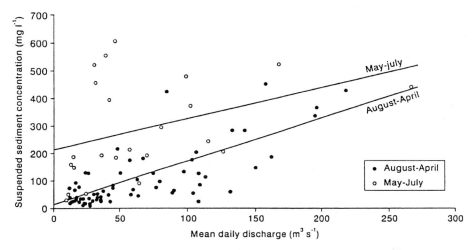

Figure 1. Comparison of early wet season (May to July) suspended sediment concentrations in the Nam Mae Kok River at Ban Tha Ton (catchment area 2980 km²) with those at other times of the year (August to April), showing the higher concentrations at low and moderate discharges during the first rains of the onset of the monsoon.

3. The Institutional Problems of Erosion and Land Degradation Monitoring

Many agriculture and forestry departments recognise that soil erosion is a problem and have set up soil conservation branches to examine the magnitude of the problem. Many aid agencies have incorporated soil erosion studies into their projects developing better land management systems. Research institutes and universities have often set up yet more investigations. However, the perspectives of such investigations are often limited. Classically, the paradigm of the government department investigation is a few years of detailed data collection and analysis, often with a specific problem in mind, and then the continuation of data collection and recording without any further analysis for many years. Occasionally such a database can be extremely informative. For example, the long records of plots examining the impact of shifting cultivation in Sarawak (Ng and Teck 1992; Teck 1992) clearly show that at the plot scale, traditional shifting cultivation causes little more erosion than occurs under natural forest (Table 1). Often however, the data remain buried in the archives of the department concerned and do not get used for further, wider studies. Such routinely collected data have a great potential for use in wider ranging research by other agencies collaborating with the government departments.

Table 1. Data on erosion rates under forest and shifting cultivation for Sarawak after Ng and Teck (1992) and Teck (1992).

Land Use	Location	Slope (degrees)	Period (years)	Soil loss t km⁻² y⁻¹ mean	Soil loss t km⁻² y⁻¹ range
PRIMARY FOREST	Niah F.R.	25-30	4	19	8.3-31
	Semongkok	25-30	11	24	7-77
SECONDARY FOREST					
a) logged 10 years previously	Niah F.R.	25-30	4	23	11-36
b) with hill padi	Semongkok	25-30	11	10	2-17
c) 2 month old lallang and scrub	Niah F.R.	33	3	1100	450-1800
HILL PADI/ SHIFTING CULTIVATION					
a) normal	Kg. Benuk	25-30	1	18	
b) terraced with cover	Semongkok	20	11	120	21-246
c) bush fallow	Semongkok	16-26	3	23.3	6-45
d) bush fallow	Tebedau	25	2	34	22-46
TRADITIONAL PEPPER	Semongkok	25-30	11	8944	5118-13912

Aid agencies can have excellent advice and enthusiasm, but the sustainability of their plot studies is often weak. Having established the plot and the different crop combinations, data collection is handed over to local technicians and the enthusiasts depart. These technicians become loaded with other responsibilities and the data collection falters. Records are not archived properly and are difficult to access. An example of this is an excellent set of plots near Hin Heup in the Lao PDR, where useful data on agroforestry systems have been collected, but the data quality is not being maintained (Oughton 1993; Douglas 1997).

Research institutes and universities have even greater difficulty in sustaining plot experiments as research funding is now almost entirely through short-term grants, and the necessary long-term ecological and hydrological research is not attractive to funders. Long-term monitoring can only be sustained by obtaining funding for a series of other projects. All too often data are collected only during the fieldwork period of a PhD or the three years of a research grant. Little information is then available on the impacts of rare events.

The type of monitoring is very often governed by what is in textbooks or by the advice received from western consultants. The soil erosion problems of arable soils in western Europe, the USA and Australia have attracted attention through the impacts of drought, wind erosion and declining soil fertility. They have led to a particular research paradigm that has led people to become concerned about rill development, about soil aggregate stability and about cultivated, ploughed land. The research applicable to such areas has often been transferred into forests, and then out of those countries to areas such as SE Asia without really questioning where the erosion occurs and what types of erosion are important. Most fields in SE Asia are small and are frequently separated from each other by patches of secondary vegetation or by bunds which restrict soil loss. However, in areas of recent forest clearance for agriculture, such as for temperate vegetables in the Kinabalu Highlands of Sabah (Douglas and Sinun 1997) or for maize cultivation in the western Korat Plateau of northeast Thailand (Douglas, 1997), severe erosion problems like those of the eastern United States develop. Under the forest, the bulk of the rain reaching the ground infiltrates and surface erosion only becomes a problem when bare patches of soil occur, through treefall or excessive trampling by animals, or when people remove the trees. The issue thus changes to questions of the occurrence, distribution and extent of bare soil.

In natural forests most of the erosion occurs in stream head hollows, on stream banks and in the occasional bare areas created by landslides or tree fall or trampling by animals. In disturbed forests and agricultural areas much, if not the majority of erosion is along footpaths, log haulage tracks and unsealed roads (Douglas *et al.* 1995; Ziegler and Giambelluca 1997). None of these erosion sources is normally assessed by routine investigations by the appropriate agencies. Few researchers give them the attention they deserve. The agricultural, forestry and water resource agencies normally responsible for measurements of soil loss or sediment transport usually neglect the significance of severe gully erosion on construction or mining sites. However, municipal authorities, national environmental ministries and local universities are now giving them more attention (Balamurugan 1991 a,b,c; Gupta 1992; Douglas and Wan Ruslan 1997).

4. Assessment of Erosion and Sedimentation for Water Resource Management

Major river basin authorities such as the Yellow River Commission, Yangtse River Commission and Mekong River Commission have long recognised the costs sedimentation causes for their dams and reservoirs. Most have conducted plot studies and trials of soil conservation works. But increasingly they have come to realise that integrated approaches to soil and water management are required (Liu and Zuo 1983; Chen 1992; Edmonds 1994; Yu and Wang 1999). The erosion problems of the loess plateau affecting the middle reaches of the Yellow River are related to gully erosion. The more level interfluve surfaces which dominate the area account for only about 44% of the soil loss, while around 56% of the erosion occurs on the gully walls and gully floors (Jiang *et al.* 1981) (Table 2). However, this is only because the eroding water reaches the gully margins in large quantities. Water control in the fields of the interfluves would reduce this supply and cut down the erosion (Huang 1988; Douglas 1989). Thus measurement is needed in the gullies and action on the farmers' fields on the interfluves.

Erosion in newly disturbed headwater areas is a problem even in relatively small river systems. The Sungei Liwagu rises high on Mt. Kinabalu in Sabah, but the temperate uplands at around 2000 m elevation around the mountain are being increasingly exploited for temperate vegetable cultivation. Plot studies show that the way in which the cabbages and lettuces are cultivated affects soil loss, but much of the erosion occurs along rills and gullies which dissect the contour furrows on which the vegetables are grown (Sinun and Douglas 1997). River suspended sediment measurements (Table 3) show a three orders of magnitude increase in yield compared with an undisturbed catchment nearby. Although measures are taken to plant the crops along the contour, the rill and gully erosion is unchecked.

Table 2. Influence of landform and land use on soil erosion in small catchments in the gullied hill loess area of the Yellow River basin, China (after Jiang *et al.* 1981)

type of land	main types of soil erosion	per cent of area	percent of catchment soil erosion	annual erosion rate t km²
interfluves	splash, rill, shallow gully	44-74	30-62	11600-26300
gully slopes	rill, gully, earth fall, landslide	25-52	35-62	16600-26,100
gully floors	gully	1-4	6-8	38500-70800
farmland	splash, rill, shallow gully	57-67	44-59	15800-19900
grazing and wasteland	gully, sheet	8-25	9-23	13400-16100
steep slopes	earth fall, landslide	13-21	20-25	21800-26500
roads, farmyards, gully floors	gully	4-7	7-13	28400-36200

Table 3. Characteristics and sediment yields of the upland study catchments near Mount Kinabalu, Sabah, Malaysia (after Sinun and Douglas 1997)

Catchment Name	Land Cover	Area km²	% of time discharge exceeds 0.125 m³ s⁻¹ km²	sediment yield t km² y⁻¹
Sungai Kalangaan	100% lower and upper montane forest	2.45	1.9	15.1
Sungai Silau-Silau	95.6% lower and upper montane forest, 4.4 % developed with tourist facilities (park buildings)	1.91	4.6	108.9
Sungai Ayamut	45 % lower montane forest, 55% cultivation roads and buildings	0.91	4.3	1076.4

In the more humid areas of the Yangtse and Mekong basins agricultural impacts on hillslopes are readily apparent, but the soil erosion and land degradation problems are not due to agriculture alone. Temporary exposures of soil and weathered rock permit large amounts of gully erosion as any inspection of an expanding urban and industrial area reveals. Even on the cohesive basalt soils around Pleiku in the Vietnamese Central Highlands, gully erosion due to roadside runoff and degraded areas around claypits is readily visible. In Malaysia, Kuala Lumpur has had a consistent outward movement of bare soil patches since 1960 (Fig. 2), but they have meant a continuing sediment problem for the managers of the Kelang River (Douglas 1978; Balmurugan 1991a). Gully erosion supplies most of the sediment from such areas (Table

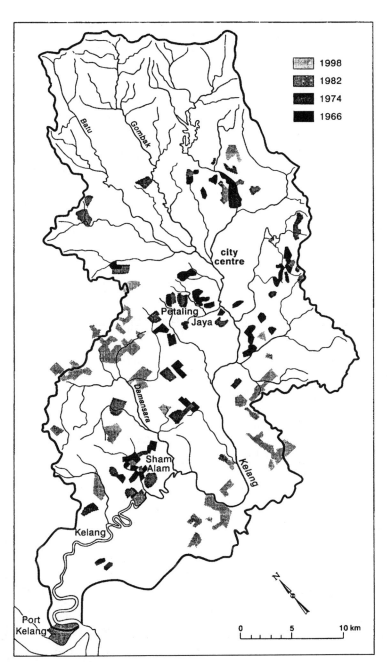

Figure 2. The outward movement of bare areas around Kuala Lumpur since 1960 based on air photo and landsat image interpretation.

4). Similar problems can be found throughout the region, but they are rarely well-documented.

Table 4. Erosion rates on a bare construction site at Mengkuang Heights in Ulu Kelang near Kuala Lumpur, Malaysia (after Mykura 1989)

	annual total eroded t y^{-1}	erosion rate per unit area t km^2 y^{-1}
Total	34,703	517,955
slope surface erosion	2,544	37,970
rill erosion	5,291	78,985
gully erosion	26,868	401,014

There seems to be a wish to deny the significance of these sediment sources in river sedimentation and to concentrate on erosion on cultivated land. Such a belief seems to have dominated the planning of erosion and sedimentation studies in the Mekong Basin (Douglas 1997). The clear evidence of forest depletion and land degradation led to a program of soil erosion measurements. Investigations in Thailand built on that country's long experience of land and water management and tackled the severe erosion in the maize growing areas of the south-western Korat Plateau. However, they were only conducted for 18 months for lack of financial support to maintain observations and because of relative hostility of the local farmers towards the field installations (Thai Project Team 1996). In Viet Nam, local hostility and maintenance problems were avoided as the plots were established at existing government meteorological and hydrological stations. The plots were well made and well cared for by local staff. Good data on the effects of crop cover on soil loss were obtained (Institute of Meteorology and Hydrology 1994). However, these plots were on the basaltic soils in the low relief area of the plateau around Pleiku, in the Central Highland tributaries of the Mekong, the Se San and Sre Pok rivers. The critical land degradation problems are not on these cohesive basaltic soils, but on the much more erodible sandstone and shale soils of the surrounding hills into which displaced traditional cultivators have been moving as migrants from other parts of Viet Nam come to the Central Highlands to cultivate coffee. No studies were undertaken of the erosion caused by mining, industrial development or urbanisation. In this Mekong program opportunities were missed because there was not a concentration on identifying the key problems and overcoming logistic difficulties to put the study areas where those problems occur. In part that failure can be laid in the hands of western advisors.

5. Timescale and Spatial Scale Issues

Even if the appropriate sites are found, the question of the period of observation remains important. The assumption is that disturbance is followed by recovery. New evidence suggests that recovery is punctuated by periods of renewed erosion triggered by extreme events (Douglas et al. 1999). Some 10 years' observations of post-logging recovery of stream sediment loads at the Danum Valley Field Centre in Sabah have shown that unusually heavy, prolonged rainfalls can induce failure of abandoned logging roads, causing landslides which send large volumes of sediment into streams. The sediment loads in subsequent storms again reached the levels which prevailed immediately after logging. Clay deposits again covered the channel beds and moderate runoff events then took several months to again reveal the natural gravel bed of the stream. Here the pulse-like nature of erosion and sediment delivery has to be emphasised. Short-term studies may not pick this up. On major rivers, such as the Segama at Danum Valley (Fig. 3), a few individual storms may account for a significant part of the total suspended sediment discharge. Observations should be maintained for as long as possible and every endeavour should be made to put the observation period in the context of the long-term rainfall record so that the range of magnitude and frequency effects investigated is understood.

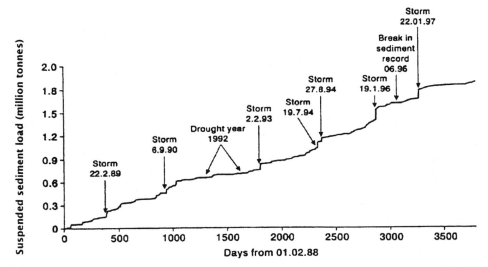

Figure 3. Cumulative curve of daily suspended sediment discharge of the Sungei Segama at Danum Valley Field Centre, Sabah, Malaysia February 1988-June 1998.

However well plot and small catchment studies are conducted, the research results are often at too small a scale for direct use in management. Researchers now endeavour to use GIS and Remote Sensing to upscale to management units and regional levels (King et al. 1998; Lee 1998). Upscaling works well if data are being used in similar situations, such as to predict erosion of maize fields across a region from a set of maize field plot studies. Care has to be taken to examine the sensitivity of erosion to environmental changes, such as the development of longer runoff paths in access track construction for timber harvesting or land clearance regimes. These changes in land surface conditions driven by human activity then have to be added to the effects of climatic change to obtain meaningful predictions about future trends in erosion. Few of the climatic change scenarios of land degradation incorporate continuing land cover modification by people.

Wider issues arise when upscaling from small catchments or hillslope studies to whole river basins. The sediment delivery ratio questions come into play (Walling 1983). Even at the plot scale, sediment deposition plays an important part in the sediment yield process (Rose and Hogarth 1998). Within small catchments, the large-scale storm period transfer of sediment into channel heads and its accumulation within channels, makes the timing of inputs from slopes differ from the downstream sediment discharge. Detailed studies of fingertip tributaries and of gullies developed during logging activities in Sabah show that immediately after disturbance sediment inputs to the headwaters are high. However, much of the eroded material remains in the channels for several weeks, or months, before finally being carried further downstream (Douglas et al. 1999). In many small channels, temporary storage of sediment behind debris dams regulates some of the sediment flux (Spencer *et al.* 1990). In major channels, within-stream sediment storage after major flows is often extremely large. These factors and the diversity of slopes and ground cover intensity make upscaling difficult.

6. Problems of Modelling and Relevance to Human Lives at the Management Scale

Complex models involving much computer power may provide more realistic predictions but there is still a gulf between the research into land degradation and the management of land systems on the ground. Two major difficulties exist: a) suitable soil parameters or erosion values are unavailable for many parts of the terrain included in geographical information systems and digital terrain models in SE Asian conditions; and b) many of the research models are difficult for managers to use, because complex local data are difficult to obtain. However, the many nomographs and tables of characteristic values available for the USLE make it easy to use, despite its inappropriateness. The physically based WEPP model has less danger than the empirical USLE and its derivatives of extrapolating the relationships on which the model is based beyond their valid range (Favis-Mortlock and Guerra 1999). Elsewhere in the tropics, WEPP has been used successfully in the northern Mato Grosso of Brazil to predict changes in field scale erosion under different future climatic scenarios, but it neglected possible changes in gullying (Favis-Mortlock and Guerra 1999). In SE Asia, many novel approaches are being developed, such as applications of the KINEROS model (Woolhiser *et al.* 1990) to erosion in agricultural and forest terrain in the hills of northern Thailand, and GIS-based modelling using remote sensing in selectively logged areas of northern Borneo (Cheesman *et al.* 2000). Being closely related to local conditions they offer greater practical utility. However, valid ground truth parameters, including a diversity of data on erosion rates, including gullying, are needed. When detailed information is available, as in the nested 0.5 km² Baru experimental catchment system in eastern Sabah (Chappell *et al.* 1999), it is possible to define the major type of landform associated with individual contributory areas and then use landform characteristics, or terrain units, to upscale to management level areas of the order of 10 to 100 km². Sometimes simple, long-term monitoring using low cost techniques such as the erosion bridge (Shakesby *et al.* 1991; Walsh and Bidin 1995) can provide the necessary data for individual landforms. Improvements in remote sensing techniques, particularly in analysis of mixed pixels in diverse humid tropical terrain, coupled with better ground truth, offer great potential for better monitoring and prediction of land degradation and erosion risks.

Many community watershed program officers in Asia argue that there is no need for further field measurements and more detailed monitoring (Sharma 1997, 1999). Instead there should be greater attention to the way people use the land and their understanding of erosion. Watershed management used to be seen as synonymous with soil conservation and erosion control, including reforestation. Today it is more synonymous with poverty alleviation and sustainable development of upland catchment areas (Sharma 1997). People rather than the natural resources have now become the foremost focus of watershed management.

When actual patterns of erosion in the fields of small farmers are examined closely, great diversity in soil loss is found. Some cultivators with permanent homes and scattered fields end up using some fields far more intensively than others for logistic reasons linked to cash cropping or off-farm employment. Thus while remote fields remain fallow, those close to settlements become severely degraded. Patterns of adoption of soil conservation measures or agroforestry techniques vary within communities and between neighbouring villages. As elsewhere, farmers in the Crocker Range of Sabah have complex family and economic circumstances that lead to some fields being degraded while other are underused (Lim and Douglas 1998, 2000). Unless the biophysical understanding of erosion is linked to these human factors, the development of measures and community action programs to combat soil erosion will be unrealistic.

7. Conclusions

The challenge to all soil erosion and land degradation scientists is to develop more purposeful studies which carry out high quality science, in the places where it matters, for long enough to be able to make reliable erosion forecasts at the management scale. The challenge for all concerned with sustainable development is to work with all stakeholders in the land to achieve land use systems that are effective. Even so, if we concentrate solely on the fields and forests, we will miss the erosion caused by industry, mining, urban development and roads of all types. There is a lot do, so our efforts have to concentrate on the real problems and on producing usable results.

8. Acknowledgments

The author is grateful to all those colleagues, students and field technicians who have worked with him in southeast Asia over the past 35 years, including particularly Low Kwai Sim, Lim Chye Lee, Jennifer Nyuk Wo Lim, Zakaria Awang Soh, Hamish Mykura, Lai Food See, Zulkifli Yusop, Wan Ruslan Ismail, Balamurugan, Waidi Sinun, Wong Wai Meng, Kawi Bidin, Jadda Suhaimi, Asman Sulaiman, Tony Greer, Mike Spilsbury, Paun Kuokon, Johnny Larenus, Jamal Mohd Hanapi, Ekwhan Toriman, Ted Murtedza, Nick Chappell, Rory Walsh and Clive Marsh. A great debt is also owed to colleagues at the Mekong River Commission, Kasetsart University, Gadjah Mada University and the Sub-Institute of Geography of Ho Chi Minh City, and to all those staff and temporary field assistants who help and support research at Danum Valley Field Centre in Sabah. The Danum Valley Management Committee, the Economic Planning Unit of the Prime Minister's Department of Malaysia, the Sabah State Secretary and the Sabah Chief Minister's Department are thanked for their permission to conduct research at Danum Valley. The collaboration of the Forestry Upstream Division, Innoprise Corporation Sdn. Bhd., in the use of the catchments is gratefully acknowledged. This is paper A/321 of the Royal Society South-East Asian Rain Forest Research Programme.

9. References

Balamurugan, G. (1991a) Sediment balance and delivery in a humid tropical urban river basin: the Kelang River, Malaysia, *Catena* **18**, 271-287.
Balamurugan, G. (1991b) Some characteristics of sediment transport in the Sungai Kelang Basin, Malaysia, *Journal of the Institution of Engineers Malaysia* **48**, 31-52.
Balamurugan, G. (1991c) Tin mining and sediment supply in Peninsular Malaysia with special reference to the Kelang River Basin, *The Environmentalist* **11**, 281-291.
Boardman, J. and Favis-Mortlock, D. (1999) Frequency-magnitude distributions for soil erosion, runoff and rainfall – a comparative analysis, *Zeitschrift fur Geomorphologie, N.F. Suppl.-Bd.* **115**, 51-70
Chappell, N.A., McKenna, P., Bidin, K., Douglas, I. and Walsh, R..P.D. (1999) Parsimonious modelling of water and suspended sediment flux from nested catchments affected by selective tropical forestry, *Philosophical Transactions of the Royal Society of London* B **354**, 1831-1846.
Cheesman, J.C., Cutler, M.C. and Douglas, I. (2000) Erosion modelling using remote sensing and GIS in eastern Sabah, Malaysian Borneo, 4th GIS Conference (in press).
Chen, Y. (ed.) (1992) *National Report of the People's Republic of China on Environment and Development. China*, Environmental Science Press, Beijing.

Ciesiolka, C.A., Coughlan, K.J., Rose, C.W., Escalante, M.C., Hashim, G.M., Paningbatan Jr., E.P. and Sombatpanit, S. (1995) Methodology for a multi-country study of soil erosion management, *Soil Technology* **8**, 179-192.

Douglas, I. (1978) The influence of urbanisation on fluvial geomorphology in the humid tropics, *Geo-Eco-Trop.* **2**, 229-242.

Douglas, I. (1989) Land degradation, soil conservation and the sediment load of the Yellow River, China: review and assessment, *Land Degradation and Rehabilitation* **1**, 141-151.

Douglas, I. (1997) *Control of soil erosion, sedimentation and flash flood hazards (basin-wide) Review and Assessment Report for Phase I (1990-1996),* Mekong River Commission, Bangkok.

Douglas, I. (1999) Sediment a major river management issue, *Rivers '99: Towards Sustainable Development, National Conference 14-17 October 1999,* Universiti Sains Malaysia, Penang, p. 2.

Douglas, I. Greer, T., Sinun, W., Anderton, S., Bidin, K., Spilsbury, M.J., Suhaimi J. and Sulaiman, A. (1995) Geomorphology and Rainforest Logging Practices, in D.F.M. McGregor and D.A. Thompson (eds.), *Geomorphology and Land Managment in a Changing Environment*, Wiley, Chichester, pp. 309-320.

Douglas, I., Kawi Bidin, Balamurugan, G., Chappell, N.A., Walsh, R.P.D., Greer, T. and Sinun, W. (1999) The role of extreme events in the impacts of selective tropical forestry on erosion during harvesting and recovery phases at Danum Valley, Sabah, *Philosophical Transactions of the Royal Society of London* B **354**, 1749-1761.

Douglas, I. and Wan Ruslan Ismail (1997) Urban erosion and sedimentation on Pulau Pinang, Malaysia, in B.W. Webb (ed.), *Erosion and Sediment Yield: Global and Regional Perspectives: Poster Report Booklet*, IAHS, Exeter, pp. 59-61.

Edmonds, R.L. (1994) China's environment: problems and prospects, in D.J. Dwyer (ed.), *China: The next decades*, Longman, Harlow, pp. 156-185.

Eiumnoh, A., Shrestha, R.P. and Baimoung, S. (1997) Remote sensing and GIS application in watershed land use planning: a case study of Mae Nam Chi sub-watershed, northeast Thailand, *Proceedings of the FORTROP'96: Tropical Forestry in the 21st Century 25-28 November 1996*, Kasetsart University, Bangkok, **6**, pp. 261-272.

Favis-Mortlock, D.J. and Guerra, A.J.T. (1999) The implications of general circulation model estimates of rainfall for future erosion: a case study from Brazil, *Catena* **37**, 329-354.

Fors, T. and Chantaviphone Inthayong (1994) *Erosion risk map of the Lao PDR: National Level*, Integrated Resources Centre, Watershed Management Section, Department of Forestry, Vientiane.

Gupta, A. (1992) Floods and sediment production in Singapore, in A. Gupta and J. Pitts (eds.), *Physical Adjustments in an Urban Landscape: the Singapore Story*, Singapore University Press, Singapore, pp. 301-326.

Huang Bingwei (1988) River conservancy and development of the North China Plains and loess highlands: strategies and research, *Great Plains Quarterly* **6**, 218-224.

Institute of Meteorology and Hydrology (1994) *Final Report: Control of Soil Erosion and Measurement of Sediment Flow at the Sesan and Srepok Catchments*, The Hydrometorological Service of SRV, Institute of Meteorology and Hydrology, Hanoi.

Jiang, D.Q., Qi, L. and Tan, J.S. (1981) Soil erosion and conservation in the Wuding River Valley, China, in R.P.C Morgan (ed.), *Soil Conservation, Problems and Prospects*, Wiley, Chichester, pp. 461-479.

King, D., Fox, D.M., Le Bissonnais, Y. and Danneels, V. (1998) Scale issues and a scale transfer method for erosion modelling, in J. Boardman and D. Favis-Mortlock (eds.), *Modelling soil erosion by water*, NATO ASI Series I, Global Environmental Changes, **55**, Springer, Berlin, pp. 201-212.

Lee, J.J. (1998) Cross-scale aspects of EPA erosion studies, in J. Boardman and D. Favis-Mortlock (eds.), *Modelling soil erosion by water*, NATO ASI Series I, Global Environmental Changes, **55**, Springer, Berlin, pp. 191-199.

Lim J.N.W. and Douglas, I. (1998) The impact of cash cropping on shifting cultivation: the Dusun in Sabah, Malaysia, *Asia Pacific Viewpoint* **39**, 315-326.

Lim J.N.W. and Douglas, I. (2000) Land management policy and practice in a steepland agricultural area: a Malaysian example, *Land Degradation and development* **11**, 51-62.

Liu Changming and Zuo Dakang (1983) The impact of South-to-North Water Transfer upon the Natural Environment, in A.K Biswas., Zuo Dakang, J.E. Nickum and Liu Changming (eds.), *Long-Distance Water Transfer: A Chinese case study and international Experiences*, Tycooly International, Dublin, pp. 169-179.

Mykura, H.F. (1989) Erosion of Humid Tropical Construction Sites, Kuala Lumpur, Malaysia, Ph.D. Thesis (Manchester).

Ng, T.T. and Teck, F.H. (1992) Soil and nutrient losses in three different land use systems on hill slopes, *Proceedings of the Technical Sessions 29th Research Officers' Annual Conference 1992*, Research Branch, Department of Agriculture, Kuching, pp. 37-55.

Oughton, G.A. (1993) *Control of Soil Erosion, Sedimentation and Flash Flood Hazards: A Report Submitted to the Mekong Secretariat*, National Office for Nature Conservation and Watershed Management, Lao PDR, Vientiane.

Rose, C.W., Coughlan, K.J. and Fentie, B. (1998) The Griffith University erosion system template (GUEST), in J. Boardman and D. Favis-Mortlock (eds.), *Modelling Soil Erosion by Water*, NATO ASI Series I, Global

Environmental Changes, **55**, Springer, Berlin, pp.-399-412.

Rose, C.W. and Hogarth, W.L. (1998) Process-based approaches to modelling soil erosion, in J. Boardman and D. Favis-Mortlock (eds.), *Modelling Soil Erosion by Water*. NATO ASI Series I, Global Environmental Changes, Vol. 55, Springer, Berlin, pp. 259-270.

Shakesby, R.A., Walsh, R.P.D. and Coelho, C.O.A. (1991) New developments in techniques for measuring soil erosion in burned and unburned forested catchments, Portugal, *Zeitschrift für Geomorphologie N.F. Supplbd* **83**, 161-174.

Sinun, W. and Douglas, I. (1997) Geomorphic and Hydrologic response to Humid Tropical Steepland Montane Forest Disturbance on Gunong Kinabalu, Sabah, Malaysia, in B.W. Webb (ed.), *Erosion and Sediment Yield: Global and Regional Perspectives: Poster Report Booklet*, IAHS, Exeter, pp. 100-102.

Sombatpanit, S. (1992) Soil Conservation in Thailand, *Australian Journal of Soil and Water Conservation* **5**, 14-18.

Spencer, T., Douglas, I., Greer, A.G. and Sinun, W. (1990) Vegetation and fluvial geomorphic processes in South East Asian tropical rainforest, in J.B. Thornes (ed.), *Vegetation and Erosion*, Wiley, Chichester, pp. 451-469.

Teck, F H. (1992) Soil loss experimental plots 1992, *Annual Report, Research Branch, Department of Agriculture, Sarawak*, pp. 167-8.

Thai Project Team (1996) *Study on Erosion and Sedimentation and its Control, Case Study: Lam Phar Phloeng Watershed, Amphoe Pakthongchai, Nakhon Ratchasima Province, Part 1: Hydrometeorological Data Collection & Analysis January - December 1995*, The Team, Bangkok.

Van Lynden, G.W.J. and Oldeman, L.R. (1997) *The Assessment of the Status of Human-Induced Soil Degradation in South and Southeast Asia*, International Soil Reference and Information Centre, Wageningen.

Walling, D.E. (1983) The sediment delivery problem, *Journal of Hydrology* **65**, 209-237.

Walsh, R.P.D. and Bidin, K. (1995) Channel head erosion in primary and logged rain forest in Sabah, in *The International Association of Geomorphologists Southeast Asia Conference, Singapore 18-23 June 1955, Programme with Abstracts*, p. 79.

Watthanasuk, Manus (1976-77) *A preliminary investigation of soil erosion using the Universal Soil Loss Equation and Satellite Photo Interpretation*, Mahidol University, Bangkok.

Woolhiser, D.A., Smith, R.E. and Goodrich, D.C. (1990) *Kineros, A Kinematic Runoff and Erosion Model: Documentation and User Manual*, U.S. Department of Agriculture, Agricultural Research Service ARS-77.

Yu Xinxiao and Wang Lixian (1999) Sustainable highland watershed management in China, in Bunvong Thaiusta, Catherine Traynor and Songkram Thammincha (eds.), *Highland Ecosystem Management. Proceedings of the International Symposium on Highland Ecosystem Managments Royal Angkhang Agricultural Station, Royal Project Foundation, Chiangmai, Thailand 26-31 May 1998*, Royal Project Foundation, Chiangmai, pp. 107-117.

Ziegler, A.D. and Giambelluca, T.W. (1997) A preliminary simulation of hydrologic change and accelerated erosion resulting from road expansion in mountainous northern Thailand, *Proceedings of the FORTROP'96: Tropical Forestry in the 21st Century 25-28 November 1996*, Kasetsart University, Bangkok, **6**, pp. 211-226.

AUTHOR

IAN DOUGLAS

School of Geography
University of Manchester
Manchester
M13 9PL
United Kingdom

I.Douglas@man.ac.uk

Part Four

Rehabilitation of Degraded Land

In Part Four the focus shifts to management: how can degraded land be rehabilitated? All bar one of the six papers come from Australia, for reasons which are not entirely clear. It might be tempting to suggest that Australians are doing more than other nations to deal with land degradation; but the increasing extent and severity of the problems in many parts of the country are a powerful counter-argument.

Despite the geographical focus, the topics covered are very diverse. Schnabel, Gonzalez, Murillo and Moreno report a preliminary evaluation of methods of improving pasture productivity and reducing soil erosion in a wooded rangeland in south west Spain. The methods included direct seeding of herbaceous species, traditional seeding of legumes, and applications of phosphate fertilisers. Their effects on pasture quality, biomass production, surface runoff and soil loss were measured using experimental plots. Their study area experienced an exceptional rainstorm which somewhat disrupted the experiments, but which produced rill erosion only on those plots subjected to traditional tillage methods.

Humphreys and Groth assess the recovery of land which had been affected by extensive sheet erosion in a region south west of Sydney, New South Wales, as a result of both grazing and cultivation for grain crops. Recovery commenced in the early 1950s, associated with increased rainfall and reduced farming intensity. The study uses space-time transformations to track subsequent changes of vegetation and soil properties (soil thickness, organic matter and chemical data), and relates the changes to the trapping of slopewash sediments behind vegetation and bioturbation by ants. The authors found that a topsoil >15 cm thick developed in <50 years, but that the trends in common soil chemical properties were much more mixed. Some properties are predicted to achieve recovery in <100 years, whereas others are predicted to take 800-1600 years. Catenary relationships are shown to be particularly imporant.

Intensive grazing by introduced stock is a common factor in land degradation in many parts of the world. John Pickard tackles the vexed question of carrying capacity for sustainable grazing in semi-arid Australia. His well researched, interesting and somewhat unconventional approach avoids the use of field-based, measured data; rather, he uses the literature and graziers' evaluations to come up with a figure of 50% of district average stocking rates as being the probable sustainable figure. Pickard now works for the New South Wales Parks and Wildlife Service in that State's pastoral country. It would be most interesting to be a witness to the discussions when he attempts to implement this finding in the region.

A very different approach - that of economics - is taken by MacLeod and Noble to the directly related question - namely the rehabilitation of degraded, semi-arid rangelands in Australia. They make the point that many landholders resist implementing remedial measures on the grounds that those measures are uneconomic to them. But MacLeod and Noble suggest that this is not necessarily the case, and mount an argument that both the underlying context and processes involved with a given degradation problem need to be considered when assessments are made of the economic value of a given restoration technique. The process needs to be well understood to assist in clearly defining the scope and nature of the problems and for seeking opportunities for low-cost options to resolve them. Managers also need a clear understanding of the objectives of rehabilitation. More attention needs to be given to taking positive action in specific cases rather than trying to solve land degradation problems in general over extensive areas.

The final two papers in this Part look at two (again highly contrasting) techniques which are or could be used in managing land degradation. The first, by Richard George and Donald Bennett, describes the results of two detailed surveys which show how airborne geophysics (electromagnetics, magnetics and radiometrics) can assist the management of secondary, dryland salinity problems in Western Australia. Both study catchments were inspected by participants in the main field trip which formed part of the Perth COMLAND conference. In the second paper, Raper, Guppy, Argent and George discuss the use of a water balance catchment model in farm and catchment planning, also in Western Australia's agricultural region. A groundwater model is used to predict the landscape-scale effects of changes in vegetation type and other water management practices. The authors state the the benfits of the model are that: (1) farmers' knowledge is included in the model calibration process; (2) farmers value the opportunity to participate in the process and learn from it, and (3) water management practices likely to succeed at the catchment scale are tested by the model simulations. Limitations of the model are discussed briefly. This paper complements those by Lobry de Bruyn and Marsh *et al.* in Part Three but is perhaps more directly linked to management than the latter papers.

CHAPTER 15

DIFFERENT TECHNIQUES OF PASTURE IMPROVEMENT AND SOIL EROSION IN A WOODED RANGELAND IN SW SPAIN

Methodology and Preliminary Results

S. SCHNABEL, F. GONZÁLEZ, M. MURILLO and V. MORENO

1. Abstract

Soil conservation measures are investigated from a representative farm in open evergreen woodlands with silvo-pastoral landuse (*dehesas*) in Extremadura. Different techniques of pasture improvement were applied, investigating their effect on pasture quality, biomass production, soil loss and surface runoff. The techniques included direct seeding of autochthonous herbaceous species, phosphate fertilisation and traditional seeding of leguminous species. The experimental scheme consisted of eight treatments and one control, each equipped with two closed soil erosion plots. The area was grazed by sheep in a controlled manner.

An exceptional rainstorm event occurred just one week after carrying out the pasture preparations in October 1997, producing heavy rill erosion only at those sites which had been subject to traditional tillage. Cross sectional measurements of the rills revealed a mean soil loss of 100 t ha^{-1} at the ploughed sites. Only a few runoff events were registered during the period following the completion of the erosion plots (February 1998 – August 1999). Mean total soil loss produced by sheetwash amounted to 0.16 t ha^{-1}. Large differences of runoff and soil loss were observed within the same treatment, probably due to small-scale variations of factors which have yet to be studied. Therefore additional measurements are now being carried out in each plot, including rainfall, soil moisture and soil density. The preliminary data indicate the necessity of treatment replication.

2. Introduction

2.1. THE DEHESAS

A research project on soil conservation in open evergreen woodlands was initiated in autumn 1997 in the south-western part of the Iberian Peninsula. An agro-silvo-pastoral landuse system is widespread, the *dehesas* and *montados*, in the Spanish and Portuguese languages, respectively (Fig. 1). The *dehesas* cover an area of approximately 30 000 km^2 (Diáz *et al.* 1997) and similar systems can be found in other Mediterranean countries (Le Houerou 1987). The vegetation consists mainly of two layers, a tree cover of evergreen oak species and a herbaceous cover. The dominant trees are holm oaks (*Quercus rotundifolia*) and cork oaks (*Q.*

239

A.J. Conacher (ed.), Land Degradation, 239–253.

suber). It is a semi-natural vegetation, produced by thinning out the tree cover, clearing the shrubs, and the evolution of a herbaceous pasture cover, favoured by the introduction of livestock (Montoya Oliver 1988). Furthermore, trees are pruned every 8 to 10 years in order to increase acorn production. The sweet fruits, especially of the holm oaks, constitute an important food for the pigs which are kept in the *dehesas*.

Dehesas are an adaptation to the edaphic and the climatic conditions of the area, where schists, granites and quartzites are the dominant rocks. Soils are generally acidic (pH 5.0 – 6.5) and have a loamy silt or loamy sand texture (CSIC 1970). They are shallow and have a low organic matter content. Soils are generally poor in nutrients, with low contents of plant available phosphorous and calcium (CSIC 1970; Dorronsoro Fernández 1992). Climate is Mediterranean with Atlantic and continental influences; in comparison to the Mediterranean coast, summer temperatures are higher and winter temperatures lower. The annual rainfall distribution shows a strong minimum during the summer months and a rainy season lasting from October until April. Mean annual precipitation typically ranges between 400 and 600 mm.

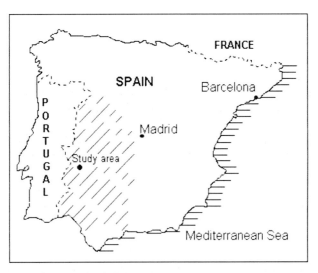

Figure 1. Location map. The shaded area indicates the distribution of the dehesas in Spain.

One of the most characteristic features of these wooded pasturelands is a multiple exploitation system, which has existed for many centuries. Different animals are kept, mainly sheep, pigs, cattle and goats. The trees are used for cork, firewood and charcoal production, but also serve as fodder during periods when pasture is scarce. Cultivation is limited to areas with deeper soils. Traditionally, cereals were cultivated in a four-year rotation cycle, whereas today two tendencies can be observed; either cultivation has been abandoned, or different crops have been introduced and used for fodder.

Traditionally, transhumance was practised, involving the seasonal movement of livestock over large distances in order to take advantage of the pastures in more humid areas during summer. Nowadays it has been almost completely abandoned, with continuous grazing being one consequence. As a result, grazing pressure is increased during summer when herbaceous production is almost zero, leading to a reduction of the vegetation cover and an increased risk of soil degradation and erosion (Schnabel 1997). Furthermore, an increase in livestock numbers has been reported mainly due to the subsidies paid by the European Union (Donázar *et al.* 1997). It has been shown that soil loss in the *dehesas* is very small under a low grazing density, but increases strongly with higher stocking rates (Schnabel 2000). This is in accordance with studies from semi-arid areas in the United States as summarised by Holechek *et al.* (1989).

The shallow soils developed on schist have a low infiltration capacity giving rise to high runoff coefficients during moderate to heavy rainstorms (Ceballos 1999). Under drought conditions the soil erosion risk is strongly increased due to the sparse ground cover (Schnabel 1997). Furthermore, the Water Drop Penetration Time Test (Letey 1969) indicates the importance of water repellency on runoff generation during droughts (Cerdà *et al.* 1998). Surface hydrophobicity is higher beneath *Q. rotundifolia* canopies than in the open spaces and is probably related to decomposing litter containing water repellent substances (Cerdà *et al.* 1998).

2.2. OBJECTIVES

Dehesas play in important role in Europe for wildlife conservation because they support a wide diversity of plant and animal species (Diáz *et al.* 1997). Conservation is thus associated with the maintenance of this landuse system, that is with pastoral exploitation. To guarantee the productivity of the dehesas constitutes therefore an important goal, not only from an economic, but also from an environmental point of view. The present project focusses on the investigation of different techniques of pasture improvement and their effects on soil erosion, runoff production and pasture evolution. The main objectives are to increase pasture quality and pasture production, in such a way that soil loss and surface runoff are reduced. Where the herbaceous cover is degraded seeding of pasture species may be a positive improvement measure. Tthe seeding technique used is crucial with respect to soil erosion. Traditionally, the hillslopes are ploughed completely, which constitutes a high soil erosion risk. Seeds cannot just be spread out on the soil surface because they are washed away by the first rains or collected by ants. Therefore different tillage systems were applied.

A mixture of autochthonous species and varieties was used in this investigation because they are well adapted to local conditions (Holechek *et al.* 1989). The use of local species and varieties for the regeneration of degraded pastures in the study area has been investigated by González *et al.* (1994) and González (1994). In addition, the study included planting fodder shrubs in rows and the application of phosphate fertilisers. The present research is justified by the absence of studies about the relationships between pasture improvement measures and soil erosion in the Mediterranean.

3. Study Area

Research was carried out on a farm which is representative of the most widespread type of *dehesas* in southwest Spain, characterised by shallow soils developed on schist, a tree cover of *Quercus rotundifolia* and sheep grazing. The experimental station is located in the private estate Las Cañas, in the Spanish region of Extremadura, 44 km southwest of the city of Cáceres and at an altitude of 300 m a.s.l. (Fig. 1, Plate 1). The area is part of a widespread peneplain, an upper Miocene erosion surface, developed on Prepaleozoic and Paleozoic material of the Hesperic massif. The undulating slopes have gradients of 5-6°. The soils are shallow, with depths ranging from 20 to 30 cm, and can be classified as Leptosols (FAO 1990). Their texture is sandy loam with a clay content of 7.4%. Organic matter content of the upper 20 cm is 1.4%, with most being concentrated in the top 5 cm of the soil. Phosphorous content is negligible with a cation interchange capacity of 8.7 meq/100g. Soils are moderately acid with a pH of 5.3.

For a general characterisation of the climate in the study area data from the meteorological station in Cáceres are used, because closer stations only provide shorter and incomplete data sets. Figure 2 indicates good agreement of monthly rainfall amounts measured at Las Cañas and at Cáceres. Mean annual temperature is 16°C, with mean maximum and minimum temperatures of 32.6°C and 4.5°C during July and January, respectively. The hydrological year starts in September, when the first rainfall is usually registered after the summer dry period. Mean annual precipitation is 514 mm, with a coefficient of variation of 30.2%. Table 1 presents the main characteristics of annual and monthly rainfall, including the precipitation amounts corresponding to different frequencies, which are used for classification: < 20% - very dry; 20 – 40% dry; 40 – 60% normal; 60 – 80% humid, and > 80% very humid. Annual and interannual variability of rainfall is high, and prolonged droughts are a common feature (Schnabel 1997).

Plate 1. Las Cañas Experimental Station. Note open evergreen woodland of *Quercus rotundifolia*, erosion plot and exclosure cage.

Rainfall intensities are lower compared with the Mediterranean coast: the 60-minute maximum intensity and maximum daily rainfall amounts are 25.8 mm h^{-1} and 61.6 mm, respectively, with a return period of 10 years (Schnabel 1998).

Tree density is 33 trees/ha. In the farm and outside the fenced experimental area the grazing density is 3.5 sheep/ha. The herbaceous cover in September of 1997 was only 10%. Given that the previous year registered high rainfall amounts, the low ground cover indicates an excessive stocking rate.

4. Methods

The experimental scheme consists of nine different treatments carried out in October of 1997, each with a surface area of 0.5 ha, located on an east facing slope (Fig. 3). Except for Treatment 9 (T9), which is the control, they were fenced and subject to controlled grazing, with an equivalent of 2 sheep per hectare. The area was grazed during several weeks at the beginning of April and in June. In the unfenced area the stocking rate was 3.5 sheep/ha. The treatments are summarised in Table 2. At T1, T2, T4, T7 and T8 a mixture of three indigenous leguminous species was introduced (*Trifolium subterraneum*, *T. glomeratum*, *Ornithopus compressus*). At T7 lupin seeds were added. At T6 forage shrubs were planted along small ridges with a narrow furrow immediately behind. Following the contour line the ridges were established with a spacing of 10 m. Three of the treatments were completely tilled with a disc plough (T2, T3, T7). In T1 and T8 direct seeding was carried out. In T4 strip tillage was applied, consisting of 2.5 m wide bands following the contour with alternating untouched

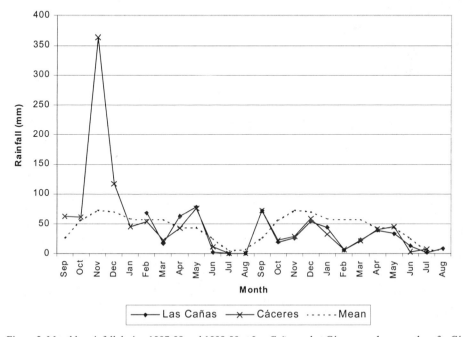

Figure 2. Monthly rainfall during 1997-98 and 1998-99 at Las Cañas and at Cáceres, and mean values for Cáceres.

Table 1. Monthly and annual rainfall characteristics at the city of Cáceres (1907 – 1996, n=90).

Month	Mean of (mm)	Coefficient of variation (%)	Maximum (mm)	Frequency (%)			
				20	40	60	80
Sep	26.3	99.7	139.1	3.8	13.3	25.2	45.1
Oct	55.6	87.2	236.3	17.6	34.7	55.0	80.5
Nov	69.4	70.5	242.1	27.8	50.4	71.3	101.2
Dec	69.4	90.4	283.8	22.1	33.7	63.7	117.5
Jan	57.7	93.3	268.2	18.2	35.0	52.0	95.7
Feb	56.4	87.5	294.0	13.2	31.1	58.8	89.8
Mar	57.1	72.4	172.6	20.8	42.6	61.5	87.2
Abr	42.7	69.3	148.5	17.2	28.0	44.6	68.3
May	43.1	74.5	157.0	14.3	27.9	46.9	65.0
Jun	25.4	102.1	132.0	5.0	13.1	24.3	40.0
Jul	4.5	150.3	30.3	0.0	0.3	2.7	8.7
Aug	6.7	174.1	60.4	0.0	0.3	3.0	12.1
Year	514.3	30.2	980.9	367.0	461.2	527.0	653.8

natural pasture and ploughed soil with introduced seeds. T5 consisted of the natural (fertilised) herbaceous cover. T3 was completely tilled, resembling the widely used clearing of hillslopes in order to reduce shrub encroachment. Granda and Prieto (1992) demonstrated that phosphate fertilisation over several years increased pasture production considerably. Therefore, calcium super phosphate was applied (200 kg ha^{-1} per year) at each treatment except the control (T9).

Seed establishment was determined in January of 1998 by counting the plants using 25 x 25 cm quadrats. Pasture cover and composition were determined by two trained persons following the dry-weight-rank method at monthly intervals (Mannette and Haydock 1963; Martín Bellido et al. 1982). Vegetation cover was estimated using a 50 x 50 cm frame, along two transects in each treatment. In addition, cover was determined in each runoff plot, with

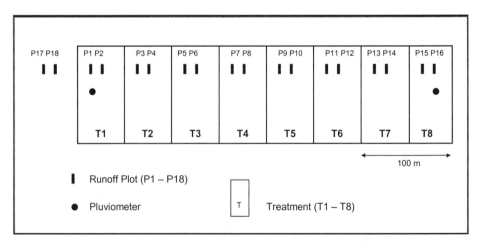

Figure 3. Experimental design at Las Cañas (the location of the runoff plots is simplified).

Table 2. Summary description of the treatments in Las Cañas Experimental Station (--- no seeding or no tillage).

Treatment. Name	Seeding	Tillage system
T1. Direct seeding I	*Trifolium subterraneum, T. glomeratum, Ornithopus compressus*	Direct seeding
T2. Traditional seeding	*T. subterraneum, T. glomeratum, Ornithopus compressus*	Disc ploughing
T3. Ploughed	---	Disc ploughing
T4. Bands	*T. subterraneum, T. glomeratum, Ornithopus compressus*	Contour strips
T5. Natural pasture	---	---
T6. Shrubs	Forage shrubs (*Cytisus scoparius, C. baeticus, Retama sphaerocarpa*	Contour ridges
T7. Lupinus	*Lupinus. sp., T. subterraneum, T. glomeratum, Ornithopus compressus*	Disc ploughing
T8. Direct seeding II	*T. subterraneum, T. glomeratum, Ornithopus compressus*	Direct seeding
T9. Control	---	---

estimates being made for four equal parts of its area. Pasture composition was determined using a 50 x 50 cm quadrat along two transects per treatment. Biomass production was measured twice per year and at two sites in each treatment. Samples were taken from 1 m^2 areas, covered previously by exclosure cages in order to avoid grazing (Plate 1).

Two erosion plots were installed in each treatment (Plate 2), located in the middle slope. Straight slope sections with similar gradients and away from tree canopies were selected. It was intended that site characteristics would be similar, in order to relate runoff and soil loss to the type of treatment. Previous analysis showed that soil properties (texture, organic matter, cation exchange capacity and nutrients) within the fenced area were similar. The experimental character of the investigation justifies the use of bounded plots (Morgan 1986). The plots have a width and length of 2 and 10 m, respectively, which are dimensions frequently used in Europe and hence permit comparison (Brandt and Thornes 1996). The collecting devices consist of a 2 m wide trough, where most of the coarse sediment is retained, a 1:5 slot divisor, and a final tank. The equipment has a maximum capacity equivalent to 55 L m^{-2} of runoff. Sampling was carried out after each rainstorm event, collecting the sediment retained in the first trough, and taking a one litre sample from the divisor and, if necessary, also from the tank. Samples were oven-dried at 105°C. Each divisor was calibrated in the field after its installation. The slots were adjusted with a file until the amount of water passed to the final tank was one fifth of the total water applied, with a maximum error of +/-0.2%. Several days after carrying out the soil preparations and the seeding, heavy rainfall started, causing a delay to the installation of the plots, which was completed in February 1998. Plots P17 and P18 (*Control*) in the unfenced area were installed in June 1999, for which reason no data are presented here.

Rainfall was measured with two automatic tipping bucket rain gauges with a resolution of 0.1 mm and 5 minute readings. Data are available since January 1998. An exceptional rainstorm occurred during the 5th of November, 1997, producing heavy rill erosion at three treatments. Soil loss was estimated volumetrically using cross sectional measurements. Cross-sectional areas were determined along the rills and at intervals of 2.5 to 5 m. This was done by measuring rill depths at each cross-section at intervals of 5 or 10 cm, depending on rill width (in the case of widths greater than 50 cm, intervals of 10 cm). Topsoil bulk density was determined at five

Plate 2. View of runoff plot consisting of collector, divisor, and tank.

points in each treatment after the seeding preparations, using sample rings with a diameter and height of 5 cm. Mean bulk density in the areas affected by rill erosion was 1.29 g cm^{-3}.

5. Results

A high variation of rainfall was registered during the study period. The hydrological years 1997-98 and 1998-99 were very humid and very dry, with amounts of 855.8 mm and 339.3 mm, respectively. Figure 2 presents the monthly precipitation of Las Cañas and Cáceres, together with the long-term mean. During November of 1997, 363.8 mm of rain were registered, which constitutes the historical maximum of monthly precipitation. Almost all months of the first study year showed above average rainfall. In contrast the second year registered very low rainfall during autumn and spring, with an accumulated water deficit of 222 mm. During the 5th and the 6th of November heavy rainfall occurred in Extremadura with 24-hour maximum amounts of 120 to 130 mm. This event has a recurrence interval of 200 years (Schnabel 1998), and was preceded by approximately 100 mm of rain (Table 3). Intensities and amounts of the rainfall events observed since completion of the runoff plots are presented in Table 3. During this period the number of moderately and highly intense rainstorms was below average.

Germination of the seeded leguminous species can be considered as unsatisfactory, with only 71 plants/m². The treatments T1 (*Direct Seeding I*) and T4 (*Bands*) presented slightly better results, with 97 and 81 plants/m², compared with the sites with traditional seeding (T2, T7), which showed a germination of 65 and 68 plants/m², respectively. The poor success rate of the ploughed sites is probably related to heavy soil erosion caused by the rainstorms following the pasture preparations, which washed away part of the seeds. However, the second site where direct seeding was carried out (T8) showed the worst result, with only 45 plants/m². The reason for the generally bad germination is not clear, considering that there was sufficient soil moisture available and no adverse temperature conditions were observed.

Table 3: Rainfall characteristics, soil loss and runoff of events (mean of 16 erosion plots), registered from March 1998 until August 1999 (I5, I60 – 5, and 60 minute maximum rainfall. 24h - 24 hour maximum rainfall. PTOT - total amount of event). * - Exceptional rainfall of November 1997.

Date	Event	I5	I60	24h	Ptot	Soil loss	Runoff
		(mm h⁻¹)	(mm h⁻¹)	(mm h⁻¹)	(mm)	(g m⁻²)	(L m⁻²)
16/04/98	1	28.8	5.2	13.0	25.0	0.5	0.1
13/05/98	2	7.8	4.9	12.4	23.3	0.6	0.1
19/05/98	3	11.8	7.4	14.9	14.3	0.0	0.0
29/05/98	4	41.6	10.1	22.5	22.6	1.1	1.1
20/09/98	5	36.5	7.2	15.6	15.6	0.0	0.0
21/09/98	6	72.8	21.5	26.1	26.1	5.8	3.8
22/09/98	7	26.0	3.7	12.4	25.5	0.4	0.0
3/11/98	8	13.0	4.0	24.9	25.1	2.0	0.2
29/12/98	9	10.4	4.4	19.2	19.2	0.0	0.0
31/12/98	10	20.8	12.4	30.2	30.2	4.2	5.6
9/01/99	11	11.8	2.8	7.8	7.8	0.0	0.0
17/01/99	12	7.8	3.3	8.6	8.6	0.2	0.0
21/01/99	13	6.5	4.4	25.6	26.2	0.2	0.9
10/03/99	14	7.8	4.4	10.0	12.3	0.1	0.0
24/03/99	15	4.0	2.9	5.2	5.2	0.0	0.0
1/04/99	16	10.4	6.0	6.4	6.5	0.0	0.0
16/04/99	17	9.1	3.3	5.1	5.1	0.0	0.0
28/04/99	18	38.5	10.0	15.0	21.6	0.5	0.6
30/04/99	19	16.8	3.8	8.0	8.2	0.0	0.0
5/05/99	20	14.4	4.8	12.2	12.2	0.4	0.0
6/05/99	21	9.6	2.0	5.2	5.2	0.0	0.0
17/05/99	22	8.4	3.0	4.5	4.8	0.0	0.0
18/05/99	23	4.8	2.7	4.4	4.8	0.0	0.0
TOTAL						15.8	12.6
2/11/97*		28.8	15.4	60.4	60.4		
3/11/97*		72.0	23.0	38.0	38.0		
5/11/97*		103.2	36.6	135.2	148.0		

Figure 4 presents the development of the vegetation cover as a mean of all treatments from January 1998 to August 1999, comparing the areas inside and outside the erosion plots. The values are similar, which means that the plot boundaries do not prevent sheep from grazing within the plots. Mean vegetation cover for the complete period was 46%. During the first year the maximum was observed in April 1998 with 64.5%. The absence of rain from June onwards together with grazing was responsible for the decrease, reaching the minimum in August with 29.1%. During the second year the maximum ground cover of 61.1 % was registered in March, followed by a decrease due to low precipitation in spring. It has to be kept in mind that pasture cover is expected to be higher during the second year than during the first year, because the initial vegetation cover was only 10% due to higher stocking rates prior to the experiment and because of the treatment measures themselves, carried out at the beginning of the study period. The deficient rainfall during the second year was clearly responsible for the lower ground cover, and also for low biomass production. During 1998-99, dry matter production amounted to 474 kg ha^{-1}, as compared with 1301 kg ha^{-1} during the first year. Water availability during autumn and spring is crucial for pasture production, and it was strongly deficient in 1998-99.

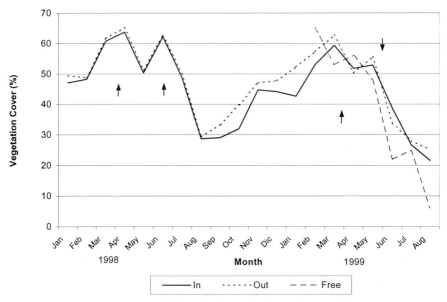

Figure 4. Mean vegetation cover within and outside the runoff plots (fenced area), and in the area with free grazing (not fenced). The arrows indicate the time of grazing by sheep.

Mean vegetation cover of each treatment and for the observation period varied between 30.8% at T8 (*Direct Seeding II*) and 65.4 % at T6 (*Shrubs*) (Table 4). Low values were also observed at T4 (*Bands*) and T3 (*Ploughed*). Cover values greater than 50% correspond with treatments T2, T5 and T1.

The exceptional rainstorm of November 1997 occurred just one week after carrying out the pasture preparations. The storm generated rills only at the three sites which had been subject to traditional tillage. Average soil loss was estimated to be 100.5 t ha^{-1}. Treatment T2 showed approximately three times higher erosion values than T3 and T7 (Table 5). Analysis of the geometric dimensions of the rills reveals that their cross-sectional areas are considerably greater in T2, which is mainly related to their greater width. Depth values are similar. Due to the shallow soils incision is limited by the C horizon and further growth of the rills is by widening. The higher soil loss of T2 can be explained by the concave slope plane curvature, which concentrates overland flow. Some of the rills of T3 enter the area of T2. Although soil loss between T3 and T7 is fairly similar, the rill dimensions are not. At T3 the rills are on average larger, whereas at T7 rill density is higher.

Table 4: Total soil loss and runoff of each erosion plot (Std.Dev. – standard deviation, Coef.Var. – coefficient of variation) and mean values of each treatment for the period March 1998 until August 1999. Veg.cover refers to mean vegetation cover inside the plots.

Treatment	Plot	Runoff (L m⁻²)	Soil loss (g m⁻²)	Runoff (L m⁻²)	Soil loss (g m⁻²)	Veg.Cover (%)
T1	P1	1.3	2.9			
Dir. Seed.I	P2	10.7	11.5	6.0	7.2	50.9
T2	P3	4.3	7.5			
Trad. Seed.	P4	0.9	3.1	2.6	5.3	61.9
T3	P5	8.1	16.6			
Ploughed	P6	12.9	34.4	10.5	25.5	38.9
T4	P7	28.8	14.8			
Bands	P8	18.3	28.8	23.5	21.8	32.8
T5	P9	26.0	22.9			
Pasture	P10	2.5	2.1	14.3	12.5	55.9
T6	P11	19.2	10.9			
Shrubs	P12	28.9	15.7	24.1	13.3	65.4
T7	P13	4.1	11.1			
Lupins	P14	5.9	24.9	5.0	18.0	42.6
T8	P15	12.4	16.9			
Dir. Seed.II	P16	16.6	29.3	14.5	23.1	30.8
Mean		**12.6**	**15.8**	**12.6**	**15.8**	**47.4**
Std.Dev.		9.6	10.0	8.1	7.5	
Coef.Var (%)		76.2	62.9	64.7	47.2	

Table 5: Results of the survey of rills generated during November 1997 at treatments T2, T3 and T7. Included is total soil loss, number of cross sections and dimensions of the rills (Std. Dev. – standard deviation).

Treatment		T2	T3	T7
Soil loss (t ha⁻¹)		182.1	65.0	54.5
Number of cross sections		129	70	93
Maximum	Mean	16.3	16.4	14.9
Depth (cm)	Median	16.0	16.1	15.2
	Std. Dev.	4.0	3.3	3.4
	Minimum	7.9	9.7	5.2
	Maximum	32.5	28.0	22.0
Width (cm)	Mean	125.8	77.4	55.2
	Median	120.0	75.0	45.0
	Std. Dev.	52.9	29.1	27.0
	Minimum	25.0	35.0	15.0
	Maximum	300.0	180.0	150.0
Area (cm²)	Mean	1193.0	827.9	496.2
	Median	1050.0	757.0	426.5
	Std. Dev.	677.0	376.8	294.1
	Minimum	122.0	239.5	103.5
	Maximum	3914.0	2008.0	1458.0
Mean Depth (cm)	Mean	9.0	10.5	8.8
	Median	9.0	10.4	8.6
	Std. Dev.	3.0	2.1	2.1
	Minimum	3.0	6.0	3.7
	Maximum	17.0	16.0	13.8

Table 3 represents the runoff and soil loss amounts as a mean of all plots for each event together with the rainfall data from March 1998 until August 1999. During this period 12 rainstorms produced runoff, but only on five occasions with amounts in excess of 0.5 L m^{-2}. The total runoff was 12.6 L m^{-2}, which represents only 2.2% of total precipitation. Maximum soil loss of 0.06 t ha^{-1} was produced by a rainstorm in September 1998, and corresponds to the event with the highest intensity during the cited period and a total rainfall amount of 26.1 mm (Table 3). Only one more event was registered with a similar soil loss (0.04 t ha^{-1}), caused by a precipitation of 30.2 mm, with a maximum 5-minute intensity of 20.8 mm h^{-1}. Total mean soil loss was 0.16 t ha^{-1}.

Total soil loss and runoff for each erosion plot are shown in Table 4. Runoff varied between 0.9 and 28.9 L m^{-2} at P1 and P12, respectively. Minimum and maximum soil losses were 0.02 and 0.34 t ha^{-1} at P10 and P6. Variation of runoff was higher than that of soil loss, with a coefficient of variation of 76.2% as compared with 62.9 %. The differences of runoff and sediment loss between plots belonging to the same treatment were on average 8.4 L m^{-2} and 0.12 t ha^{-1}. Large differences of runoff were observed at T9, T4 and T1 (Table 4). The relation between runoff and soil loss using the plot data (n=16) was weak (R^2=0.22, p=0.07). No relation could be detected between vegetation cover and runoff (R^2=0.00), whereas soil loss and vegetation cover showed a weak negative relationship (R^2=0.20, p=0.08).

The correlation coefficients improve by carrying out the analysis with the treatment values (mean of pairs of plots, n=8). The highest correlation coefficient corresponds to the negative relationship between soil loss and vegetation cover (R^2=0.66, p=0.01). However, the correlation coefficients between runoff and sediment loss or runoff and percentage cover are not significant.

Figure 5 illustrates soil loss and runoff, together with mean vegetation cover for each treatment. Sediment loss varied between 0.05 and 0.26 t ha^{-1}. The minimum value corresponds to *T2 (Traditional Seeding)* and the maximum to T3 (*Ploughed*). *Traditional seeding* is also the treatment with the lowest runoff and the second highest value of vegetation cover. Low soil loss and runoff were also registered at *Direct Seeding I*. Maximum values of overland flow were generated at unploughed sites, such as *Shrubs* and *Bands* (24.1 and 23.5 L m^{-2}), whereas the completely ploughed treatments registered on average only 6.0 L m^{-2}.

6. Discussion and Conclusions

Analysis of the vegetation cover indicates that plot boundaries did not interfere with grazing livestock. Functioning of the runoff plot devices was satisfactory. Rainfall amounts were highly variable, producing large differences in vegetation cover and biomass production. Before the plots were completed an exceptional storm was registered giving rise to heavy rill erosion. In contrast, during the period when the runoff plots were operating, rainfall amount and intensity of the events were low in comparison with the long-term mean.

Soil loss due to rill erosion was of the same order of magnitude as values reported from other Spanish areas (Faust 1995; De Alba 1997). Some of the rills generated in Las Cañas can be classified as ephemeral gullies, following the definition used by Poesen and Govers (1990). Poesen *et al.* (1996) point out the importance of ephemeral gullying as compared with sheetwash erosion in Mediterranean environments. In the case of Las Cañas, rill erosion produced in the ploughed treatments during one event amounted to 100 t ha^{-1}, whilst sheetwash

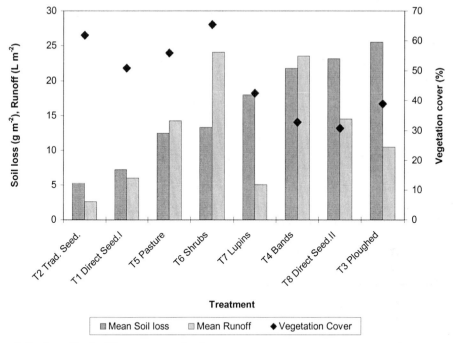

Figure 5. Total runoff and soil loss as a mean of each treatment.

during the period from March 1998 to August 1999 produced only 0.16 t ha^{-1}, which is a difference of four orders of magnitude. This illustrates one limitation of runoff plots, which are too small to monitor rill or ephemeral gully erosion.

A limitation of data interpretation of rill erosion is related to the experimental design. The study sites received runoff water from the untreated upper slope, where no rills were formed. In order to study the effect of pasture improvement techniques on rill erosion and ephemeral gullying, the treatments need to cover larger areas, including the entire slope. A further problem is related to the critical conditions for rill generation. The exceptional rainstorm registered in Las Cañas has a recurrence interval of 200 years, though rills may be formed by events of lower magnitude and frequency.

The data set is still too limited to explain the effect of the pasture improvement measures on vegetation cover, runoff and soil erosion, especially when considering the large differences between plots with the same treatment. The variation of runoff and sediment loss between different treatments was not much higher than the differences observed between the plots with the same treatment. This spatial variability is possibly related to small-scale variations of precipitation or soil properties. Therefore, soil moisture measurements are now being carried out regularly at five fixed points in each erosion plot with TDR (Time Domain Reflectometry) equipment. Precipitation is also being measured in all plots after each event using manual rain gauges. Soil physical properties will be analysed in more detail. The strong temporal variability of rainfall, and hence pasture evolution, points to the need for replication of the treatments. At the present study area measurements will continue for at least another two years.

However, the results indicate that traditional ploughing constitutes a high erosion risk during exceptionally intense rainstorms. In contrast, several months after the treatments were carried out and under low to moderately intense storms (recurrence interval < 1 year), soil losses at the ploughed sites were no higher than the losses registered at the sites with conservation treatments. Furthermore, during this period runoff generation was lower at the sites with traditional tillage, indicating higher infiltration capacity.

7. Acknowledgments

This investigation was carried out within the project IPR98B020 funded by *the Junta de Extremadura, Consejería de Educación y Juventud y Fondo Social Europeo*, Spain.

8. References

Brandt, C.J. and Thornes, J.B. (eds.) (1996) *Mediterranean Desertification and Land Use*, John Wiley & Sons, Chichester.

Ceballos, A. (1999) *Procesos hidrológicos en una pequeña cuenca hidrográfica bajo explotación de dehesa en Extremadura*, Universidad de Extremadura, Cáceres.

Cerdà, A., Schnabel, S., Ceballos, A. and Gómez Amelia, D. (1998) Soil hydrological response under simulated rainfall in the dehesa land system (Extremadura, SW Spain) under drought conditions, *Earth Surface Processes and Landforms* 23, 195-209.

CSIC (Consejo Superior de Investigaciones Científicas) (1970*): Suelos, estudio agrobiológico de la provincia de Cáceres*, Centro de Edafología y Biología Aplicada de Salamanca, Salamanca.

De Alba, S. (1997) Metodología para el estudio de la erosión en parcelas experimentales: Relaciones erosión-desertificación a escala de detalle, in J.J. Ibáñez, B.L. Valero and C. Machado (eds.), *El paisaje mediterráneo a través del espacio y del tiempo*, Geoforma Ediciones, Logroño, pp. 259-293.

Diáz, M., Campos, P. and Pulido, F.J. (1997) The Spanish dehesa: a diversity in land-use and wildlife, in D.J. Pain & M.W. Pienkowski (eds.), *Farming and Birds in Europe*, Academic Press, San Diego, pp. 178-209.

Donázar, J.A., Naveso, M.E., Tella, J.L. and Campión, D. (1997) Extensive grazing and raptors in Spain, in D.J. Pain and M.W. Pienkowski (eds.), *Farming and Birds in Europe*, Academic Press, San Diego, pp.117-149.

Dorronsoro Fernández, C. (1992) Suelos, in J.M. Gómez Gutiérrez (ed.*), El libro de las dehesas salmantinas*, Junta de Castilla y León, Salamanca, pp. 71-121.

FAO (Food and Agriculture Organization) (1990) *Soil Map of the World 1:5.000.000, revised legend*, FAO, Rome.

Faust, D. and Herkommer, M. (1995) Rill erosion in lower Andalusia, Zeitschrift für Geomorphologie Supplementband 99, 17-28.

González, F. (1994) Variedades españolas de trébol subterráneo. Origen, identificación y recomendaciones para su uso, *Monografía del Servicio de Investigación y Desarrollo Tecnológico, vol. 84*, Badajoz, Spain.

González, F., Olea, L., Moreno, V., Paredes, J., Prieto, P., Paniagua, M. and Santos, A. (1994) Recuperación de pastos de áreas degradadas en la dehesa de Extremadura, *Memoria SIA 94*, pp. 57-60.

Granda Losada, M. and Prieto Macía, P.M. (1992) La fertilización fosfórica como una mejora permanente en los pastos de la dehesa, *Revista Albear*, 0, 18-21.

Holechek, J.L., Pieper, R.D. and Herbel, C.H (1989) *Range Management, Principles and Practices*, Prentice-Hall, Englewood Cliffs, New Jersey.

Le Houerou, H. N. (1987) Indigenous shrubs and trees in the silvopastoral systems of Africa, in H. A. Steppler and P. K. R. Nair (eds.), *Agroforestry: a Decade of Development*, International Council for Research in Agroforestry, Nairobi, pp. 139-156.

Letey, J. (1969) Measurement of contact angle, water drop penetration time, and critical surface tension, *Proceedings of the Symposium on Water-Repellent Soils, 6-10 May 1968*, University of California, Riverside, pp. 43-47.

Mannette, L.T. and Haydock, K.P. (1963) The dry-weight-rank method for the botanical analysis of pasture, *Journal of the British Grassland Society* 18, 268-278.

Martín Bellido, M., López Carrión, T., Martín Javato, J., Moreno Cruz, V. and González Crespo, J. (1982) El método de los rangos para la evaluación de la disponibilidad de materia seca en pastos naturales y mejorados, *Anales del INIA, Serie Agricultura* 17, 77-89.

Montoya Oliver, J.M., Mesón García, M.L. and Ruiz del Castillo, J. (1988) *Una dehesa testigo*, Instituto Nacional para la Conservación de la Naturaleza, Madrid.

Morgan, R.P.C. (1986) *Soil Erosion and Conservation*, Longman, Essex.

Poesen, J.W. and Govers, G. (1990) Gully erosion in the Loam Belt of Belgium: Typology and control measures, in, J. Boardman, I.D.L. Foster and J.A. Dearing (eds.), *Soil Erosion on Agricultural Land*, John Wiley & Sons, Chichester, pp. 513-530.

Poesen, J.W., Vandaele,K. and Van Wesemael, B. (1996) Contribution of gully erosion to sediment production on cultivated lands and rangelands, in *Erosion and Sediment Yield: Global and Regional Perspectives* (Proceedings of the Exeter Symposium, July 1996), IAHS Publ. n°. 236: 251-266.

Schnabel, S. (1997) *Soil erosion and runoff production in a small watershed under silvo-pastoral landuse (dehesas) in Extremadura, Spain*, Geoforma Ediciones, Logroño.

Schnabel, S. (1998) La precipitación como factor en los procesos hidrológicos y erosivos. Análisis de datos de Cáceres capital, *Norba Revista de Geografía* **10**, 137-153.

Schnabel, S. (1999) Extreme events and gully erosion, *Proceedings of the International Seminar on Land Degradation and Desertification*, International Geographical Union, Lisbon 1988, pp.17-26.

Schnabel, S. (2000) Land degradation of open woodlands with silvo-pastoral exploitation, *Third International Congress of the European Society for Soil Conservation*, 28 March – 1 April 2000, Valencia, Spain.

AUTHORS

S. SCHNABEL[1], F. GONZÁLEZ[2], M. MURILLO[2], & V. MORENO[2]

[1]corresponding author
Dpto. de Geografía
Universidad de Extremadura
Avda. de la Universidad
10071 Cáceres, Spain

sschnab@geot.unex.es

[2]Servicio de Investigación y Desarrollo Tecnológico
Junta de Extremadura
PO box 22
06080 Badajoz, Spain

CHAPTER 16

LAND RECOVERY FOLLOWING EXTENSIVE SHEET EROSION AT MENANGLE, NSW

G.S. HUMPHREYS and B. GROTH

1. Abstract

An extended dry period commencing in 1934 and culminating in the most severe drought on record in the 1940s also coincided with extensive sheet erosion on agricultural land that had already experienced one and perhaps two prior erosional episodes. By 1949 about a quarter of the 4.7 km² study catchment consisted of bare land in which saprolitic subsoil material was exposed at the surface. Recovery commenced in with the onset of a wet period in the early 1950s which has continued to the present day. During this time the bare areas decreased to a third of the 1949 level. The establishment of a grass cover over these formerly bare areas is associated with the build-up of topsoil. This was examined by comparing characteristics of three sites representing 0, 25 and 45 years of recovery. During this time topsoil thickened and became lighter in texture and darker in colour. These changes correspond to a decrease in bulk density and an increase in organic matter. The build-up in topsoil depth conforms to a single logarithmic function which indicates that it may take about 1500 years to achieve 33 cm thick topsoil as found at a control site. Trends in soil chemical data are variable with less than half of the properties showing a significantly consistent trend. Of these, organic matter and base saturation at a depth of 10-20 cm, and available Na and K at 0-10 cm indicated full recovery to the control-site level within 100 years. In contrast, organic matter at 0-10 cm and total N are predicted to take >800 years.

Recovery proceeds when soil is trapped behind grass shoots at the upslope boundary between bare and vegetated areas. The sediment is supplied from beneath the armoured surface by ants and also from micro-rills that incise into the exposed soil. This material is washed downslope creating micro-alluvial fans. Hence recovery proceeds by further erosion and sediment transport. It is this lateral component to soil recovery that might explain the high rates of topsoil build-up identified in this study.

2. Introduction

The recovery of soils following natural and human-induced disturbances is well understood as a concept but less so in practice if published trends on recovery are used as a guide (for example, Bockheim 1980; Mellor 1984; Huggett 1998). Several issues contribute to limiting our understanding of recovery. The first involves devising a suitable sampling framework that adequately accounts for spatial variability and changes over a sufficient period of time.

A.J. Conacher (ed.), Land Degradation, 255–273.

One common approach is directed to short time frames of up to about five years (for example Chan *et al.* 1997; Barber and Navaro 1994; Harrison and Shackleton 1999; Ross *et al.* 1994), in which there is an emphasis on measuring labile constituents of the soil. In contrast, longer time frames, say >100 years, need to contend with changes in importance of other relevant factors which can impinge on cause and effect relationships. In this context intermediate time-runs may be more useful since the controlling factors are often better understood and the trends are beyond the indicator value of short time frame variables, but outside the complications imposed on the longer time frames. Nevertheless, despite these perceived advantages both the intermediate and longer time frames often rely on an ergodic approach where space is substituted for time, which introduces another source of variation that can be curtailed, such as by careful sampling of geomorphic units (Conacher and Dalrymple 1997), but not eliminated. However, there remains another issue that is much more fundamental than those discussed so far. Any study of soil recovery not only requires the selection of relevant attributes and sample sites but also a consideration of the model of soil formation used as a template to assess recovery (Huggett 1998). Recently, Paton *et al.* (1995) argued that topsoil mobility is a key part of soil formation and that pedogenesis is best understood if due consideration is afforded to those mechanisms that lead to topsoil mobility. They considered that in many situations topsoil mobility occurs by the combination of rainwash (a term that includes slopewash, rainsplash and sediment rafting, and employed here as a milder or non-destructive version of sheet erosion) and bioturbation. If this viewpoint is applied to the issue of recovery then it might be expected that the same mechanisms play an important role. The present contribution seeks to (i) examine soil recovery at Menangle following a period of severe land degradation; and (ii), explore whether lateral processes of sediment transport are important to soil recovery over an intermediate time frame.

3. Study Area

Foot-Onslow Creek is a small tributary of the Nepean River, with a catchment area of 4.7 km^2 located within the Razorback Range, approximately 70 km south west of Sydney, Australia (Fig. 1). The entire catchment is underlain by shales (claystones and siltstones) of the Bringelly Shale, an upper member of the Triassic Wianamatta Group (Herbert 1979). Graded hillslopes, mostly between 5 and 20°, predominate and merge into late Pleistocene and Holocene clayey valley fills. Landsliding occurs on steeper slopes in the district (Blong and Dunkerley 1976) but are not common in the study catchment, and many valley fills contain active gully systems, including Foot-Onslow Creek. Soil associations commonly consist of reddish texture contrast soils on ridgetops and midslopes, with yellow varieties towards the valley floors. These soils classify as Alfisols (Soil Survey Staff 1998) or Lixisols (FAO/ISRIC 1994). Under low vegetation cover these soils are noted for their high to very high erodibility status (Johnstone and Hicks 1984). Mean annual rainfall is 737 mm (1878-1996) but there is considerable variation (standard deviation = 223 mm), with pronounced droughts in the early 1900s and late 1930s/early 1940s (Pickup 1976), which correspond with ENSO events outlined in Allen (1988).

4. Research Strategy and Methods

4.1. LAND USE HISTORY

Widespread sheet erosion in the area was recognised by Beirne (1952) and Pickup (1976), but the timing and formative events were addressed in a cursory manner only. To overcome this a comprehensive evaluation of various historical data was undertaken. This involved accessing archival items such as early maps and reports as well as local history and oral

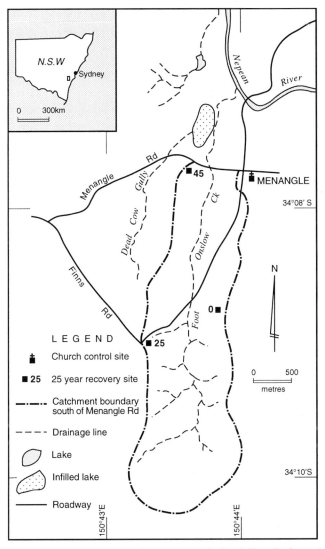

Figure 1. Location of the study area and recovery sample sites in Foot-Onslow catchment

accounts to establish a detailed land use history. Particular attention was directed to identifying events which may have contributed to erosion, such as the timing of various grazing and cropping practices. This information is reported below under the heading of 'Land degradation'.

The extent of recovery between 1949 and 1970 was depicted by Pickup (1976) as a sequence of small scale maps showing the extent of bare land determined from the interpretation of aerial photographs (Table 1). For the present study this exercise was updated to 1996 and expanded to include the lower reaches of the catchment to Menangle Road (Fig. 1) to enable a clearer examination of trends. The air photos vary in scale from about 1:13 000 to 1:27 000 and hence maximum mapping resolution is about 0.1 ha. To avoid differences in estimates between Pickup (1976) and the present study the entire time sequence was remapped. A comparison between the separately obtained results showed a very close correspondence. The lower portion of the catchment was not included in the 1996 photo run and is therefore not portrayed. However, an estimate of the amount of bare ground over this portion was obtained by ground truthing to complete the data set.

Table 1. Airphoto coverage details over Foot-Onslow catchment

Year	Series	Run No	Film No	Photo No	Date	Colour*	Scale
1949	Camden	4	547	5127-30	31 Mar	B&W	1:20000
		5		5096-9			
1956	Camden	36	241	5036-8	7 Aug	B&W	1:17000
		37		5040-3			
		38		5111-3			
		39		5114-6			
1961	Cumberland	52	1045	5075-7	26 Jun	B&W	1:13000
		53		5069-71	"		
		54	1068	5180-3	7 Aug		
		55	1047	5042-5	26 Jun		
1965	Cumberland	31	1413	5085-7	7 Oct	B&W	1:21000
		32	1414	5076-9	22 Oct		
1970	Cumberland	28	1907	5157-60	6 Jul	B&W	1:14000
		29		5122-4			
		30		5062-4			
1975	Douglas Park	4	2341	132-6	12 Oct	B&W	1:17000
		5		198-200			
1978	County of Cumberland	28	2714	234-6	14 May	B&W	1:17000
		29		254-6			
1984	Wollongong	7	3411	065-7	8 Oct	C	1:17000
		8		088-90			
1990	Wollongong	7	3758	028-30	5 Oct	C	1:16000
		8	3754	098-100			
1994	Wollongong	4	4178	117-9	4 Jan	C	1:27000
		5		140-2			
1996	BHP		1789	437-9	4 Dec	C	1:20000

* B&W, black and white. C, colour.

4.2. SITE SELECTION

Sample sites for recovery analysis (Fig. 1) were selected by identifying areas on the air photos which were under active sheet erosion in 1949, but have since become revegetated, and remained vegetated to the present day for each of the recovery intervals: 0, 25 and 45 years. [Here, 0 years recovery means that the land remained bare between 1949 and 1996 (constantly exposed B-horizon), 25 years recovery means that the land remained bare between 1949 and 1970 but since the early 1970s has maintained a grass cover, and 45 years recovery that the land was bare in 1949 but since the early 1950s it has maintained a grass cover.] While many suitable sites were identified for each of the recovery intervals, the three sites were selected for their similarity in terms of landscape position (upper midslope), slope angle (approximately 10°) and post sheet erosion land use histories (dairying pre-1940s until mid 1980s, then very light grazing until the present). The grounds of the St James Church (Fig. 1), located on the catchment boundary, provided the control site for comparison against the three recovery sites. This is the only location, within the district, which does not seem to have been affected by direct agricultural pursuits nor used as a cemetery.

4.3. RECOVERY INDICES

Several physical and morphological properties of the soil were tested at each of the four sites, including field texture, colour (Munsell), horizon thickness and bulk density, which was measured from oven dried (48 hrs @ 105˚C) 100 cm^3 samples extracted form pit faces with a 6 cm diameter metal ring. Soil chemical analysis was undertaken on air dried fine earth fractions (<2 mm) sub-sampled from thoroughly mixed bulk samples >500 g taken from pit faces and analysed using standard methods described in Blackemore et al. (1987) and included: pH (1:2.5 H_2O); available (exchangeable) K, Na, Mg and Ca from a 1M neutral ammonium acetate extraction; cation exchange capacity (CEC); base saturation (BS); available P (Olsen extraction); total P (nitric/percloric acid digestion); total N (Kjeldahl digestion), and organic matter (from organic carbon via a Walkley-Black oxidation). At each of the three recovery sites, samples were analysed at 5 cm intervals to 20 cm and sometimes deeper depending on topsoil thickness. The results were subsequently averaged at 0-10 and 10-20 cm intervals after it was established that this provided sufficient resolution for the soil depths obtained after 25 years. Three profiles were averaged at the 25 and 45 year site but only one profile at the 0 year site, which was similar to the subsoil in a control site. Because of this and cost constraints the 0 year site was not sampled further. The control site was sampled at 0-5 and 5-10 cm, and at 10 cm intervals thereafter to 60 cm. Sampling at the control site was from a core as pit excavation was not permitted, although three cores were examined.

4.4. RECENT SPATIAL PATTERN OF RECOVERY

This issue was investigated by examining more closely an area spanning an active erosion scar and a recently revegetated zone at the 0 y site. To assist in this a 5 m wide by 40 m long plot was mapped in detail to record the distribution of surface features as erosional and depositional micro-geomorphic units following an approach based on Valentin (1991). The mapped area extended from the catchment divide (0° slope) and over a convex slope unit ending within a recent revegetation growth area at the upper portion of the midslope zone.

5. Land Degradation: the Record in Foot-Onslow Catchment

5.1. LAND USE HISTORY

Foot-Onslow Creek lies within the boundaries of Camden Park Estate, where large scale agriculture first commenced in Australia. The former Estate has experienced a variety of farming practices of varying intensity for almost two centuries. Free roaming wild cattle may have entered the area in 1788, but intensive European farming practices did not commence until 1807 when sheep were introduced (Wrigley 1980). Initial production in the first decade was low, but an increase in wool demand in subsequent years caused a boom in sheep numbers to a maximum of 24 000 head in 1835. Wool production continued for a number of years, but with the onset of drought in the early 1840s and the discovery of more suitable sheep grazing areas in western New South Wales, the industry quickly declined until it ceased to be of significance by about 1860 (Jervis 1940). A period of grain cropping followed, with wheat dominating from 1840 until 1880 when an outbreak of stem rust disease forced growers to switch to a mixture of other grains, including oats, barley, rye and millet (Atkinson 1988). The success of mixed grain production, combined with an increased market demand for milk, led to a transition from grain to dairy production in the early 1880s. Dairying continued well into the 20th century, but production declined steadily in the late 1970s and early 1980s, eventually ceasing in 1985. Since then land usage has reverted to low intensity activities involving limited cropping and light grazing (there are about 200 head of mostly cattle and horses in the catchment today).

5.2. EROSION POTENTIAL

The landuse history noted above is used as a temporal framework to assess the potential for erosion, combining land use practices thought to be conducive to erosion and probable erosivity (Fig. 2). The latter is based on a continuous rainfall record from 1878 and expressed here as a running five year mean following Pickup (1976). Three time periods of high erosion fluxes are postulated. The first is the grain period between 1840 and 1880, when it is thought that most of the catchment was ploughed. The practice of this time was to plough in an upslope-downslope manner, perpendicular to the contour (Atkinson 1988). Remnant furrows preserving this orientation are evident on 1949 air photos on some ridge crests in the area (Graham 1980).

The second and third erosion phases are primarily associated with prolonged drought conditions that occurred during the dairy production period (1880-1985). Analysis of the rainfall records shows that between 1901 and 1910, annual rainfall did not exceed the lower limit of the first standard deviation (514 mm) on six occasions within the ten year period. With the end of the drought in the early 1910s, higher rainfall conditions continued for the next two decades, until the onset of the most severe drought on record between the years 1934 and 1949. During this period, annual rainfall averaged only 563 mm. The extent of erosion at the end of this period is indicated in the 1949 aerial photographs. This shows that

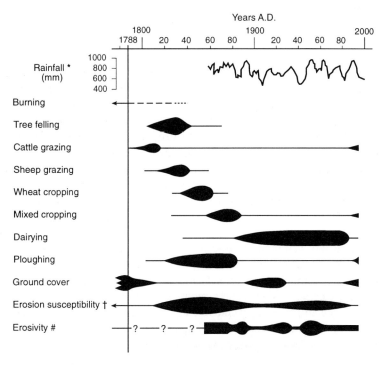

Figure 2. Relative impact of land use practices and rainfall on erosion potential in Foot-Onslow catchment: The 5 year running mean for rainfall has been extended back to 1859 using Sydney rainfall records (r2=0.82 from comparison between stations). The height of the black bands for other attributes indicates relative importance of the attribute and not relative importance between attributes at a point in time. A thin black line indicates presence of an activity, where as no line indicates absence or not important.

about 25% of the catchment was bare with most of the affected areas located on ridgetops and upper to midslope positions. Subsequent inspection and interviews with long term residents of the area, indicated that the bare areas had been subjected to sheet erosion which was so aggressive in places that the entire topsoil had been removed, exposing the highly dispersible, reddish clayey saprolitic subsoil (Pickup 1976 and reconfirmed in this study). Hence the available evidence indicates very active erosion prior to 1949, probably before 1880 and possibly between 1901 and 1910.

6. Soil Recovery Analysis

6.1. SPATIAL RECOVERY FROM SHEET EROSION 1949-1996

The results confirm the trend identified by Pickup (1976), who showed a net decline in the proportion of land affected by sheet erosion between 1949 and 1970 from 25 to 12%. This trend continued to 8% by 1996 (Fig. 3). The variability in the trend appears to relate to the timing of the photo run and whether this coincided with the end of a wetter or drier phase. Thus the 1956 result of 8%, which was not reached again until 1996, was at the end of the most sustained wetter phase since recording commenced in 1878 and was half the result in the next period of 1961 (16%), which was at the end of a drier phase. Thereafter the trend steadies, even though major droughts occurred in the early 1970s, 1980s and 1990s; but none of these matched the magnitude of the 1940s drought. The photos also indicate that recovery generally proceeds in an upslope direction, and ground inspection indicates that it involves a mixture of exotic, improved pasture grasses, especially *Paspalum districhum*, *Trifolia repens* and *Chloris gayana*.

6.2. TEMPORAL SOIL RECOVERY 1949-1996

Distinct changes in the physical properties between the 0, 25 and 45 year recovery profiles are evident (Fig. 4). At 0 y the soil retains its degraded state: saprolitic material (formerly a B_{tw} horizon), capped by an armour layer of platy ironstone gravel (up to 3-8 cm in size). By 25 y a topsoil has developed, overlying the armoured saprolite. Field texturing shows that the surface textures (0-10 cm) at the 25 y and 45 y sites have become noticeably lighter than the 0 y site, from initially medium/heavy clay (0 y) to sandy clay (25 y) to sandy clay loam (45 y) respectively. Not surprisingly, corresponding values of bulk density have decreased from 1.60 g cm^{-3} (0 y) to 1.28 g cm^{-3} (25 y) to 1.19 g cm^{-3} (45 y), perhaps indicating a net loss of fines and/or the addition of organic matter which increases from 2 to 5.3-5.9 % (Table 2). Munsell colour has also significantly darkened, from an initial red (2.5YR 5/6) to light reddish brown (5YR 6/3) at 25 y, through to brown (7.5YR 4/4) at 45 y. This trend is only partly accounted for by increased organic matter, which is slightly higher in the 25 year site in comparison to the 45 y site. It is possible that the type of organic matter differs, since more favourable soil microbial activity might be expected given the greater age and slightly higher pH after 45 y (Table 2). The build-up in the soil and change in texture corresponds to distinctive changes in soil morphology and hence in classification from a uniform clay (Uf1.42) to a gradational soil (Gn3.71) to a duplex (Dy2.11) or texture contrast soil from 0, 25, 45 years in the terms of Northcote (1979). This sequence corresponds to Entisols, Inceptisols and Alfisols (Soil Survey Staff 1998) and Leptosols, Cambisols and Lixisols (FAO/ISRIC 1994).

Topsoil thickness increased dramatically from 0 cm to 13 cm at 25 y, and to 18 cm after 45 y recovery. This represents a very high nominal increase of up to 5.2 mm y^{-1} but this value should be corrected for changes in bulk density. Compared to an initial BD of 1.6 these topsoil thickness are equivalent to 10.4 and 13.4 cm at 25 and 45 y respectively, which provides a corrected maximum topsoil production rate of 4.2 mm y^{-1}.

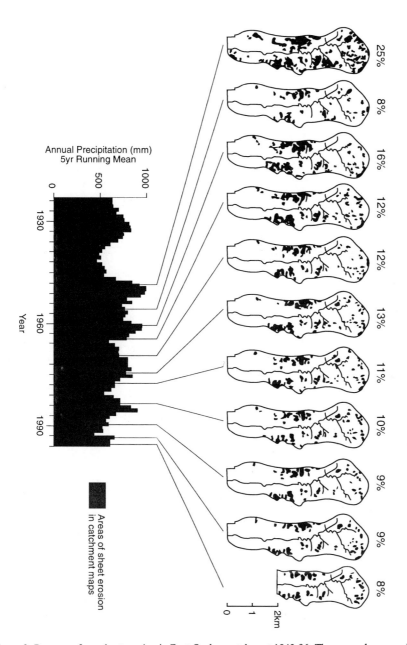

Figure 3. Recovery from sheet erosion in Foot-Onslow catchment 1949-96. The mapped area excludes bare areas associated with gullies, roads, buildings etc.

Table 2. Soil chemical and other properties at the recovery and control sites

(a) The recovery sites

Property		0 y		25 y		45 y	
		0-10 (cm)	10-20 (cm)	0-10 (cm)	10-20 (cm)	0-10 (cm)	10-20 (cm)
OM	(%)	2.0	1.2	5.9	4.5	5.3	4.4
pH		5.7	5.4	5.7	5.8	6.0	5.9
total N	(%)	0.1	0.08	0.23	0.12	0.24	0.14
avail P	(μg/ml)	3.5	4.8	4.8	4.7	4.0	3.6
total P	(μg/g)	405	573	418	384	475	470
avail Na	(me/100g)	0.40	0.29	0.22	0.23	0.19	0.21
avail K	(me/100g)	0.37	0.23	1.04	0.82	1.18	0.77
avail Ca	(me/100g)	8.5	5.2	5.8	5.4	7.4	7.6
avail Mg	(me/100g)	4.05	2.81	3.06	2.89	2.65	2.49
Ca/Mg		2.1	1.9	1.9	1.9	2.8	3.1
CEC	(me/100g)	17.0	14.0	15.2	13.6	14.3	14.5
BSat	(%)	81	57	67	69	77	75
C/N		11.6	8.7	14.9	15.7	12.8	14.5
BDens	(g/cm³)	1.60	1.63	1.28	1.42	1.19	1.43

(b) The control site; Sᵗ James Church

Property		0-5 (cm)	5-10 (cm)	10-20 (cm)	20-30 (cm)	30-40 (cm)	40-50 (cm)	50-60 (cm)
OM	(%)	11.6	5.8	4.1	3.9	2.2	1.8	2.0
pH		6.1	6.4	6.0	5.9	6.0	5.8	5.9
tot N	(%)	0.5	0.26	0.18	0.14	0.10	0.08	0.06
avail P	(μg/ml)	17.0	8.0	4.0	3.0	3.0	2.0	2.0
total P	(μg/g)	958	na	688	na	538	572	502
avail Na	(me/100g)	0.14	0.17	0.20	0.25	0.63	0.76	0.89
avail K	(me/100g)	1.51	1.01	0.59	0.31	0.24	0.19	0.23
avail Ca	(me/100g)	14.8	10.5	7.4	5.4	6.2	5.3	5.2
avail Mg	(me/100g)	4.96	4.46	4.09	3.83	5.50	5.65	5.66
Ca/Mg		3.0	2.4	1.8	1.4	1.1	0.9	0.9
CEC	(me/100g)	25.0	19.0	17.0	14.0	17.0	17.0	17.0
BSat	(%)	87	86	74	68	76	72	71
C/N		13.5	12.9	13.2	16.1	12.8	13.1	19.3

In general, the increase in soil depth conforms to a single logarithmic function of the form:

$$S = a + b \log_{10} t \qquad (1)$$

where S is the uncorrected soil depth in cm and t is the recovery period in years. This relationship has been found useful by a number of investigators (Bockheim 1980; Trustrum and De Rose 1988). The Menangle solution of

$$S = 10.291 \log_{10} t - 0.19 \qquad (2)$$

is similar to Trustrum and De Rose (1988) solutions of

$$S = 16.7 \log_{10} t - 9.5 \qquad\qquad (3)$$

$$\text{and } S = 16.16 \log_{10} t - 12.4 \qquad\qquad (4)$$

which were derived from landslide scar sites in mudstone hill terrain at Taranaki, New Zealand. In particular, Equation 3 provides a better prediction of the Menangle results than

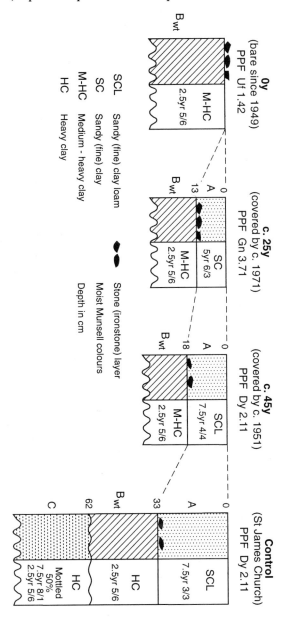

Figure 4. Changes in selected soil properties since 1949. The control site provides a further comparison

Equation 2, which is possible since the former is based on seven time periods over 80 years compared with three over 45 years at Menangle. Nevertheless, despite this similarity it is probably not a generalised solution, for there is no compelling theory to support any particular function.

A similar approach can be applied to trends in soil chemical parameters, but this is also hampered because only three sampling periods were used. In general the results are mixed, but several of the indicators display strong positive trends with respect to soil recovery. They include organic matter, pH, total P, total N, available K, available NA, available Ca, available Mg and Base Saturation in at least one of the two soil depth classes analysed (0-10 and 10-20 cm). To improve the interpretation, these soil chemical data were assessed in terms of three properties; consistency and significance of the trend, and advantage to general plant growth (Table 3). Trend consistency refers to whether each successive value continues to increase (or decrease). An exception to this occurs if the 25 y value is slightly greater (for an increasing trend) than the 45 y value, provided that the 25 y value is much greater than the 0 y value. Trend significance refers to the net incremental change based on the rating classes employed by NSW government agencies (Hazelton and Murphy 1992). If the change spans two or more rating classes the trend is significant, provided that the trend is generally favourable to plant growth. A rating of 1 indicates that all values fall within an existing class and is treated as not significant. Favourableness to plant growth refers to levels deemed suitable for cereal crops and improved pastures in NSW (Hazelton and Murphy 1992). The only attributes that satisfy these criteria include organic matter at both depths, available K and Na, and total N at 0-10 cm and Base Saturation at 10-20 cm (Table 3). Assuming that Equation (1) is applicable then the recovery period (t_r) for these control dependant variables (y) is obtained by rearranging Equation (1) for t:

$$t_r = 10^{\,(y-a)/b} \qquad\qquad\qquad (5)$$

This shows that the time to recovery is mostly between 20 and 80 years (Table 3) with the OM and total N in the upper 10 cm requiring much more time (>800 years). The difference between the 0-10 and 10-20 cm OM recovery periods possibly reflects on the inappropriateness of the single logarithmic equation (Equation 1) in this situation, even though the OM and N values better match the estimated soil depth recovery period of about 1600 years if the original soil was 33 cm thick. Of course the pre 1800 topsoil depth at all sites is unknown, but even altering the depth of the 'control' will not change the order of magnitude difference in recovery period found between different attributes at Menangle. Alternatively, the mixed chemical results may be simply an expression of chaotic behaviour during the early period of soil evolution (Huggett 1998).

6.3. LATERAL TRANSPORT OF SEDIMENT IN RECOVERY

As noted above the trend in the spread of grass cover over bare ground since 1949 proceeds in an upslope direction, especially for sites on the upper hillslope (Fig. 3). Mapping revealed three dominant zones along the transect; a source (or deflation) zone, a transportation zone and a depositional zone (see Fig. 5). The source zone (0-8 m from the crest) is located on the upper portion of the transect. It consists of remnant B-horizon saprolite covered by a gravel pavement of planar shaped iron indurated shale fragments (b-axis up to 50 mm). This gravel layer probably developed from the residual material left behind during the removal of the fine earth fraction from the original upper soil profile. Vegetation is absent or very sparse and there is little microtopographic relief. Saprolite material also forms the surface of the transportation zone (8-28 m from the crest). However, the surface microtopography is somewhat variable with the upper portion characterised by a more hummocky surface consisting of flat-topped vegetated 'islands', which have a slightly higher relief (usually 2-3 cm) than the surrounding unvegetated surface. A lag gravel pavement is present but the frequency of shale fragments is much lower than occurs in the source zone. Further downslope, the relative abundance of these sub-units decreases and is replaced by the development of rills and micro-alluvial fans. The area below this point, the deposition zone (28-40 m from the crest), consists entirely of micro-alluvial fans consisting of eroded, reworked upslope material. The surfaces of the fans are partially covered by vegetation, mostly grasses, and the fans thicken to 4 cm towards the base of the transect, possibly indicating a longer period of establishment and/or more suitable conditions for growth.

Table 3. Assessment of soil chemical trends

Property	Depth (cm)	Trend	No of rating classes	Suitable for plant growth*	45 y value as % of control	Usability (predicted recovery period)
				Indice interpretation		
OM	0-10	increasing	3	good,imp	61	yes (827y)
	10-20	"	3	good,imp	>100	yes (24y)
pH	0-10	minor increase	1	good	96	no
	10-20	"	1	good	98	no
tot N	0-10	increasing	2	reasonable	63	yes (1543y)
	10-20	increasing	1	reasonable	78	no
avail P	0-10	fluctuates	1	deficient	32-38	no
	10-20	decreasing	1	deficient	>90	no
total P	0-10	increasing	1	reasonable	53	no
	10-20	fluctuating	1	reasonable	68	No
avail Na	0-10	decreasing	2	good	82	yes (83 y)
	10-20	decreasing	1	good	95	No
avail K	0-10	increasing	2	good,imp	>100	yes (68 y)
	10-20	fluctuating	just 2	good,imp	>100	Marginal
avail Ca	0-10	fluctuating	1	low	46-58	No
	10-20	increasing	1	low	100	No
avail Mg	0-10	decreasing	1	high	56	No
	10-20	fluctuate	1	high	61-71	No
CEC	0-10	decreasing	1	poor	65	No
	10-20	fluctuating	1	poor	80-85	No
BSat	0-10	fluctuate	1	satisfactory	77-89	No
	10-20	increasing	2	satisfactory	100	yes (51y)
C/N	0-10	fluctuating	1	poor	97	N
	10-20	fluctuating	3	poor	>100	N
Soil depth		increasing	-	-	55	(1608)

*imp = improving

From these observations the following sequence of detachment, transport and deposition processes is implied. Following deposition of micro-alluvial fans, vegetation growth occurs. This acts as a positive feedback to increase surface roughness, which in turn reduces the velocity of turbid runoff originating from the upslope source zone. This promotes deposition of suspended and bedload particles in an upslope direction. The process is further assisted in that deposition increases base level and thus reduces slope angle and runoff velocity, thus promoting further deposition and hence aggradation of a surface layer. Continued favourable conditions encourage the spatial expansion of vegetative cover in an upslope direction, gradually reducing the area of sheet erosion and initiating recovery of the underlying soil.

6.4. THE CONTRIBUTION OF BIOTURBATION TO RECOVERY

In many areas dominated by the gravel pavement surface (primarily the source zone and to

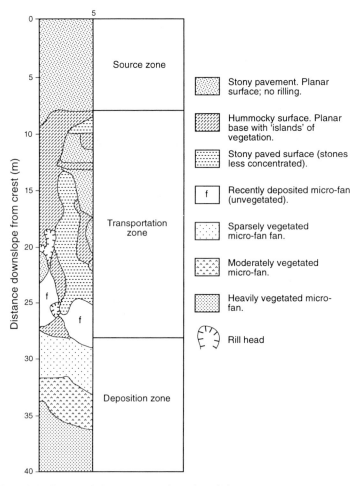

Figure 5. Surface morphology map spanning a degraded area to a recovery zone

a lesser extent the transportation zone), the concentration of gravel is so great that it armours the surface and protects it from further erosion. Hence a negative feedback is expected, since the source of sediment supply is reduced which would curtail recovery. However, this does not occur for two reasons; (i) incipient rilling in the transportation zone, as noted above, which incises the armour layer, and (ii) the work of mesofauna, especially ants, in mining soil from beneath the gravel and depositing the spoil on the surface as mounds. Two ant species are particularly prevalent across the catchment, especially on degraded areas where vegetation cover is low. The meat ant (*Iridomyrmex purpureus*) constructs large mounds (2-3 m circumference), and the sugar ant (*Camponotus consobrinus*) constructs smaller (~20 cm diameter), ephemeral, funnel shaped mounds. Both of these species are opportunistic and readily invade degraded land. For the meat ant an important factor is the access to open sites to maximise radiant energy input into mounds, but for the sugar ant the lack of competitors may be the more important factor. The meat ant mounds at the site are typical for the species (Cowan *et al.* 1985), whereas mound construction by the sugar ant is not prevalent in undisturbed sites (Humphreys 1994).

These observations imply that recovery is initiated through the revegetation of newly deposited micro-alluvial fans. However, recovery cannot proceed or is curtailed if source material for these fans is limited by armouring of upslope areas. Mounding by ants overcomes this constraint by providing a constant supply of material which is deposited over the armoured layer. Rainwash then redistributes and sorts the mounded sediment which is then trapped behind grass tufts, twigs and other surface obstructions in the deposition zone (Fig. 6). The constant reworking of new topsoils may account for a coarsening of texture via net export of clay (Paton *et al.* 1995). Overall, a self-organising system is in operation (Huggett 1998).

7. Discussion

The results of this study indicate that recovery occurs via several related parameters, including suitable rainfall since 1949, the spread of grasses, and the build-up of a more fertile topsoil. In conventional soil conservation practice the benefit of increasing vegetation cover is normally perceived as protecting the surface to reduce soil loss and hence provide conditions conducive to recovery. The current study indicates that what appears to happen differs from this commonly accepted viewpoint. The mechanisms leading to recovery at Menangle involves the entrapment of soil. This entrapped soil is derived from ongoing erosion by rainwash from existing bare surfaces, micro-rilling, and also sediment provided by mesofauna mounds. That is, recovery stems from further erosion in that the products of erosion are used to build a topsoil. This combination is similar to those reported in rangelands (Ludwig and Tongway 1995), whereby water and/or wind transported sediment, seeds and nutrients are trapped behind barriers to provide a site that is conducive to plant growth. This applies to parallel bands of vegetation on gentle slopes in rangelands referred to as vegetation arcs and also to smaller features known as litter dams on steeper slopes (Eddy *et al.* 1999). On the hillslopes at Menangle the patches of vegetation downslope of the bare areas act as barriers. From a management perspective, creating barriers across the degraded sites could enhance recovery. These could consist of fences, piles of plant cuttings or even strips of planted vegetation, depending on what was deemed suitable based on

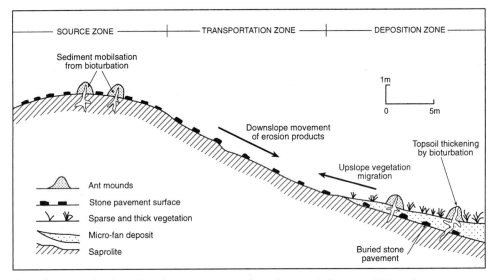

Figure 6. Suggested pathway for soil recovery based on the surface morphology map of Figure 5

criteria such as cost effectiveness and environmental integrity (Ludwig *et al.* 1997). Regardless of the selected medium, however, a suitable strategy for determining spacing between the barriers is required and several approaches are outlined in soil conservation manuals (for example, Morgan 1995).

The recovery sequence outlined in this study identifies the importance of lateral movement and the roles of rainwash and bioturbation. In essence the recovery recorded at Menangle represents the re-establishment of a topsoil which because of continued bioturbation conforms to a biomantle (Johnson 1990). The speed of this recovery may be considered high but it is very similar to the results of others. As noted above the Menangle results of topsoil depth fit very closely the findings of Trustrum and De Rose (1988) on deforested hillslopes in New Zealand. In a subsequent study on forested hillslopes Trustrum *et al.* (1990) reported even higher rates of soil build-up. These studies attest to the potential for rapid soil development which concurs with the lateral transport model outlined in Paton *et al.* (1995) but not with traditional approaches based on leaching mechanisms (Bockheim 1980; Foss and Segovia 1984). A similar constraint applies to the use of chronofunctions which are often advocated as a method for examining soil formation over time (Bockheim 1980; Mellor 1984; Huggett 1998). At present the theoretical basis for accepting any of the widely used chronofunctions such as linear, exponential, power and asymptotic equations is not well established and this provides a major constraint to their use other than as an exercise in trend fitting.

Within the Menangle area three periods of potential erosion have been identified, corresponding to particular combinations of drier conditions and land use practices: 1840-80; early 1900-1910, and 1934-1949. Only the last period is reasonably well documented. Nevertheless, it is not known if the extent of sheet erosion reached a maximum at the time of the first air photographs in 1949 or at some earlier period. There is some evidence to suggest that erosion was both more extensive and at a higher rate during the first period. For example, there is a strip of post settlement alluvium along the axis of Foot-Onslow Creek. Some of it seems to date back to at least the early 1900s and probably much earlier, judging from the condition and style of buried remnant fence posts and bridges. Such issues can only be resolved via additional investigations. Nevertheless, we note that elsewhere in NSW extensive changes to river style and sediment load have occurred following initial disturbance, but there are lag effects between apparent defining events and responses (Eyles 1977; Wasson and Galloway 1986; Brooks and Brierley 1997). Hence, determining precise cause and effect relationships may prove difficult.

8. Conclusion

This study sought to account for the recovery of soil following a period of extensive sheet erosion. The results show that a topsoil >15 cm thick with a field texture distinctly coarser than the source saprolite (sandy clay loam compared to medium clay) develops in <50 years. The speed of soil build-up is similar to that found in other studies in different terrains. In comparison, the trends in common soil chemical properties are much more mixed with some properties predicted to achieve recovery, that is, to match a control value, in <100 years whilst others are predicted to take 800-1600 years. However, there remains an important caveat to these predictions: at present the theoretical basis for accepting linear or various non linear expressions is not well founded and provides a fundamental constraint to the application of chronofunction analysis.

The study also indicates that the pathway to recovery involves the lateral transport of soil particles/aggregates over short distances and their entrapment in vegetated areas. A combination of rainwash and bioturbation plays a key role. This finding is similar to the landscape functioning model for rangelands (Ludwig and Tongway 1995) and also to the model of soil formation outlined by Paton *et al.* (1995).

9. Acknowledgments

We wish to thank the Hawkesbury Nepean Catchment Management Trust for their financial support for this research, Mr Roy Lawrie NSW State Agriculture for advice on the soils, Ms Sarah West for advice and assistance with historical records, and Messrs Harry Finn, Neville Jeffery and other local land owners for sharing their knowledge on landuse practices.

10. References

Allan, R.J. (1988) El Nino Southern Oscillation influences in the Australian region, *Progress in Physical Geography* **12**, 313-348.

Atkinson, W.A. (1988) *Camden*, Oxford University Press.

Barber, R.G. and Navarro, F. (1994) The rehabilitation of degraded soils in eastern Bolivia by subsoiling and the incorporation of cover crops, *Land Degradation and Rehabilitation* **5**, 247-259.

Beirne, K.G. (1952) Soil erosion in the Camden district, *Journal of the Soil Conservation Service of New South Wales* **8**, 123-131.

Blakemore, L.C., Searle, P.L. and Daly, B.K. (1987) *Methods of Chemical analysis of soils*, New Zealand Soil Bureau Scientific Report.

Blong, R.J. and Dunkerley, D.L. (1976) Landslides in the Razorback Range area, New South Wales, Australia, *Geografiska Annaler* **58A**, 139-147.

Bockheim, J.G. (1980) Solution and use of chronofunctions in studying soil development, *Geoderma* **24**, 71-85.

Brooks, A.P. and Brierley, G.J. (1997) Geomorphic response of lower Bega river catchment disturbance, *Geomorphology* **18**, 291-304.

Chan, K.Y., Bowman, A.M. and Friend, J.J. (1997) Restoration of degraded vertisols using a pasture including a native grass (*Astrebla lappacea*), *Tropical Grasslands* **31**, 145-155.

Conacher, A.J. and Dalrymple, J.B. (1977) The nine unit landsurface model: an approach to pedogeomorphic research, *Geoderma* **18**, 1-154.

Cowan, J.A. Humphreys, G.S., Mitchell, P.B. and Murphy, C.L. (1985) An assessment of pedoturbation by two species of mound building ants, *Camponotus intrepidus* (Kirby) and *Iridomyrmex purpureus* (F. Smith), *Australian Journal of Soil Research* **22**, 98-108.

Eddy, J., Humphreys, G.S., Hart, D.M., Mitchell, P.B. and Fanning, P.C. (1999) Vegetation arcs and litter dams: similarities and differences, *Catena* **37**, 57-73.

Eyles, R.J. (1977) Birchams Creek: the transition from a chain of ponds to a gully, *Australian Geographical Studies* **15**, 146-157.

FAO/ISRIC (1994) *World Reference Base for Soil Resources*, FAO/ISRIC, Wageningen and Rome.

Foss, J.E. and Segovia, A.V. (1984) Rates of soil formation, in R.G. LaFleur (ed.), *Groundwater as a Geomorphic Agent*, Allen & Unwin, Boston, pp. 1-17.

Graham, O.P. (1980) The Erosional Development of a Gully in the Razorback Area, N.S.W., unpublished B.A. Honours Thesis, Macquarie University, Sydney.

Harrison, Y.A. and Shackleton, C.M. (1999) Resilience of South African communal grazing lands after removal of high grazing pressure, *Land Degradation and Development* **10**, 225-239.

Herbert, C. (1979) *The Geology and Resource Potential of the Wianamatta Group*. Geological Survey of New South Wales, Bulletin No. **25**.

Huggett, R.J. (1998) Soil chronosequences, soil development and soil evolution: a critical review, *Catena* **32**, 155-172.

Humphreys, G.S. and Mitchell, P.B. (1983) A preliminary assessment of the role of bioturbation and rainwash on sandstone hillslopes in the Sydney Basin, in R.W. Young and G.C. Nanson (eds.), *Aspects of Australian Sandstone Landscapes*, University of Wollongong, Wollongong, pp. 65-80.

Humphreys, G.S. (1994) Bioturbation, biofabrics and the biomantle: an example from the Sydney Basin, in A.J. Ringrose-Voase and G.S. Humphreys (eds.), *Soil Micromorphology: Studies in Management and Genesis*, Elsevier, Amsterdam, pp. 421-436.

Jervis, J. (1940?) *The Story of Camden. A Modern Farming Community Closely Allied with the Earliest Australian History*, Published by Arthur G. Gibson, Printed by Cumberland Newspapers Ltd., Parramatta.

Johnson, D.L. (1990) Biomantle evolution and the redistribution of earth materials and artifacts, *Soil Science* **149**, 84-102.

Johnstone, D. and Hicks, R.W. (1984) *Urban Capability Study Camden Park Development Area*, New South Wales Department of Agriculture.

Ludwig, J.A. and Tongway, D.J. (1995) Spatial organisation of landscapes and its function in semi-arid woodlands, Australia, *Landscape Ecology* **10**, 51-63.

Ludwig, J., Tongway, D., Freudenberger, D., Noble, J. and Hodgkinson, K. (1997) *Landscape Ecology Function and Management*, CSIRO, Melbourne.

Mellor, A. (1984) Soil chronofunctions on neoglacial morraine ridges, Jostedalsbreen and Hotunheimen, southern Norway: a quantitative pedogenic approach, in K.S Richards, R.R. Arnett and S. Ellis (eds.), *Geomorphology and Soils*, Allen & Unwin, London, pp. 289-307.

Morgan, R.P.C. (1995) *Soil Erosion and Conservation*, Longman, Essex.

Paton, T.R., Humphreys, G.S. and Mitchell, P.B. (1995) *Soils: A New Global View*, UCL Press, London.

Pickup, G. (1976) Geomorphic effects of changes in river runoff, Cumberland Basin, NSW, *Australian Geographer* **13**, 188-193.

Ross, D.J., McQueen, D.J. and Kettles, H.A. (1994) Land rehabilitation under pasture on volcanic parent materials: changes in soil microbial biomass and C and N metabolism, *Australian Journal of Soil Research* **32**, 1321-1327.

Soil Survey Staff (1998) *Keys to Soil Taxonomy*, USDA, Washington DC.

Trustrum, N.A. and De Rose, R.C. (1988) Soil depth-age relationship of landslides on deforested hillslopes, Taranaki, New Zealand, *Geomorphology* **1**, 143-160.

Trustrum, N.A., Blaschke, P.M., De Rose, R.C. and West, A.W. (1990) Regolith changes and pastoral productivity declines following deforestation in steeplands of North Island, New Zealand, *Transactions 14th International Soil Science Congress*, Kyoto, Japan, 125-130.

Valentin, C. (1991) Assessing the space and time variations of the surface features and the cultivation profile, *Asialand Workshop on the Establishment of Experiments for the Management of Acid Soils*, International Board for Soil Research and Management, Bangkok, Thailand, IBSRAM Tech. Notes No. **5**, 105-128.

Wasson, R.J. and Galloway, R.W. (1986) Sediment yield in the Barrier Range before and after European settlement, *Australian Rangeland Journal* **8**, 70-90.

Wrigley, J.D. (1980) *A History of Camden, New South Wales*, Prestige Duplicating Service.

AUTHORS

G.S HUMPHREYS[1] AND B. GROTH[1,2]

[1]Dept. Physical Geography, Macquarie University, Sydney, 2109
ghumphre@laurel.ocs.mq.edu.au

[2]NSW Dept. Land & Water Conservation, PO Box 3935 Parramatta, NSW 2124
bgroth@dlwc.nsw.gov.au

CHAPTER 17

SAFE CARRYING CAPACITY AND SUSTAINABLE GRAZING: HOW MUCH
HAVE WE LEARNT IN SEMI-ARID AUSTRALIA IN THE LAST 170 YEARS?

JOHN PICKARD

1. Abstract

The paper reviews Australian literature on estimating carrying capacity on rangelands from
1823 to 1998. Graziers assume that current stocking rates (numbers of domestic stock per unit
area) are identical with the sustainable carrying capacity. It is argued that in virtually all cases,
current stocking rates in semi-arid Australia exceed the carrying capacity of the landscape.
Despite this, these stocking rates are considered acceptable by both graziers and government
agencies. Empirical research by graziers shows that lowering stocking rates by up to 60%
increases incomes and improves the vegetation. These lowered stocking rates are probably
close to safe carrying capacities and hence sustainable grazing.

Two pragmatic approaches to estimating the stocking rate of 'sustainable grazing' are
suggested. The first identifies benchmark properties where the vegetation community of interest
is rated to be in excellent condition. By obtaining the history of stocking rates, an estimate of
safe carrying capacity is possible. The second is simpler: use 50% of the district average
stocking rate for the community. This figure will probably be very close to the 'sustainable
grazing' stocking rate. The '50% of district average' approach is objective, based on ecological
theory and empirical research, and is legally defensible.

2. Introduction

One of the perennial and most vexed questions facing pastoralists and other managers of
grazing land is the deceptively simple: 'how many stock can the land safely carry?'. Despite
this simplicity, we still do not appear close to an unambiguous answer. One indication of this
is the discrepancy between estimates by graziers and those by range administrators in an
unpublished analysis of stocking rates in advertisements. An allied problem has been use of
carrying capacity as a measure to determine rents on many rangelands. The methods used have
been far from transparent, the results frequently at variance with other estimates, and even more
frequently, misunderstood and misapplied by many range scientists, government agencies and
conservationists. Although recent research by Johnston *et al.* (1996a, b) suggests some
convergence of estimates by scientists and pastoralists, this may be a consequence of the nature
of the particular study and the summer-dominated forage growth in the study area.

For most of the past 200 years of European pastoralism in Australia, the rangeland industry
has been underpinned by a basic tenet: that there is a 'safe' carrying capacity. Recently this has
been replaced with more sophisticated views which better account for the inherent variability

A.J. Conacher (ed.), Land Degradation, 275–289.
© 2001 *Kluwer Academic Publishers. Printed in the Netherlands.*

of rainfall and pasture production. Similarly, a considerable amount of recent research and grazier attention has focussed on 'total grazing management' which incorporates all herbivores: domestic, feral and native. The result can best be described as 'adaptive stocking' where "stock numbers fluctuate in response to the wide range of seasonal conditions" (Marree Soil Conservation Board 1997:113). This is essentially a form of nomadic pastoralism where the stock are moved locally or regionally in response to the conditions. Robson (1994) specifically proposed a radical variation on this as an alternative to fixed land tenure in western New South Wales. However, despite these advances, I suggest that many graziers, range scientists and range administrators, and certainly many conservationists still approach stocking rates with an implicit assumption that they should match some fixed but sustainable carrying capacity.

This paper reviews a number of approaches to estimating safe carrying capacity in Australian rangelands from 1826 to 1998. The focus is on one of the basic tenets of our rangeland industry: that there is a 'safe' carrying capacity. I then briefly examine some ecological theory of carrying capacity before suggesting two pragmatic approaches to estimating safe carrying capacity. At the end of the paper, some of the issues raised by the articles are discussed.

3. Assessing/Estimating Carrying Capacity in Australia, 1826-1998

This section documents a range of approaches to assessing or estimating carrying capacity in Australia over the last 170 years (Table 1). Many show a consistent basis in practical, often hard-won empirical experience by graziers. Also, there has been essentially no real improvements in accuracy of estimates despite a considerable amount of research by government agencies and range scientists. I describe one or two in some detail (e.g. rental carrying capacity, and the methods used in the Western Lands Lease Management Plans) because these are so widely misunderstood.

Table 1. Summary of Australian experience with estimating carrying capacity

Year	Who?	Basis of estimate	Sustainable?	Source
1826 - 1997	Grazier / pastoralist	Experience	Yes - too high	Atkinson 1826, Marsh 1842, McMichael 1984, "Crowfoot" 1917, Lamond 1927, Anon 1997
1949+	Government	Previous stocking rates (for rent)	Not relevant	Pickard 1994
1968, 1983	Scientist	Benchmark property	Probably	Condon 1968, Walker 1983
1986, 1988	Grazier / pastoralist	Deliberate lowering of stocking rate	Yes	Purvis 1986, Morrissey and O'Conner 1988
1996 1996a,	Scientist	Biomass and percent utilisation	Maybe	Johnston et al. 1996b
1997	Graziers	Adaptive stocking	Possibly	Marree SCB 1997
1998	Legislation	Wishful thinking	"in the D-G's opinion"	NSW Native Vegetation Conservation Act
1998	Scientist	Benchmark properties 50% of district average	Yes (?) Yes (?)	Pickard 1998

3.1. REGULATING STOCK NUMBERS (1826)

In July 1822 James Atkinson took up two grants near Bong Bong, on the tablelands south of Sydney. Within a few years, he was cultivating running stock over several thousand hectares and his farm was almost entirely fenced. He was a highly respectable farmer in the colony when he left Sydney on 17 February 1825 to return to England (Fletcher 1975). Atkinson could speak authoritatively on grazing in the colony and on his return to England in 1825 he grew tired of answering "innumerable questions and enquiries respecting the present state of New South Wales, and especially of its Agriculture and Grazing" (Atkinson 1826:v). His opinions are not necessarily free from bias. Fletcher (1975:16) considered that Atkinson was too harsh in some of his judgments: "He [Atkinson] had an inbred intolerance of poor husbandry which caused him to over-react against anything which fell short of his ideal". Thus the material quoted below can be taken as direct criticism of many of the smaller graziers who had little capital, and probably saw no benefit from understocking.

Atkinson described how overstocking in the early years was easily solved by simply moving further inland and squatting on more land:

> Very few of the stock owners have sufficient land to support the whole of their stock, and are therefore obliged to have recourse to the unoccupied tracts in the interior. When any person finds himself overstocked - and very few make the discovery, or, at any rate, will take measures to remedy the evil, until their cattle are half-starved - they go into the interior, or *bush*, as it is termed, beyond the occupied parts of the country, ... (p. 64, italics in original).

His observations, on what was 170 years later to be called 'minimum stocking' (Pickard 1994a) are particularly interesting. Summarising, Atkinson favoured fencing to allow better management and breeding but he cautioned that all graziers must

> ... take care to understock their land; but the reverse is too often the case, many stocking their pastures to the extent of what they will bear in plentiful seasons; and the consequence is, that in scarce times they are short of food, and frequently perish from want, ... Every person ought to regulate his quantity of stock by the number his land will support in the worst seasons; he would then always have plenty, and escape the deplorable consequences attendant upon overstocking.

> Plentiful times frequently occur, when the land will bear, for many months, or even two or three years together, five times as much stock as it will at others: unless at these times, therefore, sufficient stock is kept to consume the grass, a great deal of feed will be wasted, and the pastures much injured; but then, unless the farmer provides artificial means, he will have no recourse against the return of scarcity; *his only safe plan, therefore, is to keep his land constantly understocked* (pp. 68-69, italics added).

Atkinson's description clearly shows that the empirical knowledge of good stock management has been in Australia for almost 200 years.

3.2. "…NOT AMPLE ROOM ON HIS RUNS; …" (1842)

Two decades later, M.H. Marsh gave evidence at the Select Committee on Immigration. Most of his evidence concerns the finances and profitability of sheep enterprises, but he comments indirectly on stock numbers:

> The calculations [of profit based in 15 000 sheep of which 7000 are breeding ewes] are also wholly inapplicable to a sheep farmer who has not ample room on his runs; … I am convinced that the indigenous grasses, with fair treatment, do not deteriorate in quantity and quality, but when a run has once been overstocked, and the grasses pulled up by the roots by the sheep, it is impossible to say how long it may be before the country can recover, … (Marsh 1842).

At this time, many squatters who wanted to carry more stock simply occupied additional land and expanded their stations. But this was frequently only possible by droving out to the limits of other squatters and taking up new land. Marsh's comments are directed at squatters who are hemmed-in and unable to simply expand.

3.3. SMALL AND LARGE LANDHOLDERS STOCK AT DIFFERENT RATES (1844)

McMichael (1984) presented summaries of the area and stock carried on the four largest and four smallest landholdings in each of the 14 squatting districts in 1844. These are derived from official returns sent from the colony to the British government. The squatting districts are rather ill-defined (Figures 4 and 5 in King 1957) and run from what is now southern Queensland to western Victoria but at variable distances inland.

There was no consistency in assessing stocking rates in the various districts. Small and large landholders in six districts have generally similar assessments but the remaining eight are wildly disparate. These differences are not related to rainfall gradients, but may be due to the duration of settlement of the various districts. This will not be explored further in this paper, suffice it to say that small and large landholders in both the same and different districts assessed stocking rates differently.

3.4. "THE QUESTION OF CARRYING CAPACITY" (1917)

In one of a series of articles to *The Pastoral Review*, an experienced grazier writing under the pseudonym "Crowfoot" described his empirical method of determining carrying capacity.

> In my opinion the only safe plan is to find out over a series of years what a piece of country will carry comfortably, and stock a little lighter than that capacity, eating off closest in the spring months the parts that are most likely to be devastated by bushfires should such occur. Protect any open plains or broken country by good breaks, and always leave plenty of plants to seed. … It is a mad policy to eat off everything while it is growing, as many do for fear of droughts and fires; by understocking good nutritious plants grow, many of them drought-resisting, that die right out under overstocking. Under the methods I have enunciated, together with plenty of water, the grazier will see

most droughts out without serious loss, and in many cases without any ("Crowfoot" 1917: 862-863).

3.5. "CARRYING CAPACITY" (1927)

Henry G. Lamond (1885-1969) was a regular columnist for *The Pastoral Review* over several decades. He was very highly regarded by graziers and pastoralists across Australia (Walsh 1993). In 1927 he wrote an article on carrying capacity in which he describes two approaches. The first involved an almost casual glance at the landscape then asserting a carrying capacity. The second was based on local experience of seasons and stocking rates.

> Without attempting to lay down any hard-and-fast rules, without attempting to dogmatise, and dealing only with conditions which I know, I have placed before you my views of carrying capacity. My idea is that *carrying capacity should be based on the limit of stock the country will carry in healthy condition from wet season to wet season, and that estimate should be based only on country which is available* (Lamond 1927:335) [italics added].

Lamond's article engendered considerable correspondence. One letter from the anonymous WLA in Queensland (16 August 1927, p. 787) deserves to have the last word:

> With regard to safe stocking, it seems to me an impossible thing. ... I would suggest taking an average of good and fair seasons, and stocking accordingly, and when a drought comes along - as it undoubtedly will - grin and bear it.

While his suggestion sounds sustainable, this Queensland grazier accepted stock death during drought. Despite grazier rhetoric, stock rarely die from lack of water during droughts, they die from lack of feed. And if there is insufficient feed, then the grazing can scarcely be called 'sustainable'.

3.6. WESTERN LANDS LEASE RENTAL CARRYING CAPACITY (1949+)

Rents for 'leases for the purpose of grazing' (usually called 'western lands leases') in the semi-arid rangelands of the Western Division of New South Wales are determined according to S19C of the Western Lands Act (1901). Basically, rent is charged on average carrying capacity over the previous decade. The Western Lands Act does not provide a definition of 'carrying capacity', apparently assuming that it is self-evident. S19C(2) requires that the previous 10 years must be considered, suggesting that there is or should be some mechanism of feedback for determining carrying capacity. The mechanism is not specified, but from other research into the history of stocking rates (Pickard 1990, 1991, 1994b), I have concluded that what is normally called 'rental carrying capacity' is determined empirically as follows.

In the late 1940s when the carrying capacity of many thousands of leases was determined, the Western Lands Commission compiled all available data on stocking rates of each lease, namely the numbers of all domestic stock which had been carried in previous years, particularly the 1930s and early 1940s. Using these data, each lease was examined on its own and a carrying capacity determined. The data used were not restricted to the lease, but also came from properties, nearby leases, shearing returns, and returns to government. Detailed examination of a single lease reveals what happened with reappraisal. The original 1949 assessment was accepted each decade (Pickard 1991). There is nothing in the files to support any decision: the original value was simply rolled over in an unskilled clerical operation. Checking several hundred files revealed an identical pattern.

Thus, stocking rates in the 1930s are the basis for rental carrying capacity in the Western Division today. While other information has been incorporated, for example estimates of carrying capacity from Western Lands Lease Management Plans / Property Resource Plans (see below), there is no basis in ecosystem theory for rental carrying capacity. Indeed, rental carrying capacity was primarily seen as a problem of valuation rather than a problem of ecosystem management. This is still true today, and is one of the major problems with the current rental system. For valuation,

> ...the *area* [*sic*] necessary to run a unit of stock is of comparative, and not absolute, importance, and within fairly wide limits, the actual stock numbers may be used in the valuation process ... When the grazing is upon natural pastures it will be shown that consideration must be given to the extent of land (sheep or beast area) which will run one head of stock, year in and year out... (Murray *c.* 1965:319-320).

Murray does not describe how to determine carrying capacity, but offers detailed suggestions on how to verify it using the number of stock carried by the grazier.

Returning to the mechanism of determining rental carrying capacities, it is quite clear that previous stocking rates are the basis for determining carrying capacity. At no time was there any real attempt (other than occasional property inspections) to examine the productivity of the ecosystem and determine whether the stocking rates were reasonable.

3.7. "ESTIMATION OF GRAZING CAPACITY ON ARID GRAZING LANDS" (1968)

Condon (1968) developed a more objective system for estimating grazing capacity (which he did not define) by using multiplicative modifiers applied to a standard system

> ... from which all other grazing capacities are determined. This should be an area of known and proven grazing capacity,... intermediate in most of the factors concerned.

The unit chosen as a base in western New South Wales (NSW) was a red sandy loam brown acid soil, of level to slightly undulating topography, carrying open mulga scrub with woolly butt (*Eragrostis eriopoda*) or corkscrew grass (*Stipa variabilis*) pastures of moderate to good palatability. Long-term stocking records show this class of country to have a grazing capacity of ... [1 sheep to 5.1 ha] at an average annual rainfall of ... (254 mm). (Condon 1968, p. 115).

The standard land class was in the Bourke-Wanaaring district (Walker 1983). The modifying factors include soils (fertility, moisture, erodibility), topography, tree density, drought forage, pastures, condition, rainfall, and barren areas.

The rating values were developed from records of stocking during a period (1927-1946) when there were background populations of kangaroos (at the rate of approximately one kangaroo per two sheep), small populations of feral goats and unknown, but probably fairly large, populations of rabbits (R.W. Condon, pers. comm., Walker 1983:5).

It is clear that, like rental carrying capacity, Condon's method was based on previous stocking rates. There is no evidence that the approach was ever used by graziers or land administrators.

3.8. "THE WESTERN LANDS LEASE MANAGEMENT PLANNING SCHEME" (1983)

In 1965 the Western Lands Commission requested assistance from the Soil Conservation Service of New South Wales in maintaining productivity of the grazing lands. The Service initiated the Western Lands Lease Management Plan (WLLMP) Scheme in 1967. Subsequently the plans were renamed Property Resource Plans (PRPs). The objectives of the Scheme are to prepare a management plan of each property to guide lessees in developing sustainable management and to assess the carrying capacity of each property. Initially, the carrying capacity data were deliberately removed from copies of plans sent to each grazier, but since the mid 1980s the estimates were supplied to the graziers.

The PRP assessment procedure is based on the earlier work by Condon (1968) and is fully described by Walker (1983). Properties are mapped into land classes, each with relatively constant topography, soil, vegetation and hence assumed constant carrying capacity. Each class is then compared with a standard land class with known long-term stocking rate (and hence assumed known long-term carrying capacity). The carrying capacity of the new land class is determined by applying multiplicative rating scales for topography, soils, vegetation, tall shrubs and trees, drought forage, pastures, condition and rainfall. The original rating tables of Condon were updated in Walker (1983). The final number is assessed carrying capacity (dry sheep equivalents per 100 ha) which "can be safely supported by each land class" (Walker 1983). These are summed for the area of each land class in each paddock on the property, and then for the property as a whole.

> This calculated carrying capacity is one which should enable landholders to carry stock 12 months into a drought without causing damage to soils or pastures. If stocking is much above the recommended level the ability to carry stock into a drought will be reduced and serious damage may occur on erodible soil types (Walker 1983:5).

Thus there is no suggestion that the calculated carrying capacity is sustainable beyond 12 months into a drought. This implies that the assessments are higher than sustainable carrying capacity. Without showing the actual evidence, Walker provides some support for the assessments: but again, it is with actual stocking rates.

The assessments of carrying capacity in the PRPs were added to the data on stocking rates in Rental Reappraisal Files (described above), but I have seen no evidence that the rental carrying capacity was ever modified solely because of the PRP assessment. Similarly, I have seen no evidence that graziers ever adopted this approach.

3.9. "NURTURING THE LAND" (1986)

Bob Purvis of *Atartinga Station* in the Northern Territory faced a dilemma when he inherited his property: continue managing in the same style as his father and achieve the same results (country in poor condition, financial losses in most years) or do something different. He decided to reduce his stocking rate. His income increased, so he reduced his stock numbers further and his income again increased. This continued until he was carrying well below previous numbers and the district averages (Purvis 1986).

The relevance of this empirical approach is the response of the landscape: the country improved. As he reduced stock numbers, all of the indicators of good condition improved, the indicators of poor condition declined. Purvis concluded that he was approaching the safe carrying capacity of his landscape. It was 30-50% lower than previous stock numbers.

3.10. "28 YEARS OF STATION MANAGEMENT" (1988)

In a similar, but quite independent approach, Morrissey and O'Connor (1988) found that the property *Annean* in semi-arid Western Australia made considerably more money with only 30% of the previous stock numbers. At the same time, the country improved. They concluded that they were approaching the safe carrying capacity: one-third of the district average.

3.11. "OVERGRAZING: PRESENT OR ABSENT?" (1991)

The term 'overgrazing' is used very widely, but it is rarely defined. Wilson and MacLeod (1991) defined it quite narrowly as a change in vegetation structure and a concomitant reduction in animal production. They were unable to find any evidence of reduced production, and concluded that overgrazing was not very widespread.

However, their definition is very restricted and entirely production-oriented, ignoring other values of the landscape. If, for example we include loss of biodiversity, then the answer would be substantially different (Landsberg *et al.* 1997). These apparently tangential observations are relevant, because a central tenet of a safe carrying capacity is that it is sustainable. And despite difficulties of definition, 'sustainable' does imply long-term maintenance of landscape attributes other than simple production. Consequently, Wilson and MacLeod's study actually adds little to our understanding of sustainable stocking rates and safe carrying capacity.

3.12. "OBJECTIVE 'SAFE' GRAZING CARRYING CAPACITIES FOR SOUTH-WEST QUEENSLAND" (1996)

A special issue of *The Rangeland Journal* (volume 18, part 2, pp. 191-391) is devoted to 'Grazing management'. Many of the papers are relevant to the concept of 'sustainable grazing', but most demonstrate that we are indeed seeking a Holy Grail.

Of particular relevance are two papers by Johnson *et al.* (1996a, b) in which they develop and test a model for southwest Queensland. The model seems to extend Christie's (1984) notion of a safe level of utilisation, here set at 15-20% of annual forage. Available biomass is calculated on a rainfall model using annual precipitation and a rainfall use efficiency factor. The predicted levels of stock numbers are remarkably similar to those assessed by experienced local graziers. Although the test reveals that the model is accurate, it is essentially limited to areas where rainfall is the primary determinant of forage. While this may the case over large areas of Australia's rangelands, a significant proportion is less regular than the summer-dominated rainfall in the test area.

The close relationship between the model prediction and graziers' estimates is of some interest. While Johnston *et al.* take this as supporting their model, it could equally well be taken as supporting the accuracy of grazier experience. Thus we are still unable to develop a standard for comparison other than graziers' stocking rates. This suggests a fundamental sterility in the research approaches.

3.13. ADAPTIVE STOCKING AT MARREE (1997)

Many arid rangelands are simply not suited to set stocking rates: rainfall is too variable and pasture production too ephemeral or sparse. In these, 'adaptive stocking' is more appropriate:

> Stock numbers will fluctuate in response to the wide range of seasonal conditions, from small breeding herds during prolonged drought, through normal sustainable levels, to larger herds in growth periods following major rain or flooding. ... With an adaptive system, both the condition of the largely ephemeral pastures and the condition of the stock have to be considered. ...

Adaptive stocking has been aided and can be carried out economically, because of the availability of transport and roads which allow stock to be both removed for sale or brought in for herd increases relatively rapidly (Marree Soil Conservation Board 1997:113).

Superficially this appears to be sustainable. However, elaboration of the approach in the District Plan suggests otherwise:

> high [grazing] pressure immediately about watering points cannot be totally avoided, and there will necessarily be a 'sacrifice' area of land in degraded condition. Whether the loss of land condition is limited to a few tens of metres about the water as it should be, or extends to cover a significant proportion of the paddock, depends on the nature of the land unit the water is in and the number of stock using the water (p. 114).

The Plan suggests using PVC pipe to relocate watering points in land less susceptible to heavy grazing, and to spread the grazing load. Fencing is important for controlling stock distribution and the Plan recommends fencing out different land systems.

Several elements of the Plan suggest that it may not always be sustainable. First, the Plan accepts large piospheres (sacrifice areas) which are the norm in most of arid and semi-arid Australia. Second, fencing along land system boundaries has been suggested for decades by extension agencies. The fact that it still needs to be suggested indicates that it will continue to be the exception. Third, the Plan rightly points out that there are times where the strategy must take second place to other factors (economic and financial). Finally, the Plan makes only passing reference to any landscape value (such as biodiversity) other than forage for stock.

In total, these indicate that the 'adaptive' strategies may work when conditions are appropriate, but as soon as times turn bad, many factors will mitigate against sustainability. The almost complete dismissal of landscape values other than forage is inconsistent with any reasonable definition of 'sustainable'.

3.14. SUSTAINABLE GRAZING IN THE DIRECTOR-GENERAL'S OPINION (1998)

Recent legislation in New South Wales no longer refers to 'safe carrying capacity'; instead it uses a new term 'sustainable grazing', which appears to be grazing at levels equal to the safe carrying capacity. 'Sustainable grazing' is exempt from the requirements of the *Native*

Vegetation Conservation Act because it is regarded as not destroying native vegetation. 'Sustainable grazing' is clearly defined in the Act (cl5(4)) as "the level of grazing that, in the opinion of the Director-General [of the Department of Land and Water Conservation], the vegetation is capable of supporting without resulting in a substantial long-term modification of the structure and composition of the vegetation".

The problem with the definition is that it provides no guidance to the Director-General, apparently Parliament believed that the Director-General could achieve what had never been achieved in the history of rangeland management in Australia. Worse, the Act implies that the opinion of the Director-General could be defended successfully in court.

4. Defining a Chimera: What is a Safe Carrying Capacity?

Some 170 years ago, Atkinson provided a clear rule of thumb: stock for the worst seasons. A century later, Lamond provided what he believed to be a clear definition of (safe) carrying capacity. But as it relies on a subsidiary definition of 'healthy condition' of the stock, it essentially ignores the landscape. This is a fundamental divergence between graziers/ pastoralists on the one hand, and range scientists/conservationists on the other.

Over the last century, range scientists in Australia (and their earlier counterparts) have focussed on the condition of the land (to cite but a few examples, the quite explicit views in Foran *et al.* 1990 and Ludwig *et al.* 1997, or the approaches of Condon 1968, refined by Walker 1983; Christie 1984; and various sections in Harrington *et al.* 1984). In contrast, many (most?) graziers implicitly assume that their stocking rates are equivalent to the carrying capacity and have been safe and correct for several decades (for examples see Palmer 1991; Pickard 1993). A very few notable exceptions have been Purvis (1986) and Morrissey and O'Connor (1988), who quite explicitly set out to empirically determine a stocking rate that would generate maximum economic return. Although this not necessarily the same as a safe carrying capacity, Purvis' descriptions of the condition of his land indicate that he was approaching this target.

From these few examples, it is clear that the industry and the scientists are using two dissimilar, indeed, non-commensurate, measures of health: stock *versus* land. Range scientists have spent a lot of effort relating stocking rate to climate and yields (examples: Reid and Thomas 1973; Roshier and Barchia 1993). An underlying, and usually unacknowledged, assumption is that such research will help define the elusive safe carrying capacity (usually assumed to be equal to stocking rate). A similar theme pervades extension literature (Simpson 1992; Walker 1992). This attitude reached its nadir in the assertion by the Working Party on the Effects of Drought Assistance Measures and Policies on Land Degradation (1988) that we should determine for each property the "long-term economic optimal stocking rate". This ignores the well-known and demonstrated failure of closer settlement (see for example the Royal Commission 1901). It also ignores the less well-known failure of the home maintenance area concept (Young 1985; MacLeod 1990; Passmore and Brown 1992, but note Condon's 1982 view that closer settlement caused improvement in land condition) in the Western Division and south-western Queensland. Finally, it assumes some stability in commodity prices and operating costs. Reality suggests otherwise.

Over the past century, stocking rates have varied regionally (Squires 1981; Noble 1997) and on individual properties (Pickard 1990). Estimates of carrying capacity (often based on stocking rates) have similarly varied (Pickard 1991). The suggestion of a "long-term economically optimal stocking rate" is flawed both theoretically (Bartels *et al.* 1993) and practically. The magic number simply does not exist. Thus any such attempts would be a waste of time, effort and money.

Even if there is a fundamental difference in approaches and attitudes of graziers and scientists, does it matter? Does the 'healthy animal' approach of a grazier give a stocking rate significantly different from the 'healthy land' approach used by a range scientist to estimate carrying capacity? Before addressing this, it is necessary to summarise some of the theory of carrying capacity in semi-arid ecosystems.

5. Some Elementary Theory of Carrying Capacity, and Results from Africa and USA

A considerable amount of Australian range and ecological research is directed at determining carrying capacity: sheep, cattle, kangaroos, tourists, four-wheel drive vehicles, total population. In every case, there is a fundamental assumption that there is some threshold number. At the threshold, the population is sustainable in the long-term. Higher levels will lead to degradation in the environment, lower levels can be increased if desired. Despite the interest in determining the magic number, there are important questions about the very concept of carrying capacity.

Several chapters (for example Bartels *et al.* 1993) in the recent book by Behnke *et al.* (1993) suggest that the concept is a chimera in rangelands at disequilibrium (such as in Africa and almost certainly, Australia). Bartels *et al.* list many definitions of carrying capacity, and all are relevant to the definition of 'sustainable grazing' in the Native Vegetation Conservation Act. However, they are all very difficult to determine with any consistency.

There are two main groups of definitions and approaches. One, from theory looks at the maximum number of individuals that can be supported on an area. In ecological theory this point on a logistic population growth curve is defined as K. This implies "that carrying capacity is the total resources available divided by the minimum maintenance requirement of each individual". However, maximum sustainable yields occur at half the carrying capacity predicted by the logistic model (Bartels *et al.* 1993:90).

The other approach, from range management in USA, is more confused. There are many definitions, not all commensurate. Fundamentally, maximum sustainable yields occur at about 50-75% of the resource-restricted carrying capacity (Bartels *et al.* 1993:92). Holechek and Pieper (1992) describe their method of estimating stocking rates on rangelands in New Mexico. They use a rating approach which is based on utilisation levels of standing crops of key forage species. With suitable multipliers (similar to Condon's described above) they accurately estimate actual stocking rates. Once again, the standard for comparison is what the ranchers actually carry.

Of course, many Australian range managers are not operating at the 50-70% level suggested by Bartels *et al.* They are operating at considerably higher stocking rates as demonstrated unequivocally by the pastoralists themselves (see the examples of Purvis 1986, and Morrissey and O'Connor 1988 described above).

In summary then, carrying capacity is a flawed concept with varying definitions, but there is some consistency that sustainable stock numbers occur at about 50% of maximum numbers.

6. Two Pragmatic Approaches to Defining a Chimera

The Australian literature reviewed above suggests two pragmatic approaches to defining 'sustainable grazing'. There is a logical inconsistency in showing that sustainable carrying capacity is a chimera and then proposing yet more methods for finding it. However, the concept is so deeply rooted in the grazing industry that better approaches are warranted, and the newer alternatives of 'adaptive stocking' are still being developed.

6.1. APPROACH 1: BENCHMARK PROPERTIES: EXCELLENT VEGETATION CONDITION IS DUE TO LIGHT STOCKING

Condon (1968) was the first to codify the 'benchmark property' approach. The basis for this is that vegetation in excellent condition is usually (always?) associated with light stocking. The method is simple, but time-consuming. Skilled staff with considerable field experience deliberately seek properties with vegetation communities considered to be in excellent condition. They then ask the graziers for their explanations of the excellent condition of the community. In most cases, the reason will be light stocking rates. The most critical information therefore are details of the stocking history of the property. This type of information is accumulated for all of the vegetation communities under consideration. When sufficient data are available, it is possible to determine a stocking rate which is associated with excellent condition in each of the communities. This stocking rate is the upper limit of safe carrying capacity and sustainable grazing.

The are several advantages to this method: The association of excellent vegetation condition and light stocking is intuitively obvious and widely acknowledged. The approach rests on the management styles of graziers who are usually regarded as better managers. The link between stocking rates and vegetation condition is strong enough that any stocking rate determined in this way could be defended in court.

6.2. APPROACH 2: REDUCE THE GRAZIERS' STOCKING RATES BY 50%

The empirical results of Purvis (1986) and Morrissey and O'Connor (1988) show that the commonly accepted stocking rates are far too high. These two papers are still the only real evidence for this in Australia for this. Interestingly, the results did not come from scientists, but from graziers. However, similar results have been reported in the USA (pers. comm. Professor Phil Ogden, Department of Range Science, University of Arizona, April 1993).

If we accept that these two graziers achieved sustainable grazing levels, then the only question is how to extend this more widely. I suggest a simple and pragmatic approach. Determine the district average stocking rate for the particular land system or vegetation community of interest. This is straightforward and would use the same approach as the Western Lands Commission in the late 1940s in determining rental carrying capacity (see above). Reduce the estimate by 50% and we are probably fairly close to a safe carrying capacity or sustainable grazing level.

For several reasons, I do not believe that any reasonable amount of research would substantially change the level determined in this apparently arbitrary manner. Graziers tend to stock at the highest level that will maintain good animal health. This is always more than is optimal for the environment. Reductions by Purvis and Morrissey and O'Connor by up to 60%

achieved what appeared to be sustainable grazing. There is strong evidence (quoted above) from ecosystem theory to support the 50% reduction.

This method has several advantages. It does not require skilled staff to implement, it applies across all vegetation communities, it is intellectually justifiable and has a firm base in theory, it is based (although somewhat tenuously) on grazier knowledge and experience, and it is quick to apply.

7. A Way Forward? Comparing Estimates and a Dialogue

There is considerable evidence that the whole issue of carrying capacity as a function of land condition is only of concern to range scientists and conservationists. The decades of inaction by range administrators show that they are reluctant to seriously consider land condition in semi-arid regions. Also, given operational difficulties and a long history and entrenched culture of doing nothing, there is little to suggest that the Director-General will effectively implement the sustainable grazing provisions of the NSW Native Vegetation Conservation Act.

Even if this was not the case, what *are* the techniques used by graziers and pastoralists to assess the carrying capacity of their country? At a more prosaic level, there is also the problem of the units of carrying capacity. Scientists prefer to use 'dry sheep equivalents', graziers use 'flock sheep'. There is no simple conversion from one to the other. Thus, any measure of carrying capacity may not, in itself, be particularly useful.

Does it matter that there is a fundamental difference in approaches and attitudes of graziers and scientists? Does the 'healthy animal' approach of a grazier give a stocking rate significantly different from the 'healthy land' approach used by a range scientist to estimate carrying capacity? Is the '50% reduction' approach as accurate as any of the others? Are the differences consistent within and between range types? Can one predict the others? The obvious questions pour out.

One possible solution could be arranged in conjunction with one of the Biennial Conferences of the Australian Rangeland Society. A tour could be organised where both graziers and scientists stop at a number of sites in different range types where the carrying capacity / stocking rate is described by both a grazier and a range professional. At the end of the tour, all the assessments are compared, and then used to establish a *dialogue* (two individuals talking *with* each other on the *same* topic) rather than the present *duologue* (two individuals talking *past* each other on *different* topics).

In the meantime, how many of us recognise that we have not really progressed very far since 1826? All of the science and research does not seem to have improved the accuracy of our estimates of safe carrying capacity. If you disagree, please join the dialogue. But the imperative is to provide clear evidence that *graziers* (rather than researchers) actually use the 'objective' methods you describe.

8. Acknowledgments

I thank the field staff of the Western Lands Commission (Frank McLeod, Geoff Cullenward, Ted Lowe, Peter Spencer, Rawleigh Smith, Eric McCormick) who introduced me to estimates of rental carrying capacity that sparked my initial interest in safe carrying capacity. Comments by Jim Noble (CSIRO Division of Wildlife and Ecology) and David Eldridge (NSW Department of Land and Water Conservation) clarified my thoughts and arguments. Ian Hannam (NSW Department of Land and Water Conservation) contracted with me for a report on "specifying sustainable grazing" that allowed me to complete my review of the literature and see the issues more clearly. Of course, I am quite happy to retain sole responsibility for the views expressed in this paper.

9. References

Atkinson, J. (1826) *An Account of the State of Agriculture and Grazing in New South Wales,* J. Cross, London (Facsimile edition, edited by B.H. Fletcher (1975), Sydney University Press, Sydney.

Bartels, G.B., Norton, B.E. and Perrier, G.K. (1993) An examination of the carrying capacity concept, in R.H. Behnke Jnr, I. Scoones and C. Kerven (eds.) *Range Ecology at Disequilibrium. New Models of Natural Variability and Pastoral Adaptation in African Savannas*, Overseas Development Institute, London, pp. 89-103.

Behnke, R.H. Jnr, Scoones, I. and Kerven, C. (eds.) (1993) *Range Ecology at Disequilibrium. New Models of Natural Variability and Pastoral Adaptation in African Savannas*, Overseas Development Institute, London.

Christie, E.K. (1984) Production and stability of semi-arid grassland, in D. Parkes (ed.) *Northern Australia. The Arenas of Life and Ecosystems on Half a Continent*, Academic Press, Sydney, pp. 157-171.

Condon, R.W. (1968) Estimation of grazing capacity on arid grazing lands, in G.A. Stewart (ed.) *Land Evaluation*, Macmillan of Australia, Melbourne, pp. 112-124.

Condon, R.W. (1982) Pastoralism, in J. Messer and G. Mosley (eds.) *What Future for Australia's Arid Lands?* Australian Conservation Foundation, Hawthorn, pp. 54-60.

"Crowfoot" (1917) Pastoral problems. The question of carrying capacity, *The Pastoral Review* (15 September 1917) **27**, 862-863.

Düvel, G.H. and Scholtz, H.P.J. (1992) The incompatibility of controlled selective grazing systems with farmers' needs, *Journal of the Grassland Society of South Africa* **9**, 24-29.

Fletcher, B.H. (1975) Introduction, in J. Atkinson (1826) *An Account of the State of Agriculture and Grazing in New South Wales,* J. Cross, London (Facsimile edition, edited by B.H. Fletcher (1975), Sydney University Press, Sydney), pp. 5-22.

Foran, B.D., Friedel, M.H., MacLeod, N.D., Stafford Smith, D.M. and Wilson, A.D. (1990) *A Policy for the Future of Australia's Rangelands*, CSIRO Division of Wildlife and Ecology, Canberra.

Harrington, G.N., Wilson, A.D. and Young, M.D. (eds.) (1984) *Management of Australia's Rangelands*, CSIRO, Melbourne.

Holechek, J.L. and Pieper, R.D. (1992) Estimation of stocking rate on New Mexico rangelands, *Journal of Soil and Water Conservation* **47**, 116-119.

Johnston, P.W., McKeown, G.M. and Day, K.A. (1996a) Objective 'safe' grazing carrying capacities for south-west Queensland Australia: Development of a model for individual properties, *The Rangeland Journal* **18**, 244-258.

Johnston, P.W., Tannock, P.R. and Beale, I.F. (1996b) Objective 'safe' grazing capacities for south-west Queensland Australia: model application and evaluation, *The Rangeland Journal* **18**, 259-269.

King, C.J. (1957) An outline of closer settlement in New South Wales. Part 1. The sequence of the land laws 1788 - 1956, *Review of Marketing and Agricultural Economics* **25**, 1-290.

Lamond, H.G. (1927) Carrying capacity, *The Pastoral Review* (14 April 1927) **37**, 334-335.

Landsberg, J., James, C.D., Morton, S.R., Hobbs. T.J., Stol, J., Drew, A. and Tongway, H. (1997) *The Effect of Artificial Sources of Water on Rangeland Biodiversity,* CSIRO Australia Wildlife and Ecology, Lyneham.

Lange, R.T., Nicolson, A.D. and Nicolson, D.A. (1984) Vegetation management of chenopod rangelands in South Australia, *Australian Rangeland Journal* **6**, 46-54.

Ludwig, J., Tongway, D., Freudenberger, D., Noble, J. and Hodgkinson, K. (eds.) (1997) *Landscape Ecology Function and Management. Principles from Australia's Rangelands,* CSIRO Australia, Collingwood.

MacLeod, N.D. (1990) Issues of size and viability of pastoral holdings in the Western Division of New South Wales, *Australian Rangeland Journal* **12**, 67-78.

Marree Soil Conservation Board (1997) *Marree Soil Conservation District Plan,* Marree Soil Conservation Board, Marree.

Marsh, M.H. (1842) Evidence of M.H. Marsh to the Select Committee on Immigration (*Votes and Proceedings of Legislative Council of New South Wales)*, reproduced in McMichael (1984) pp. 263-264.

McMichael, P. (1984) *Settlers and the Agrarian Question, Foundations of Capitalism in Colonial Australia,* Cambridge University Press, Cambridge.

Morrissey, J.G. and O'Connor, R.E.Y. (1988) 28 Years of Station Management, Paper presented to 5th Biennial Conference of the Australian Rangeland Society, Longreach, Queensland, June 1988.

Murray, J.F.N. (*c.* 1965) *Principles and practice of valuation,* Commonwealth Institute of Valuers, Sydney.

Noble, J.C. (1997) *The Delicate and Noxious Scrub. CSIRO Studies on Native Tree and Shrub Proliferation in the Semi-arid Woodlands of Eastern Australia,* CSIRO Australia Wildlife and Ecology, Lyneham.

Palmer, D. (1991) Western New South Wales - a miracle of recovery, *Australian Journal of Soil and Water Conservation* **4**, 4-8.

Passmore, J.G.I. and Brown, C.G. (1992) Property size and rangeland degradation in the Queensland mulga rangelands, *The Rangeland Journal* **14**, 9-25.

Pickard, J. (1990) Analysis of stocking records from 1884 to 1988 during the subdivision of Momba, the largest property in semi-arid New South Wales, *Proceedings of the Ecological Society of Australia* **16**, 245-253.

Pickard, J. (1991) Land management in semi-arid environments of New South Wales, *Vegetation* **91**, 191-208.

Pickard, J. (1993) Western New South Wales - increased rainfall (and not miracles) leads to recovery, *Australian Journal of Soil and Water Conservation* **6**, 4-9.

Pickard, J. (1994a) Land degradation and conservation in the semi-arid zone of Australia: grazing is the problem ... and the cure, in C. Moritz and J. Kikkawa (eds.) *Conservation Biology in Australia and Oceania,* Surrey Beatty, Chipping Norton, pp. 131-137.

Pickard, J. (1994b) Rents for Grazing Leases, Western Division, Spatial and Temporal Patterns 1949 -1993, Unpublished consultancy report for Western Lands Commission.

Pickard, J. (1998) Specifying Sustainable Grazing, Unpublished consultancy report prepared for Land and Vegetation (Access) Unit, Department of Land and Water Conservation.

Purvis, J.R. (1986) Nurture the land: my philosophies of pastoral management in central Australia, *Australian Rangeland Journal* **8**, 110-117.

Reid, G.K.R. and Thomas, D.A. (1973) Pastoral production, stocking rate and seasonal conditions, *Quarterly Review of Agricultural Economics* **26**, 217-227.

Robson, A.D. (1994) The case for multiple use reserves in woody weed country, Paper presented to 8th Biennial Conference of the Australian Rangeland Society, Katherine, Northern Territory, June 1994.

Roshier, D.A. and Barchia, I. (1993) Relationships between sheep production, stocking rate and rainfall and commercial sheep properties in western New South Wales, *The Rangeland Journal* **15**, 79-93.

Royal Commission (1901) *Royal Commission to Inquire into the Condition of the Crown Tenants of the Western Division of New South Wales,* Legislative Assembly of New South Wales, Sydney. 2 volumes.

Simpson, I. (ed.) (1992) *Rangeland Management in Western New South Wales,* NSW Agriculture, Orange.

Squires, V. (1981) *Livestock Management in the Arid Zone,* Inkata Press, Melbourne.

Walker, P.J. (1983) The Western Lands Lease Management Planning Scheme, *Technical Bulletin* **23**, Soil Conservation Service of New South Wales, Western Region.

Walker, P.J. (1992) Managing for drought, *Department of Conservation and Land Management Technical Paper* **1**.

Walsh, G. (1993) *Pioneering Days. People and Innovations in Australia's Rural Past,* Allen and Unwin, St Leonards.

Whalley, R.D.B. and Lodge, G.M. (1987) Use of native and natural pastures, in J.L. Wheeler, L.J. Pearson and G.E. Robards (eds.) *Temperate Pastures: their Production, Use and Management.* CSIRO, Melbourne, pp. 121-135.

Wilson, A.D. and MacLeod, N.D. (1991) Overgrazing: present or absent? *Journal of Range Management* **44**, 475-482.

Working Party on the Effects of Drought Assistance Measures and Policies on Land Degradation (1988) *Report of the Working Party on the Effects of Drought Assistance Measures and Policies on Land Degradation,* Australian Government Publishing Service, Canberra.

Young, M.D. (1985) The influence of farm size on vegetation condition in an arid area, *Journal of Environmental Management* **21**, 193-203.

AUTHOR

JOHN PICKARD *

Graduate School of the Environment
Macquarie University NSW 2109

* Current address
National Parks and Wildlife Service of NSW
PO Box 2111 Dubbo NSW 2830 Australia

email: john.pickard@npws.nsw.gov.au

CHAPTER 18

RECONSIDERING THE ECONOMIC SCOPE FOR REHABILITATING
DEGRADED AUSTRALIAN SEMI-ARID RANGELANDS – ISSUES OF CONTEXT,
PROCESS AND INTEGRATION.

N.D. MACLEOD and J.C. NOBLE

1. Abstract

The degradation of Australia's semi-arid rangelands is generally recognised to be both
widespread in scale and ongoing. This degradation is seen to be the source of significant
economic and social losses. However, concerted action by landholders to address this issue with
on-ground works on any real scale has been restrained. This restraint may be partly due to a
general perception that the characteristics of many restoration technologies would render them
uneconomic for adoption by private landholders. It may also be due to confusion as to what the
objectives of restoring rangeland landscapes should ultimately be.

These issues are explored from the perspective of individual land mangers as opposed to the
broader community. It is suggested that, in some cases, rangeland restoration options may be
more economic than previously judged. Both the underlying context and processes involved
with a given degradation problem need to be considered when assessments are to be made of
the economic value of a given restoration technique. A clear vision for a given rangeland
landscape is also argued to be a precursor for exploring integrated approaches to restoration
management that might offer realistic potential for implementation.

2. Introduction

The degradation of Australia's semi-arid and arid rangelands is widespread (Woods 1983).
Despite advances in the range of restoration technologies available for landholders (Noble *et
al.* 1984; National Research Council 1990) and an increased community commitment to
conservation initiatives (e.g. Decade of Landcare, Natural Heritage Trust funding), degradation
still occurs on a significant scale (Tothill and Gillies 1992). This is of genuine concern because
these same resources underpin a significant proportion of the national income attributed to
agricultural production and hold significant alternative values of both an economic and socio-
cultural nature (ANZECC & ARMCANZ 1996). However, concerted action by private
landholders, and to a lesser extent, public land managers, to address this issue on any real scale
has been restrained. That is, despite considerable policy and community interest, the actual
level of on-ground restoration work on private rangeland holdings is still falling far short of the
magnitude of the task as suggested by the reported scale of the degradation.

A question arises as to why so little genuine interest is being shown in addressing rangeland
degradation problems, given that they are genuine economic and socio-cultural resources. One

291

answer is that many private landholders may perceive it to be infeasible or uneconomic to do much about the problem and are, therefore, prepared to accept the ongoing degradation. There is some support for this suggestion in the limited economic literature on rangeland restoration and management (Johnston *et al.* 1990). For example, while there are relatively few economic studies which have specifically examined restoration technologies from the perspective of individual managers, these have generally pointed to a poor economic performance.

Some results of a range of published economic studies conducted within a semi-arid rangelands context are presented in Table 1. These cover most of the more common rehabilitation technologies, including prescribed fire, chemical treatment of shrubs, goat browsing, blade-ploughing and waterponding. While there is a broad categorical overlap between the purpose of each of the technologies, to change the quantity and quality of herbage available for subsequent grazing, they differ significantly in the manner in which they are effected. Unlike the first four technologies which seek to reduce shrub densities in rangeland pastures, waterponding is used to revegetate scalded areas which have lost their main soil layers. With the exception of prescribed fire, and to a lesser extent goat browsing, the majority of these techniques are rated as being non-economic by conventional economic criteria – i.e. negative net present value and benefit-cost ratios less than unity against a benchmark discount rate (10%). The testing procedure and these criteria are considered in more detail in a following section. The studies were largely conducted in the 1980s and there has been limited subsequent formal interest in reviewing these technologies. The findings for each of the technologies are generally consistent with on-going opinion of their relative economic efficiency.

Table 1: Economic performance measures from a range of rangeland restoration studies (derived from MacLeod and Johnston 1990).

Restoration Technology	Author	Net Present Value @ 10% discount rate	Benefit-Cost Ratio @ 10% discount rate
Prescribed fire shrub control (western NSW)	Burgess 1987	AU$8.40/ha (US$5.80)	4.6
Chemical shrub control (western NSW)	Burgess 1986b	-AU$59.50/ha (-US$41.20)	0.2
Goats for shrub control	Davies 1986	AU$3.56/ha (US$2.55)	Not defined
Blade ploughing for shrub control (western Queensland)	Murphy 1989	-AU$33.90/ha (-US$23.75)	0.6
Waterponding for scald reclamation (western NSW)	Penman 1987	-AU$17.70/ha (-US$12.40)	0.6

However, the *process* and *context* within which the particular form of degradation occurs were not canvassed in detail in these studies. It is possible that when these considerations of process and context are factored into rangeland management decision-making, the answer would often be a more encouraging one. Moreover, the exploration of opportunities for integrating various restoration technologies into technically and economically feasible packages has not been given much consideration for Australian rangelands. There may be scope for improving the level of restoration work on the ground when this gap is redressed. These issues are now explored, drawing largely on reflections from the authors' previous research in this field.

3. Economic Evaluation of Restoration Technologies

Why do contemporary economic studies generally suggest that the potential return from restoring degraded rangelands is relatively poor? The answer may lie partly in the nature of the typical resource degradation process and partly in how the economic analyses themselves have been structured, particularly the context in which they are set.

3.1. DEGRADATION AND DISCOUNTING

Rangeland degradation processes are typically slow and occur over long periods of time, although this can be punctuated (MacLeod et al. 1993). The early stages of these processes are often hard to observe and seem to have limited immediate impact on primary and secondary productivity or financial returns. For example, grazing animals are highly selective and can often compensate for a deleterious change in the composition of herbage on offer (Ash and Stafford-Smith 1996), while forage quality and animal productivity can initially improve within a degrading rangeland pasture (Ash et al. 1995). Benefits and costs accrue from an investment at different times, especially when several different technologies and implementation options are available to choose from. For example, most of the more common restoration technologies involve the commitment of substantial resources at different times (e.g. capital works, reinforcement and maintenance treatments) and the benefits are generally captured well into the future, often taking the form of avoiding the loss of future income. Therefore, time-dependent and consistent appraisal techniques, such as cost-benefit analysis, are required to evaluate the economic value of such investments (Pearce and Turner 1990). This involves the application of discount factors to the cost and benefit streams which are associated with the particular degradation process and restoration practice to calculate economic performance indices, such as net present values and cost-benefit ratios (Table 1).

The discounting calculus highlights certain characteristics of rangeland restoration technologies which may contribute to their generally poor performance relative to many other classes of investment which are available to land managers (Pannell 1999). That is, the effect of discounting, even where modest discount rates are employed, on a sum received well into the future can be very large, whereas near-present sums are only slightly affected. For example, the present value of one dollar discounted at 5% is 95 cents and 61 cents respectively after 1 and 10 years. The costs incurred with many rangeland restoration technologies are typically high relative to the production value of many rangeland pastures, as is the case for many chemical and mechanical treatments to control shrub encroachment or soil erosion (Johnston et al. 1990). Improvements in livestock productivity are often slow to appear after treatment, particularly in cases where a period of resting from grazing is required to guarantee the effectiveness of treatments. Because the discounting process places weights most heavily against more distant future sums and least heavily against more immediate sums, the relative cost to benefit disadvantage of management intensive technologies is amplified.

This should not to be taken as a criticism of the use of discounting, although some conservationists and resource economists have explored arguments for applying zero or otherwise very low discount rates to problems involving natural resource use (Goodin 1982; Pearce and Turner 1990). It is simply highlighting an apparent disadvantage facing some technologies when decision-makers are constrained in their choices and have viable alternatives which are reflected in positive discount rates. The choice of appropriate discount rates remains controversial and its resolution is beyond the scope of the present exercise (Pearce *et al.* 1989).

Some restoration technologies which possess low cost to potential benefit characteristics which would notionally be favoured by discounting face other problems associated with their use, such as poor treatment efficiency under some circumstances. For example, for prescribed fire in semi-arid rangelands seasonal or management factors often prevent short-cycle serial application required for economic returns (Noble 1997a) and the dietary preferences of goats may preclude the key target species (Harrington 1979). The issue of poor efficacy in individual technologies is revisited in a later section on integrated management systems.

3.2. PROBLEM FRAMING AND OUTCOMES

The context of a given degradation problem is argued to be important to the economic value of its resolution. However, the specification and exploration of this context have not been major features of previous economic evaluations of rangeland restoration technologies. A major reason for this may relate to the underlying purpose of many of those evaluations, which has been to generalise and publish the findings of restoration research projects. Many of these projects have been conducted as small-scale treatments and the economic evaluations have attempted to place the work within a wider management context. A general analytical approach involves the application of partial budgeting techniques to a limited scenario, which has commonly been set as a representative paddock with limited detail provided of the broader enterprise context in which it might be located (MacLeod 1997). This is the case for the majority of the studies which are listed in Table 1. The scenarios are commonly hypothetical and do not relate to an actual paddock for which specific details of the resource endowment and its spatial pattern can be obtained. Rather, they are typically treated as if they were uniformly endowed with resources. Moreover, the scenarios rarely depict the actual processes by which the degradation to be treated is occurring or the dynamic nature of the eventual response to the treatment. It is not clear whether the degradation problem has been permanently fixed by the treatment or merely contained and is thus capable of re-emerging. These considerations would be expected to have an impact on the economic outcome of a given restoration investment.

The practice of structuring the evaluations in a non-contextual fashion is not necessarily an oversight on the part of the investigators. It may be a deliberate choice to simplify the analyses because of the recognised heterogeneity of the contexts of the various stakeholders likely to be targeted for adoption of the technology (MacLeod and Johnston 1990). Because range enterprises are of different sizes and resource endowments, and likely to have experienced different degrees of degradation, it is judged confusing or misleading to evaluate the application of a particular technology against a specific case. This approach, it might be argued, would lose context for all of the other cases which would, by definition, be different. The scenario of a general problem within a representative sub-unit (e.g. paddock), on the other hand, is seen to be sufficiently close to the general context of most stakeholders to be useful for their decision-making.

4. Accounting for Process and Context in Restoration Studies

The conclusions of many studies of rangeland restoration options suggest a vicious cycle of high and unwarranted costs associated with restoring resources with generally low pre-degradation productive value (Johnston *et al.* 1990). However, this conclusion may well relate to a fairly narrow context of gross degradation and marginal lands where broad-scale treatment of severely degraded rangelands is involved. Restoration in these circumstances may, indeed, be non-economic and the most appropriate strategy may be to avoid further degradation. In many circumstances, however, the economic picture may be brighter than these highly aggregated and simplified studies would suggest. This is when the context and process of a particular degradation problem or remedial treatment are specifically accounted for.

It is acknowledged that some rangeland degradation problems will be uneconomic to treat. This may include those problems which emerge at both a low density and over a broad-scale, particularly where treatment effectiveness and application costs are directly proportional to the treatment area (Tisdell *et al.* 1984). Such a pattern, as depicted schematically in Figure 1a, may be observed in many cases of shrub encroachment in semi-arid woodlands (Harrington *et al.* 1984) and timber regrowth in sub-tropical woodlands (Mott and Tothill 1984). However, caution is necessary in making such conclusions because a restoration treatment which may generally be viewed as marginal or non-economic for broad-scale application may, in fact, be economic when applied in certain specific contexts. Such a case, depicted in Figure 1b, may apply when the emergent degradation problem is both concentrated within the landscape and capable of rapid or broad-scale spread, or where treatment can lead to the permanent removal of the threat. Containment of gully erosion and prevention of weed outbreaks from point sources within a paddock or watercourse would be examples (Tisdell *et al.* 1984).

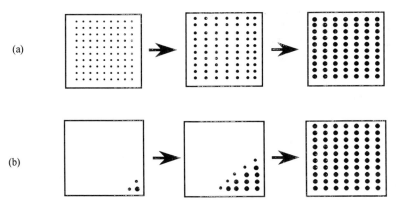

(a)

(b)

Figure 1. Two hypothetical patterns of degradation (e.g. shrub encroachment or soil erosion) are depicted that reflect different processes of expansion that result in a similar terminal state. Pattern (a) emerges uniformally across the paddock from multiple nodes and progressively increases in severity at each localised site. Pattern (b) commences from a single point and expands across the paddock over time. The economic and technical implications of managing the two cases will differ, particularly if treatment techniques and costs are proportional to the total area treated.

Caution is also required in pre-judging cases where seemingly intractable degradation problems might be resolved economically through the co-operative efforts of individual managers, exploiting economies of scale (MacLeod and Johnston 1990). An example is the co-operative efforts of several landholders using co-ordinated application of aerial ignition techniques for applying prescribed fire for eucalypt regrowth in southwestern New South Wales (Noble 1986). In this case, the issue need not be confined to questions of reducing the cost per unit area treated. The relevant technological issue may more likely be one of finding ways to cover the large spatial scales involved economically (average paddock sizes 20 000 ha) within relatively narrow windows of opportunity for treatment effectiveness (Noble 1986). This issue of exploiting economies of spatial and temporal scales will be a serious consideration for many degradation management problems which have distinct and episodic drivers and predisposing conditions (MacLeod *et al.* 1993).

In another context, the treatment of a severely degraded area might promote an opportunity to remove a constraint on the overall economic performance of the rangeland enterprise. It might also allow a more profitable activity to be included within the total enterprise. An example of the first case (removing a constraint) might apply to waterponding of severely scalded rangeland in western New South Wales. As previously noted (Table 1), this technology has been assessed by Penman (1987) as being non-economic for general application. However, when used strategically the technique may provide a source of quality forage to sensitive animal classes at critical times of the feed year, thereby raising the overall profitability of the total livestock enterprise. This may be the case where the high quality ponded pasture can provide a plane of nutrition for breeding animals sufficient to raise reproduction and survival rates to levels which would warrant a switch from wether sheep enterprises to self-replacing breeding ewe enterprises. At the time the Penman study was conducted, the gross margin differential between these two enterprises marginally favoured grazing wether sheep when these were confined to highly degraded pastures (Burgess 1986a). If stategic use of the restored pasture can be used to raise flock reproduction rates to levels which are consistent with ewes grazing more open-pastures, the order of difference is approximately 150% (Figure 2). The cited study, however, was restricted to considering marginal improvements in productivity for breeding ewes alone within the context of a single paddock.

An example of the second type (including a more profitable enterprise) is also a waterponding case from the Katherine region of the Northern Territory (Sullivan 1991a, 1991b). In this case, the reclaimed land was targeted for intensive pasture development to take advantage of the fact that it would not require timber clearing which ordinarily carries a significant up-front capital cost (Sullivan 1991a). The restored land could be sown to improved pastures for quality feed, hay or seed harvesting. This would generate an early cash flow to assist with recouping the capital and maintenance costs which contributed to much of the poor economic result observed for the western New South Wales example (Penman 1987). The estimated cost of reclamation of the scalded land was A$37/ha (US$26) when landholders used their own road-making or fire-breaking equipment, and a projected gain of approximately A$6 ha/year (US$4.20) was identified when it was returned to its former use of grazing by cattle (Sullivan 1991b). However, the gain from switching to a summer crop for harvest and sale was projected to be A$125 ha/year (US$88), allowing the costs to be recouped almost immediately. While there are potential risk and management issues involved in decisions to change the enterprise mix of a rangeland holding, the potential for gain is apparent in these two examples.

The issue of building economic assessments of restoration technologies on platforms involving hypothetical sub-units of enterprises, such as paddocks, has been highlighted within the context of ignoring the total enterprise. Another major shortcoming of this approach stems from the lack of recognition given to the spatial heterogeneity of the soil and vegetation

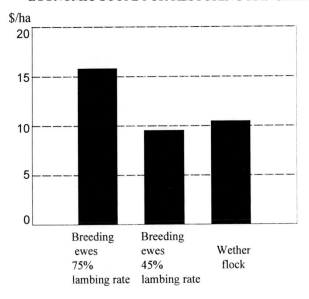

Enterprise

Figure 2. Gross margin per hectare estimates for three sheep enterprises in the Cobar region of western New South Wales. (Source: Burgess 1986a).

resources typical of real paddocks, and their inherently different productive capacities (Ash and Stafford Smith 1996). The implication of this resource heterogeneity for restoration outcomes is simply illustrated with an example of green timber management in the grassy eucalypt woodlands of southeastern Queensland. The resource mix of a single paddock drawn at random from a case study property (MacLeod and McIvor 1998), 'Philpott' - 25°36 S, 151°21 E, is presented in Table 2. The paddock resources are characterised by four land types described by locally recognised soil-vegetation associations (Smith and Dent 1993). Each land type is further categorised into one of four timber density classes according to the basal area of standing green timber present. The potential herbage productivity varies considerably between the broad land classes and within the individual tree density sub-classes. Within a decision context involving green timber control to promote future livestock production and income, the economic outcome would vary considerably according to which particular land classes are treated in the paddock and the actual density of timber involved. Highly productive land types with thick timber stands (e.g. Class B and C, timber class 3 or 4) are more likely to provide an economic response to a given timber control treatment than will less productive types with similar or lower timber densities (e.g. Class A and D, timber class 1 and 2). Any analysis which treats the paddock and property resources as being homogeneous and ignores the actual productivity differences of individual land types and existing tree densities will offer limited relevance.

5. Integrated Restoration Approaches

A major consideration for determining the economic value of rangeland restoration options, which is frequently overlooked, involves the integration of treatments which on their own are potentially ineffective or non-economic into programs which may be effective. The process of

Table 2: Herbage productivity of four land classes located in a randomly drawn paddock from a grazing property, 'Philpott', located at Mundubbera (25°36 S, 151°21 E), southeastern sub-coastal Queensland.

Land Class	Tree Density Sub-Class	Standing green timber basal area $(m^2\ ha^{-1})$	Pasture herbage total dry matter $(kg\ ha^{-1}\ yr^{-1})$	Area of land class in paddock (ha)
A. Deep sandy and	1	0	1279	10
loamy soils –	2	0-5	995	16
riparian zone	3	6-12	519	12
	4	12+	235	3
B. Non-sodic	1	0	2015	6
texture contrast	2	0-5	1350	11
soils – mid-	3	6-12	530	0
slopes	4	12+	243	0
C. Sodic texture	1	0	1560	46
contrast soils –	2	0-5	1022	23
lower slopes	3	6-12	302	8
	4	12+	147	10
D. Shallow rocky	1	0	896	5
soils in hilly to	2	0-5	830	8
mountainous	3	6-12	474	10
areas	4	12+	344	11

exploiting the emergent properties or compounding efficiency of various treatments is the central feature of integrated restoration systems, such as the Integrated Brush Management Systems and Brush Busters system used for woody shrub management in the southwestern USA (Scifres 1987; Ueckert *et al.* 1997). While less advanced in their development and application, similar integrated approaches are presently being explored to overcome the poor economic outlook for individual control technologies for semi-arid rangelands in Australia (Noble *et al.* 1999). The key principle underlying all integrated systems is the application of combinations of treatments with their order and timing set strategically set in such a way as to:

a)   compound the overall effectiveness of the individual treatments;
b)   reduce the limitations of individual or combined treatments, and/or
c)   reduce the density or magnitude of the target problem to a sufficiently concentrated or limited form to be economically treated by subsequent components of the package.

5.1. A CASE STUDY FROM THE SEMI-ARID WOODLANDS OF EASTERN AUSTRALIA

The potential value of integrated restoration approaches is illustrated drawing on the authors' experience in semi-arid woodlands where the cumulative effect of shrub encroachment has led to a substantial decline in pastoral profitability (Noble 1997b). For example, in a study of sheep enterprise profitability for western New South Wales, Burgess (1986a) assessed the economic performance of a breeding ewe flock grazing in shrub-encroached country to be approximately one half of that for open country. This magnitude of income loss when scaled up to a typical enterprise of 20 000 ha was approximately A$55 000 (US$38 500), while regional losses were estimated to be A$80 million (US$56 million) at the time the study was completed (MacLeod

1993).

A concerted program of research which commenced in the 1970s identified the major ecological and historical factors which are responsible for the changes in vegetation structure and composition of these woodlands (Noble 1997b). However, while the potential for a wide range of management options has been investigated, shrub encroachment largely remains an intractable problem. For example, while prescribed fire may offer an economic and ecologically viable method of control, it generally requires serial applications to be effective (Hodgkinson and Harrington 1985). Seasonal conditions rarely provide adequate fuel to allow this to occur at the required frequency (Noble *et al.* 1986). More input-demanding treatments, such as chemical control involving lethal levels of application and mechanical clearing, have been judged to be non-economic for broad-scale application (Burgess 1986b; Murphy 1989). While this conclusion is largely based on the context-limited analyses criticised earlier, it is consistent with the experience of both researchers and pastoralists that shrub control is a complex problem, with little likelihood of finding a single, low-input solution.

Eradication of native shrubs is neither practical nor desirable in either an economic or an ecological sense. Rather, the appropriate and most effective rehabilitation strategies should seek to restore a balance between over-storey and under-storey components within the vegetation mosaic. This will require the systematic and serial application of several control options over relatively long time periods when seasonal and economic conditions permit (MacLeod *et al.* 1993). To this end, a longer-term research program was established to develop integrated management systems for restoring productive, and sustainable, vegetation mosaics (Noble *et al.* 1999).

The previous body of research had identified the changed fire frequencies and intensities which followed European settlement and the introduction of livestock grazing as the principal cause of the shrub encroachment problem (Noble 1997b). Moreover, because the cost of implementing prescribed fire programs is relatively low compared with alternative restoration technologies (MacLeod and Johnston 1990), it should play a central role in any effective integrated management program. However, despite the optimism shown towards prescribed fire management in the 1980s (Hodgkinson and Harrington 1985), it has never been applied on any significant scale beyond the initial research trialling. This is largely due to complex interactions between the target species (many of which re-sprout post-fire), seasons and fuel dynamics (Noble *et al.* 1986). It may also be due to a limited understanding of what the fire management program is actually intended to achieve, referred to as requiring an appropriate 'vision' in the following section. For example, the actual proportion of any given area which was previously maintained as open grassland or woodland by periodic wildfire remains unclear, as much of the affected country originally comprised dense scrub interposed between open grasslands (Noble 1997b). The implicit aims of the early fire studies seem to have been to re-create an homogeneous shrub-free landscape, which is unrealistic.

To overcome the natural limitations to serial applications of fire, alternatives were sought as secondary or tertiary treatments to mimic the impact of sequential fires. The most promising strategy has involved the low concentration application of a chemical defoliant to young shrub coppice post-fire (Noble 1997b). This is consistent with both the ecological role of fire in a serial regime and the conclusions of a parametric benefit-cost analysis which suggested a threshold chemical cost for broad-area viability which was considerably below that actually required for lethal defoliation (Noble *et al.* 1991). For example, by substituting a parametric value for the actual chemical agent cost into the benefit-cost model used by Burgess (1986b), a break-even chemical application cost of approximately A\$8.50/ha (US\$5.90) was derived (Noble *et al.* 1991). This value compares with a systemic arboricide (picloram) application cost of approximately A\$75/ha (US\$45/ha) at that time (MacLeod and Johnston 1990). These values are directly related to the cost-price regime for sheep enterprises prevailing at the time and were recognised to be indicative only, but they helped to set guidelines for future screening of potential agents. An initial fire will only run in areas with sufficient fuel, effectively identifying those landscape zones offering potential for increased herbage production following secondary treatment (Daly and Hodgkinson 1996; Noble 1997b). This provided a guide to the critical landscape components which might be targeted economically in the post-fire treatment sequences.

Once an effective defoliant was isolated (glyphosate applied in the autumn to two-year-old coppice), further experiments were conducted throughout the semi-arid woodlands to identify the most appropriate strategies for broad-scale application (Noble 1997b). Aerial application of low volume chemical (0.5 kg ha^{-1} glyphosate) to very young coppice was the most effective treatment and was also consistent with the break-even chemical application cost (Noble *et al.* 1991). Present work is seeking further scope to reduce these costs through the refinement of the mix of treatments (fire, chemicals, resting from grazing), application rates and operational logistics. For example, co-operative action by several adjoining landholders over large areas could reduce aircraft procurement and operating costs through shared aircraft ferrying costs, reduced spray runs and generally exploiting economies of scale in chemical purchases and site preparation (Noble 1986). These will be explored, but the point to be made is that there is scope for integrated management strategies, based on prescribed fire followed by low-concentration defoliation of sprouting shrubs, to be economically feasible for application over extensive semi-arid rangelands. The key to success is to understand the context in which the problem is occurring and to exploit any opportunities to act on the problem. The use of prescribed fire to identify areas with productive potential within individual paddocks enables precise aerial application of secondary defoliant treatments (Noble *et al.* 1997). To do this is also consistent with having a clear 'vision' of the post-treatment landscape.

6. Clear Objectives for Landscape Management as a Prerequisite to Economic Restoration.

The importance of understanding the context within which a given restoration problem or opportunity is framed has been highlighted. Part of this framing is the spatial heterogeneity of range landscapes and the differential productivity of the resources constituting the mosaic to be managed. Many practical range managers appreciate this differential and manage their resources accordingly. However, it is also likely that many decisions on restoring degraded landscapes are being considered without a clear vision of what the landscape might actually look like in a reasonably productive or 'natural' state.

Of course, beyond important issues of conserving biodiversity, there is no inherent value in maintaining rangeland landscapes in some pre-European (or pre-Aboriginal) state, nor is it entirely clear what those states were actually like. Most of the evidence is largely anecdotal, drawn from explorers' journals and settlers' accounts, although some detailed archival material is available from various commissions, land surveys and station records (Noble 1997b). All rangeland landscapes are an artefact of human and natural forces, and an ideal pattern, while set within ecological limits, will be the outcome of human values and choice processes. Nevertheless, the notion of a natural or productive state is a useful starting point for constructing a vision of where restoration efforts should be targeted and of the desired end result. This would help eliminate the waste of scarce resources in restoring landscape elements for which restoration is inappropriate or, at best, marginally beneficial.

An example is again drawn for native shrub proliferation in semi-arid woodlands. The historical evidence suggests that these woodlands were characterised, at the time of European settlement, by a rich mosaic of vegetation and soil associations. The mosaic included dense tracts of the native tree and shrub species which constitute the shrub encroachment or woody weed problem facing these woodlands over most of the present century (Noble and Brown 1997). These mixed shrub communities obviously did not detract from what were then perceived to be highly productive grazing lands waiting to be exploited by domestic livestock. Therefore, the degradation context is more one of increasing shrub densities and, to a lesser extent, altered species compositions resulting from poor grazing and fuel management than the terms 'encroachment' and 'woody weeds' generally imply. This loose association of the problem species and the nature of the problem itself with that of agricultural weeds, may have given rise to a major block to seeking economic strategies to resolve the problem. That is, a management perception has emerged that the problem species are unnatural in the landscape and should ideally be removed or their densities substantially reduced across the entire landscape. A similar phenomenon has been identified with perennial weeds in temperate rangeland pastures (Auld *et al.* 1987) for which elimination control strategies appropriate to annual weeds within cropping systems were over-riding rational assessments of the economic nature of the problem.

In both cases of semi-arid and temperate rangelands, by ignoring the 'natural' landscape pattern, its constituent elements and their inherent differential productivity, there can be a tendency to pursue strategies which are economically wasteful or to frame restoration problems in an unrealistic manner. This is explored schematically in Figure 3, which depicts two hypothetical control options for a heterogeneous landscape (Fig. 3a) which has become degraded through an encroachment process between adjacent land classes (I and II). For example, under a complete elimination strategy (Fig. 3b), considerable resources may be diverted to removing problem species which were naturally present in similar densities on parts of the landscape (II, Fig. 3a) and which are readily suited to rapidly re-colonising that part of the treated area. Under a partial treatment strategy (Fig. 3c), only the more productive land class (I) is targeted, presumably at a lower economic cost.

7. Conclusion

While the problem of degradation of rangeland resources is both wide-scale and increasing, efforts to fix the problem on any significant scale are restrained. This restraint may, at least in part, be supported by a general perception that many restoration efforts would be non-economic to undertake. In many instances, this may be a realistic assessment. However, caution is urged against discounting too readily the potential for economically undertaking certain rangeland restoration investments, particularly by individual land managers. This may arise through unwarranted generalisations being made of particular classes of degradation problems and potential restoration techniques.

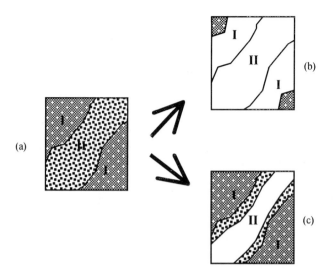

Figure 3. An hypothetical starting state involving a degraded paddock (3a) comprising two different land classes (I, II) with associated vegetation structures. The darker patches in the lower and upper quadrants (land class I) represent the original dominant vegetation. The same vegetation has encroached onto the other land class (II) over time. The two management options are to remove the majority of vegetation on both land classes I and II (3b) or to accept the original pattern and manage the encroachment on land class II only (3c). The economic and ecological outcomes associated with pursuing the alternative management strategies will invariably be different.

The key to making a realistic assessment lies in having a clear understanding of the context within which a specific restoration problem, or opportunity, is framed. At the same time, the process through which the problem is becoming manifest is important. This process needs to be well understood to assist in clearly defining the scope and nature of the problems and for seeking opportunities for low-cost options to resolve them. Integrated strategies of restoration management may offer considerable scope for achieving this objective. However, the process of finding economic solutions to degradation problems will also be enhanced when managers have a clear vision of what the ultimate objectives of restoring rangeland landscapes should be.

It is acknowledged that broad-scale restoration of degraded rangelands is generally problematic, in both an economic and an ecological sense. The ideal remains to avoid degradation in the first instance through widespread adherence to sustainable land management practices. However, much rangeland is already degraded and blanket generalisations may be made too readily concerning the specific merit of individual technologies which might otherwise be brought to bear on restoration. Some real progress will be made to reduce barriers to on-ground actions by both private landholders and the wider community if more attention is given to reviewing the potential for taking positive action in specific cases. In doing this, identifying the specific processes and context underlying these individual cases and the scope for integrating treatment options will be an important component to devising solutions.

8. Acknowledgment

Most of the individual pieces of cited research and development which have been incorporated within the integrated approach to shrub management described in the paper have been supported jointly by the International Wool Secretariat and CSIRO.

9. References

ANZECC and ARCANZ (1996) *National Strategy for Rangeland Management*, Department of Environment, Sport and Territories, Canberra.

Ash, A.J. and Stafford Smith, D.M. (1996) Evaluating stocking rate impacts in rangelands: animals don't practice what we preach, *Rangeland Journal* **18**, 216-243.

Ash, A.J., McIvor, J.G., Corfield, J.P. and Winter, W.H. (1995) How land condition alters plant-animal relationships in Australia's tropical rangelands, *Agricultural Ecosystems and Environment* **56**, 77-92.

Auld, B.A., Menz, K.M. and Tisdell, C.A. (1987) *Weed Control Economics*, Academic Press, London.

Burgess, D.M.D. (1986a) Livestock budgets – Cobar and Bourke Shires, *New South Wales Department of Agriculture and Fisheries Farm Business Notes* No. 30, Sydney.

Burgess, D.M.D. (1986b) An economic perspective on chemicals for shrub control in the Western Division, 2nd Edition, *New South Wales Department of Agriculture and Fisheries Farm Business Notes* No.31, Sydney.

Burgess, D.M.D. (1987) The Economics of Prescribed Burning for Shrub Control in the Western Division of New South Wales, *New South Wales Department of Agriculture and Fisheries Farm Business Notes* No.34, Sydney.

Goodin, R. (1982) Discounting discounting, *Journal of Public Policy* **2**, 53-72.

Harrington. G.N. (1979) The effects of feral goats and sheep on the shrub populations in a semi-arid woodland, *Australian Rangeland Journal* **1**, 334-345.

Harrington, G.N., Mills, D.M.D., Pressland, A.J. and Hodgkinson, K.C. (1984) Semi-arid woodlands, in G.N. Harrington, A.D. Wilson and M.D. Young (eds.), *Management of Australia's Rangelands*, CSIRO, Melbourne, pp. 189-207.

Hodgkinson, K.C. and Harrington, G.N. (1985) The case for prescribed burning to control shrubs in eastern semi-arid woodland, *Australian Rangeland Journal* **7**, 64-74.

Johnston, B.G., MacLeod, N.D. and Young, M.D. (1990) Economic forces and pressures on the Australian sheep-grazed rangelands: implications for future research objectives and priorities, *Australian Rangeland Journal* **12**, 91-115.

MacLeod, N.D. (1993) Economic cost of shrub encroachment in western New South Wales, in. E.S. Delfosse (ed.), *Pests of Pastures: Weed, Invertebrate and Disease Pests of Australian Sheep Pastures*, CSIRO Information Services, Melbourne, pp. 58-63

MacLeod, N.D. (1997) Case studies: a useful tool for integrating socio-economic data into grazing systems research, *Proceedings 8th International Grasslands Congress*, Winnipeg-Saskatoon, Canada 8-19 June. Volume 2, pp. 24.23-24.24.

MacLeod, N.D. and Johnston, B.G. (1990) An economic framework for the evaluation of rangeland restoration projects, *Australian Rangeland Journal* **12**, 40-53.

MacLeod, N.D. and McIvor, J.G. (1998) Managing Grazed Landscapes - Exploring Production and Conservation Tradeoffs in Sub-Tropical Woodlands, in. J. Gooday (ed.), *Proceedings of the Bioeconomics Workshop (Post Australian Agricultural and Resources Economics Conference)*, University of New England, Armidale, 22 January 1998, ABARE, Canberra, pp. 51-63.

MacLeod, N.D., Brown, J.R., and Noble, J.C. (1993) Ecological and economic considerations for the management of shrub encroachment in Australian Rangelands, *Proceedings of 14th Asian-Pacific Weed Science Society Conference*, Volume 2, Brisbane, pp. 118-121.

Mott, J.J. and Tothill, J.C. (1984) Tropical and subtropical woodlands, in G.N. Harrington, A.D. Wilson and M.D. Young (eds.), *Management of Australia's Rangelands*, CSIRO, Melbourne, pp 255-269.

Murphy, R. (1989) Economics of some pasture restoration options, *Australian Rangeland Society/Southern Maranoa Pastoralists Association, Workshop*, Bollon, 25 May.

National Research Council (1990) *The improvement of tropical and subtropical rangelands*, Board on Science and Technology for International Development, Office of International Affairs, National Academy Press, Washington D.C., pp. 139-185.

Noble, J.C. (1986) Prescribed fire in mallee rangelands and the potential role of aerial ignition, *Australian Rangeland Journal* **8**, 18-30.

Noble, J.C. (1997a) The potential for using fire in Northern Australian pastoral lands, in A.C. Grice and S.M. Slatter (eds.), *Fire in the management of Northern Australian pastoral lands*, Tropical Grasslands Society of Australia, Occasional Publication No. 8, pp. 41-47.

Noble, J.C. (1997b) *The Delicate and Noxious Scrub: CSIRO Studies on Native Tree and Shrub Proliferation in the Semi-Arid Woodlands of Eastern Australia*, CSIRO Wildlife and Ecology, Canberra.

Noble, J.C. and Brown, J.R. (1997) A landscape perspective on rangeland management, in J.A. Ludwig, D.J. Tongway, D.O. Freudenberger, J.C. Noble and K.C. Hodgkinson (eds.), *Landscape Ecology: Function and Management*, CSIRO, Melbourne, pp. 79-92.

Noble, J.C., Cunningham, G.M. and Mulham, W.E. (1984) Rehabilitation of degraded land, in G.N. Harrington, A.D. Wilson and M.D. Young (eds.), *Management of Australia's Rangelands*, CSIRO, Melbourne, pp. 171-186.

Noble, J.C., Harrington, G.N. and Hodgkinson, K.C. (1986) The ecological significance of irregular fire in Australian rangelands, in P.J. Joss, P.W. Lynch and O.B. Williams (eds.), *Rangelands: A Resource Under Siege, Proceedings of 2nd International Rangeland Congress, Adelaide,* Australian Academy of Science, Canberra/Cambridge University Press, Melbourne, pp. 577-580.

Noble, J.C., MacLeod, N.D. and Griffin, G.F. (1997) The rehabilitation of landscape function in rangelands, in J.A. Ludwig, D.G. Tongway, D.O. Freudenberger, J.C. Noble, and K.C.Hodgkinson (eds.), *Landscape Ecology: Function and Management*, CSIRO, Melbourne, pp 107-120.

Noble, J.C., MacLeod, N.D., Ludwig, J.A. and Grice, A.C. (1991) Integrated shrub control strategies in Australian semi-arid rangelands, in A. Gaston, M. Kernick, and H. Le Houerou (eds.), *Proceedings of the 4th International Rangeland Congress*, Service Central d'Information Scientifique et Technique, Montpellier, volume 2, 846-849.

Noble, J.C., MacLeod, N.D., Grice, A.C. and Jones, P. (1999) Managing mosaics in semi-arid woodlands of eastern Australia, in D. Eldridge and D. Freudenberger (eds.), *People and rangelands – building it's future. Proceedings of 6th International Rangelands Congress,* Volume 2, Aitkenvale, pp. 1024-1025.

Pannell, D.J. (1999) Explaining non-adoption of practices to prevent dryland salinity in Western Australia: implications for policy, *SEA Paper 99/08*, Agricultural and Resource Economics, University of Western Australia, Nedlands.

Pearce, D.A. and Turner, R.K. (1990) *Economics of Natural Resources and the Environment*, Harvester Wheatsheaf, Hemel Hempstead.

Pearce, D.A., Markandya, A. and Barbier, E. (1989) *Blueprint for a Green Economy,* Earthscan Publications, London.

Penman, P. (1987) An economic evaluation of waterponding, *Journal of Soil Conservation New South Wales* **43**, 68-72.

Scifres, C.J. (1987) Decision analysis approaches to brush management planning: ramifications for integrated range resources management, *Journal of Range Management* **40**, 482-490.

Smith, G.K. and Dent, D.J. (1993) Understanding and managing soils in the inland Burnett region, *Training Series QE93001*, Queensland Department of Primary Industries, Brisbane.

Sullivan, S. (1991a) Waterponding for scald reclamation (Katherine Region), *Landcare Fact Sheet,* No. 10, Conservation Commission of Northern Territory, Darwin.

Sullivan, S. (1991b) Costs and benefits of waterponding for scald reclamation, *Landcare Fact Sheet,* No. 14, Conservation Commission of Northern Territory, Darwin.

Tisdell, C.A., Auld, B.A. and Menz, K.M. (1984) Crop loss elasticity in relation to weed density and control, *Agricultural Systems* **13**, 161-166.

Tothill, J.C. and Gillies, C. (1992) The pasture lands of northern Australia: their condition, productivity and sustainability, *Tropical Grasslands Society Occasional Paper,* No. 5, Brisbane.

Ueckert, D.N., McGinty, W.A. and Addison, D.A. (1997) Brush Busters: common-sense brush control for rangelands, Dow AgroSciences LLC, Indianapolis, Indiana USA, *Down to Earth* **52**, 34-39.

Woods, L.E. (1983) *Land Degradation in Australia,* Australian Government Publishing Service, Canberra.

AUTHORS

N.D. MACLEOD[1] and J.C. NOBLE[2]

[1] CSIRO Tropical Agriculture, 120 Meiers Rd, Indooroopilly, Q 4068, Australia
neil.macleod@tag.csiro.au
[2] CSIRO Wildlife and Ecology, G.P.O Box 284, Canberra, ACT 2601, Australia
jim.noble@dwe.csiro.au

Corresponding author – MacLeod

CHAPTER 19

AIRBORNE GEOPHYSICS PROVIDES IMPROVED SPATIAL INFORMATION
FOR THE MANAGEMENT OF DRYLAND SALINITY.

RICHARD J. GEORGE and DONALD L. BENNETT

1. Abstract

Biophysical information forms the technical base for farm planning and landscape
management. Such information is required at spatial scales which are suited to the
application of the user. Within the past five years, two catchments in the South West of
Western Australia were studied to assess the contribution that airborne geophysics data
(specifically electromagnetics, magnetics and radiometrics) makes, in conjunction with
existing datasets, to the understanding and management of land degradation, principally
dryland salinity. The Towerrinning (30 000 ha) and Toolibin surveys (70 000 ha) are
located in areas of intensely weathered Archaean granitic terrain in a low gradient
landscape. At both sites over 80 bores were drilled and extensive ground geophysical
surveys were undertaken as a part of a formal process of calibration and landscape
investigation.

Our analysis suggests that the airborne systems used have the ability to map: geologic
structures critical for the management of land and water salinity at a local scale; farm-scale
changes in regolith salt-store and groundwater salinity, and local and regional-scaled
patterns of soils. When combined they provide for better salinity hazard assessment and
management decision making, not previously available from any other data sources. Finally,
when compiled in a simple to use GIS format, a wide array of people may easily access and
use the data.

2. Introduction

The clearing of native vegetation has resulted in a change to the water and salt balance of
agricultural catchments. As a result, rising groundwater levels and the mobilisation of salt
within the deeply-weathered regolith is leading to the salinisation of large areas of formerly
productive agricultural land (Ferdowsian et al. 1996). The future of Western Australia's
(WA) agricultural land, its water resources and its natural environment depends on at least
partially returning the water balance towards pre-clearing conditions (George et al. 1997).
However, at present it is not possible to do this as there are too few management options
available which are both effective and economic. As a result, land managers seek ways to
adopt existing management systems selectively to suit local (paddock-scale) conditions. The
development and implementation of these systems rely on accurate spatial data to select,
locate and then underpin prediction models which consider the effects of management

A.J. Conacher (ed.), Land Degradation, 305–318.

changes. Additionally, improved knowledge of the landscape may allow more accurate diagnosis of causes, prognosis of trends, and delineation of treatments which may improve catchment-scale management by both land owners and resource management agencies (George *et al.* 1997).

Conventional methods of identifying the spatial elements of landscape hydrology on either a paddock or regional scale include soil analysis, airphoto interpretation, regional mapping, ground geophysics and drilling. These are expensive and labour intensive, and hence are not appropriate for use in every catchment. Given the magnitude of the salinity problem, and the limitations of the currently used methods of quantification and description, it will take many decades to define accurately the landscape to the level required. While Landsat Thematic Mapper (Landsat TM) and other satellite-based systems are currently being developed as a tool for mapping the surface features (such as the expression of salinity throughout WA), there are currently no 'systematic' tools for investigating 'landscape hydrology' (in three dimensions) apart from airborne geophysics.

Nulsen *et al.* (1996) observed that up to 14 potential datasets were required for effective farm and catchment management of dryland salinity. George *et al.* (1999), at the conclusion of the National Airborne Geophysics Project, later refined these datasets in the light of recent developments in technology. The 12 core biophysical datasets required now include:

- topography and drainage patterns derived from a hydrologically sound digital elevation model (DEM) (+/- 1 m accuracy);
- orthophotos (corrected);
- hydrological data (water table depth, salinity, hydraulic properties and trends);
- multi-spectral data (satellite; Landsat TM, airborne multi-spectral and thermal scanners) and classifications (eg soil salinity maps, soil chemistry);
- magnetics (to improve geology);
- radiometrics (to improve geology and soils/landscapes);
- field geology and related interpretations;
- electromagnetics (ground and airborne; targeted application);
- meteorological data (rainfall, evaporation etc);
- soils - Land Management Units (LMUs and other classifications);
- land use (spatial and temporal), and
- cadastre and other infrastructure.

Of these biophysical datasets, only magnetics, radiometrics, airborne electromagnetics (AEM) (collectively referred to here as 'airborne geophysics') and hydrological data will not be available routinely in the Western Australian agricultural area after 2001. State-wide programs to acquire geophysical data have not been established, although about 20 catchments have been flown at various times over the past decade, with several geophysical systems. At least three companies operating in Australia have 'broad spectrum' airborne geophysical data acquisition capability, electromagnetics, magnetics and radiometrics systems. Many more companies offer magnetics and radiometrics services. In WA, the development of the SALTMAP AEM system (Pracilio *et al.* 1998) followed almost a decade of collaborative research. Our current study focuses on the application of airborne geophysics in the agricultural areas of WA, furthering the work by de Broekert *et al.* (1996). In this paper we consider the application of the SALTMAP system, making reference where possible to the recent development of the new TEMPEST (June 1999) acquisition system.

This papers aims to describe the application of airborne geophysics to the definition of landscape processes and the development of land management options. It shows that to be effective, biophysical data must be at the scale required by the manager rather than that at which State and Federal Agencies have existing data. To undertake this task, we chose two catchments (Toolibin and Towerrinning: Fig. 1) in differing terrain where extensive field analyses have been conducted. We first review the major results in terms of the utility of the three geophysical systems; magnetics, radiometrics and AEM, and then discuss the application of the three datasets for farm and catchment management.

3. Methods

3.1. LOCATION

The Towerrinning catchment is located approximately 250 km SSE of Perth in the middle of Blackwood River system (Fig. 1). The catchment has an area of approximately 12 000 ha (while a greater area was investigated; 30 000 ha), a mean annual rainfall of 500 mm and is more than 80% cleared for agriculture (rain-fed grazing and cropping). Lake Towerrinning lies at the base of the catchment and is an important social and environmental asset. Approximately 15% of the catchment is already salt-affected and it is predicted that as much as 20-25% could become saline within 30 years. Rates of groundwater rise vary from 0 to 1.0 m yr^{-1}, with the mean of about 0.3 m yr^{-1} in uplands and on the slopes (George et $al.$ 1996).

The Toolibin catchment is located approximately 250 km SE of Perth at the headwaters of the Blackwood River system (Fig. 1). The Toolibin catchment has an area of approximately 47 000 ha, a mean annual rainfall of 400 mm and is more than 90% cleared for agriculture (rain-fed cropping and grazing). Lake Toolibin, a 'Wetland of International Importance' (Ramsar Convention) lies at the catchment outlet. Approximately 6% of the catchment is severely salt-affected and it is predicted that within 60-100 years as much as 25-30% could become saline if the water balance is not altered and groundwater levels continue to rise (current average rates of 0.25 m yr1; George 1999).

3.2. GEOPHYSICAL SURVEYS

Airborne magnetics and radiometrics systems measure the magnetic and radioactive element characteristics of the earth from an elevation of approximately 20-120 m. At both catchments a magnetometer sensor mounted at the rear of the aircraft (at 60 m) measured the total magnetic field of the earth. Short wavelength variations in the magnetic field are due to variations in the magnetic mineral content of the rocks near to the surface of the ground. Longer wavelength variations are usually due to deeper sources.

Gamma-ray spectrometry (radiometrics) measures the natural gamma radiation emitted by radioactive daughter decay of three elements - potassium (K), thorium (Th) and uranium (U) - from within approximately the top 300 mm of the earth's surface. The maps prepared from radiometric surveys provide information about the soil parent materials and other

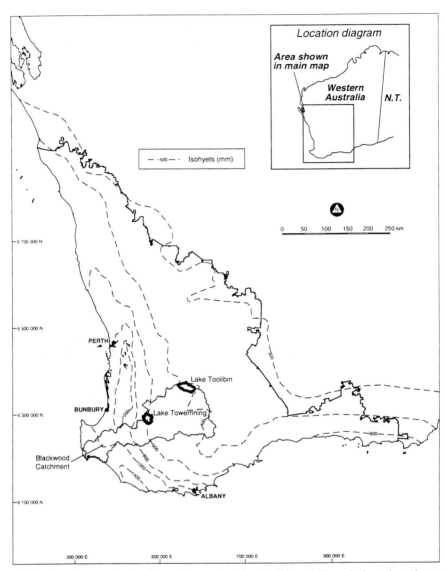

Figure 1. Location diagram showing the position of the Lake Toolibin and Towerrinning sub-catchments, the Blackwood catchment and major rainfall isohyets within the agricultural areas of WA. The majority (> 90%) of the land between the 1000 mm isohyet and the 300 mm isohyet (18 M ha-1) has been cleared of all native vegetation.

properties such as surface soil texture, weathering, leaching, depth and clay types.

SALTMAPTM is a fixed wing, time-domain airborne electromagnetic AEM system (Sattel 1998) which differs from conventional airborne AEM systems. The transmitter (Tx) is a single loop, slung horizontally around the extremities of the aircraft with an effective area of 186 m^2. The loop has a moment of 18 600 Ampere per metre squared (A m^{-2}) and the transmitter waveform is a 495 hertz (Hz) alternating polarity square wave, with a pulse width of one millisecond (ms) (Pracilio et $al.$ 1998).

The SALTMAP receiver (Rx) used in this survey consisted of one orthogonal coil mounted in a towed instrument, which is positioned 95 metres behind and 70 metres below the aircraft (nominally 50 m above the ground). The receiver has a bandwidth (the range of frequencies over which it is sensitive) of 50 Khz and samples each waveform 100 times. Each reading is stacked over 200 ms, thus five readings can be made per second - equating to approximately 12.5 metres along the ground. Additionally, the complete waveform of the transmitter is measured. Data pre-processing includes primary field stripping, conversion to step response and data binning into eight selected primary windows. Data are then reported as apparent conductivity from these windows. Typically, the groups are binned into three clusters; AC2_3 (apparent conductivity of windows 2 and 3), AC4_5, and AC7_8). Data are then modelled using a 3-Layer inversion process to determine a total conductance (product of conductivity by thickness) value of the upper resistive zone (Layer 1, L1) and the lower conductive zone (Layer 2, L2). The basement is assumed to be infinitely resistive. Line spacing for the surveys ranged from 150 to 300 m and the direction was 0-180 degrees (Towerrinning) and 020-200 degrees (Toolibin).

3.3. ANALYSIS IN TWO CATCHMENTS

The Towerrinning AEM survey was completed in March 1997. An area of over 30 000 ha was flown (Archaean granites and Tertiary sediments in a dissected landscape). Drilling and ground data collection were initiated in 1992, however collation and analysis of more recently acquired data have been completed only recently. The Toolibin survey area (Archaean granitic terrain in a low gradient landscape) covered 35 000 ha of AEM and an area of approximately 70 000 ha of magnetics and radiometrics (150 m flight line spacing). All of the Toolibin surveys were flown by October 1997. Field assessment ended in mid August 1998. At both sites over 80 bores were drilled and large areas of ground geophysical surveys were undertaken. Calibration and field testing was carried out using: drilling, soil assays, ground and downhole geophysics; airphoto interpretation, and community workshops. Both statistical and empirical methods were used to interpret the data (when compared with existing data or current interpretation of such knowledge). Statistical analysis was first undertaken using Microsoft Excel and then GENSTAT. All data (for example from ~80 drill holes) at each catchment were used to analyse the geophysical data and products derived from their interpretation. Downhole (EM39) and ground based electromagnetics instruments (EM31, EM38) were also used to calibrate the AEM systems. A detailed analysis of the Toolibin data is presented in George (1999).

4. Results

4.1. GEOPHYSICAL SURVEY RESULTS

Results from both the Toolibin and Towerrinning studies have shown that airborne magnetics have the ability to map geological structures which are not always apparent from outcrop or airphoto interpretation and to map geological structure at a paddock (field) scale (~100 m, Plate 1). The Toolibin study showed that airborne radiometrics (specifically by use of the ternary images, Plate 2) have the ability to map patterns of soil properties (clay mineralogy and texture) more effectively and efficiently than existing landscape interpretation technologies, particularly when integrated with field observations. It also assists interpreters to map at a paddock scale (~100 m) instead of existing regional scales (~1000-5000 m) and allows farmers to compare soils, map similar units within catchments or regionally, and use their own detailed knowledge to improve mapping on their farm. At Towerrinning, acquisition problems prevented the same level of analysis and application of the radiometric data.

At Toolibin, statistical analysis showed that airborne electromagnetic data acquired by SALTMAP explains *up to* 90% of the variance in salt-store at depths from 5-10 m to bedrock (Table 1). It maps regional variability in measured regolith salt-store at depth (> 10 m) which could not be derived by other measured hydrologic attribute, including terrain (topography). It maps local variations in salt-store over distances of the order of 100 m and accounts for between 60 and 90% of the variance in electrical conductivity measured by ground and downhole EM systems (Table 1). AEM also accounts for 75% of the spatial variance seen in the distribution of groundwater salinity and aids the identification of features significant for catchment management (such as resistive and conductive linear anomalies associated with sediments and dykes), and identifies those which require further investigation (Table 2, Plate 3). The Towerrinning study showed that airborne electromagnetic data acquired by SALTMAP accounts for some of the variance in salt-store at depth and maps local and regional variability in regolith salt-store, but with less statistical significance than at Toolibin. Finally, at both Toolibin and Towerrinning, AEM resolved the location of valley palaeochannel, its groundwater salinity and related sediments.

At Toolibin, SALTMAP could not reliably measure near-surface (0-5 m) salt store, or map either absolute or specific channel conductance in highly conductive terrain in the Toolibin valley (>10 Siemens across ~ 30% of catchment), although meaningful spatial relationships exist between field data and AEM in several areas. At neither catchment could AEM reliably map regolith thickness or bedrock topography at the paddock-scale.

When combined with other spatial data in a GIS environment, airborne geophysics also provides other information to assist landscape interpretation. For example it allows the user to visualise the landscape and its geological attributes, and by interpretation map the spatial distribution of salt-store which indicates likely hydrological and geological features.

4.2. CATCHMENT MANAGEMENT IMPLICATIONS

The application of airborne geophysics (particularly AEM) for farm and catchment planning is critically dependent on the availability of skilled landscape interpreters to link data about physical processes to 'on-farm' management. At present these linkages are limited, although improved understanding and more accurate biophysical data can be used to aid the assessment of risk and the impact on groundwater hydrology of proposed management systems, using groundwater models. Specifically, we considered that a better understanding was possible with the combined use of all datasets (enabling mapping of groundwater salinity and salt store) than before acquisition of the data. Significant scope exists for improvements in AEM (in particular, developing accurate regolith thickness maps). At Toolibin, recent TEMPEST data have improved basement mapping and resolution in the vertical distribution of salt (5-50 m).

All geophysical data require skilled interpretation, particularly in the absence of well-

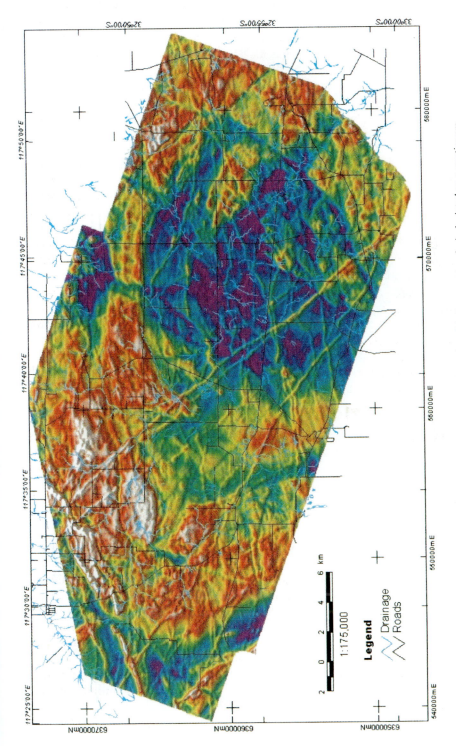

Plate 1. Toolibin magnetic image; where the magnetic intensity of the basement rocks is depicted as a continuum from low intensity (even grained granites, gneissic and deformed rocks) to basic dolerite and other intrusive dykes (from George 1998).

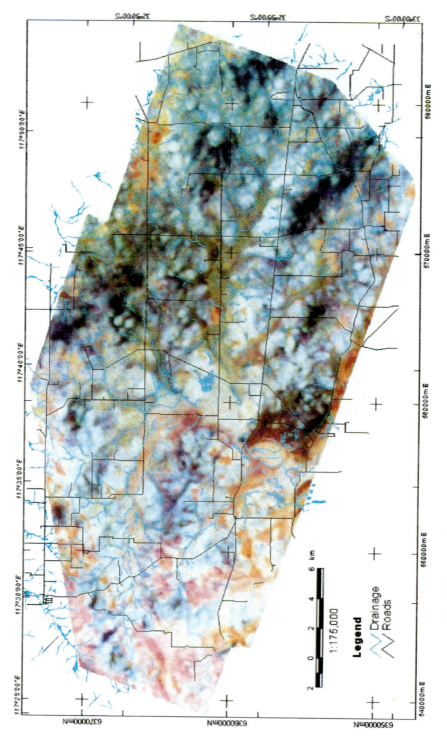

Plate 2. Ternary radiometric image; where the intensity of color is derived as a function of the relative abundance of uranium, potassium and thorium. For example, dark areas are those with low radioelement levels in all channels (eg quartz sands), while red and pink areas are those shallow soils with relatively high potassium radioelement levels (from George, 1998).

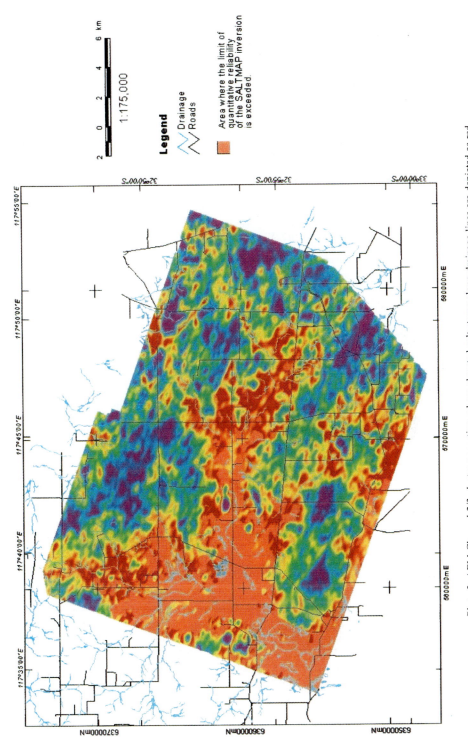

Plate 3. AEM Channel 2/3 electromagnetics; where elevated salt-stores along drainage lines are depicted as red, while those areas with little or no salt store are depicted as blue or purple. These areas occur in association with those identified as poor in all radioelements by the radiometrics image (from George 1998)

Table 1: Statistical relationships (r value) between measured attributes (drill data), AEM (apparent conductivity and conductance for 3 channels) and Terrain

	terrain (m)	AC 2_3	AC 4_5	AC 7_8	L1c (mS)	L2c (mS)	L1 C	L2 C
Regolith								
Bedrock (m)	0.072	-0.110	0.145	0.425	0.024	0.222	0.390	0.333
Sediments(m)	-0.474	0.246	0.263	0.078	-0.378	0.193	-0.245	0.301
Saprolite (m)	0.354	-0.328	-0.212	0.019	0.288	-0.321	0.513	-0.189
SWL (m)	0.824	-0.426	-0.675	-0.719	-0.305	-0.700	-0.160	-0.716
Saturated (m)	-0.565	0.183	0.636	0.874	0.281	0.729	0.501	0.809
GW Conductivity	-0.829	0.425	0.793	0.859	0.152	0.820	0.080	0.900
Salt-store EC1 :5 m/m								
EC1:5 0-bedrock	-0.651	0.462	0.682	0.766	0.439	0.754	-0.090	0.777
EC1:5 sediments	-0.707	0.135	0.641	0.820	0.440	0.749	0.592	0.813
EC1:5 saprolite	-0.594	0.324	0.695	0.716	0.373	0.797	0.484	0.769
EC 1:5 unsaturated	-0.335	0.325	0.340	0.148	0.189	0.348	0.282	0.305
EC 1:5 saturated	-0.641	0.347	0.723	0.723	0.282	0.775	0.379	0.815
Tot. EC1:5	-0.640	0.327	0.721	0.763	0.267	0.792	0.272	0.813
0-5m EC1:5	-0.446	0.235	0.382	0.312	0.275	0.433	0.292	0.411
0-10m EC1:5	-0.508	0.310	0.518	0.441	0.275	0.574	0.183	0.552
0-20m EC1:5	-0.642	0.377	0.703	0.694	0.350	0.774	0.220	0.758
0-30m EC1:5	-0.624	0.328	0.756	0.808	0.349	0.855	0.291	0.830
Borehole EM39 (mS/m)								
1-5m	-0.619	0.360	0.585	0.531	0.075	0.556	-0.128	0.646
1-10m	-0.715	0.543	0.728	0.625	0.154	0.718	-0.126	0.736
1-15m	-0.805	0.594	0.864	0.823	0.223	0.856	-0.048	0.890
1-20m	-0.777	0.551	0.894	0.908	0.198	0.916	-0.001	0.952
1-25m	-0.772	0.538	0.895	0.928	0.202	0.924	0.006	0.966
1-30m	-0.845	0.497	0.886	0.956	0.306	0.938	-0.005	0.968
1m-bedrock	-0.769	0.685	0.918	0.905	0.272	0.903	-0.016	0.936
1m-end of data	-0.769	0.520	0.890	0.947	0.293	0.941	-0.002	0.955
5-10m	-0.741	0.664	0.795	0.653	0.206	0.801	-0.112	0.755
5-15m	-0.810	0.649	0.909	0.860	0.262	0.912	-0.008	0.912
5-20m	-0.748	0.567	0.899	0.900	0.205	0.932	0.030	0.943
5-25m	-0.750	0.546	0.894	0.918	0.204	0.931	0.027	0.955
5-30m	-0.820	0.492	0.878	0.947	0.315	0.942	0.013	0.955
5m-bedrock	-0.640	0.703	0.912	0.874	-0.053	0.911	-0.026	0.914
5m-end of data	-0.751	0.526	0.887	0.936	0.293	0.944	0.006	0.944
Unsaturated zone	-0.629	0.848	0.883	0.715	0.065	0.930	0.109	0.824
Saturated zone	-0.624	0.520	0.878	0.939	-0.107	0.819	-0.150	0.931
Sediments	-0.505	0.370	0.696	0.699	-0.201	0.630	-0.138	0.746
Saprolite	-0.667	0.673	0.919	0.914	-0.089	0.892	-0.037	0.943
Ground EM								
EM31v (mS/m)	-0.807	0.764	0.817	0.551	0.150	0.750	-0.107	0.802
EM31h (mS/m)	-0.773	0.757	0.844	0.640	0.213	0.805	-0.118	0.826

AC2_3 = Apparent conductance Channel 2_3, AC4_5 = Apparent conductance Channel 4_5 AC7_8 = Apparent conductance Channel 7_8 L1c = Layer 1 conductivity, L2c = Layer 2 conductivity L1 C = Layer 1 Conductance, L2 C = Layer 2 Conductance

Table 2: Variance explained by AEM data to account for salt store (EM39, EM31) in the regolith

Response	Variance accounted	Level of significance of regression	Significant effects (in order of variance accounted)
EM39	%		
1-5(mS/m)	40.7	0.001	Terrain
1-10m	49.6	< 0.001	Terrain
1-15m	79.3	< 0.001	L2c, terrain
1-20m	92.6	< 0.001	L2c, AC45, AC23, L1c
1-25m	95.1	< 0.001	L2c, AC23, L1c
1m-bedrock	89.2	< 0.001	L2c, AC23, L1c
1m-end of data	92.7	< 0.001	L2c, AC23, L1c
2-5m (mS/)	36.4	0.001	Terrain
2-10m	49.9	< 0.001	Terrain
2-15m	76.7	< 0.001	AC45
2-20m	93.1	< 0.001	L2c, AC45, AC23, L1c
2-25m	95.3	< 0.001	L2c, AC45, AC23, L1c
2m-end of data	92.8	< 0.001	L2c, AC45, AC2 L1c
5-10m	61.8	< 0.001	L2c
5-15m	85.5	< 0.001	AC45
5-20m	94.4	< 0.001	L2c, AC45, AC23, L1c
5-25m	95.1	< 0.001	L2c, AC45, AC23, L1c
5m-end of data	91.5	< 0.001	L2c, AC45, AC23, L1c
EM31 (mS/m)	56.7	< 0.001	L2c
EM31h (mS/m)	67.9	< 0.001	AC45

L1c = Layer 1 conductivity,
L2c = Layer 2 conductivity
AC23 = Apparent conductance Channel 2_3,
AC45 = Apparent conductance Channel 4_5.

defined, conceptual models of landscape processes. In addition, interpretation of AEM and radiometrics requires field validation. Drilling and logging of these datasets are viewed as essential adjuncts for reliable interpretation of data in different geological settings. At present, while the interpretive products derived from airborne systems are of a reasonable standard, their widespread use would be unwise without assistance from local and qualified interpreters with a better knowledge of the landscape's physical processes (George *et al.* 1999). Collaborative work between the skilled geoscientists, farmers and service providers would help improve these interpretations.

Farmers preferred experienced practitioners to interpret all data. However, an initial evaluation suggests that most farmers could use the radiometrics maps (especially the ternary image; Plate 2) 'as is' to improve paddock management, and that they could try to use magnetics to help define salinity risk. AEM was also seen as an aid in defining higher risk areas when areas of high salt store were related to the locations of dykes. All accepted that colour maps of saltstore were only associated with risk if groundwaters were within, or were likely to rise into, the root-zone. Finally, farmers could see applications of AEM in better defining drill targets, by selecting areas of lower groundwater salinity (low AEM) and higher yield (magnetics - fractures and faults). However, without experienced advisers developing methods to utilise the data, or establishing practical applications from the data, the value of airborne geophysics would be diminished.

Datasets developed from airborne and other remote sensing platforms provide input data for biophysical models. Important input parameters to be added include soils (digital data from radiometrics), regolith materials and thickness (from AEM), groundwater flow system dimensions (magnetics and AEM), and groundwater salinity (from AEM). For example, radiometric data can be used as inputs to groundwater or decision support models as the soil is represented as a continuum of pixels, containing sample points taken at millions of sites, rather than from relatively selected sampling points. To compile the 1:100 000 soils map of the Toolibin catchment, 50 shallow auger holes were dug over a period of two weeks to supplement airphoto interpretation of the 46 000 ha catchment. By contrast, the radiometrics image sampled approximately 0.5 million points in less than two days. Combining the two systems creates greater accuracy and value.

Another management implication of collating spatial information is the compilation of these data on a CD-Rom, with a 'user-friendly' viewing package bringing all 12 layers of biophysical systems together for farmers, for the first time. However, digital datasets delivered to farmers without practical applications being developed, or training, were of little use or value. Other applications from this advance in technology are developing quickly as computers are more often being used by agriculture. For example, geophysical datasets are now being used as a part of precision farming systems, which when coupled with economic analyses, can aid the effectiveness of fertiliser and herbicide applications, provide maps of groundwater recharge, or simply act as a spatially accurate method of recording farm business details.

Preliminary economic analysis (George 1999) suggests that the Toolibin geophysics provided information which *may* justify the cost of collection and interpretation (A$7.00 (US$4.00) per ha). However, the analysis suggests that farmers and/or the community require a combined return on investment of between A$0.66 and $1.66 (US$0.37-0.94) ha yr^{-1} (for 20 years), to make acquisition profitable. At current costs of implementing salinity abatement programs (A$200-1000 (US$114-570) ha^{-1}), these costs appear small if the data are effective in improving decision-making. Recent reductions in acquisition prices (A\$3 (US\$1.70) per ha) may further reduce this cost.

5. Discussion

The development of catchment management strategies to combat salinity depend on the assessment of risk, the availability of management systems (with adequate information to guide their implementation), and the existence of an operating environment where the strategies can be established economically and in a socially acceptable way. High resolution spatial information forms the basis for the decision-making platform, providing information from which individuals and groups can assess risk and plan actions. Information on depth to watertable, rates of watertable rise, locations and extents of likely and existing discharge areas, all form a part of the requisite information.

At present, the available soil and geologic information is usually collated at a scale of 1:250 000, while in most catchments knowledge of salt store and regolith properties can only be obtained by referring to distant experimental sites or empirical relationships developed by previous researchers. In addition, hydrogeological data are scant. In most catchments there are few bores, and only some have any temporal data (Nulsen 1998). In the entire wheatbelt of Western Australia (which extends over some 20 million ha), only 4000 bores have time series data (rates of change), and most of these have been drilled relatively recently (< 5 years old).

While geophysical data cannot directly determine the important factors related to hydrology such as depth to the watertable and rate of rise, it can significantly improve the ability of interpreters to identify risk areas, likely recharge zones, groundwater system characteristics and areas of high and low regolith salinity. Specifically, magnetics surveys on the Archaean basement provide a map of the 'plumbing system', radiometrics the soil systems and AEM the regolith characteristics and salt store. Together in a geographic information system, the 12 biophysical datasets provide nearly all that is required to develop local-scaled conceptual models, quantify hydrologic conditions and together with a modelling tool, assess the likely impact of treatment systems.

The current lack of available geophysical data and the interpretations they allow, hampers the development of risk maps and the compilation of management plans. Plans are based on poorly-derived biophysical data, especially in cases where inadequately trained and equipped landscape interpreters are operating. For example, in many cases revegetation and engineering systems are being established without knowledge of their site requirements and likely consequences.

Cost of acquisition is the primary impediment to widespread surveying, even though recent developments have meant that geophysical data may be obtainable for a relatively low cost. We have recently seen that complementary data acquisition programs have already acquired high resolution DEMS, orthophotos and current salinity hazard maps for less than A\$0.30 (US\$0.17) ha^{-1} when delivered for the entire 20 M ha 'wheatbelt'. Estimates of prices for magnetics and radiometrics acquisition are of a similar magnitude, while the addition of AEM increases the price by a factor of almost 10. We believe that radiometrics and magnetics, together with other rapidly developing multi-spectral systems, offer the first step towards answering questions posed by the community. In this context, better biophysical data lead to both practical and cost-effective farm and catchment plans. These datasets also form part of wider opportunities for improved mineral exploration, rural and regional development and other local business opportunities. In time, AEM systems can be added to the biophysical base as applications are better developed.

6. Acknowledgments

This project was completed as a part of a wide-ranging research program on the application of airborne geophysics to farm and catchment management. At Towerrinning, much of this was part of an Australian Research Council collaborative grant with Murdoch University (C. Clarke, R. Bell), Fugro and other partners (Don Cochrane). At Toolibin, it was based largely on work undertaken within the National Airborne Geophysics Project (George 1999 and George *et al.* 1999), using data gathered by partners managed by the WA Airborne Geophysics Technical Advisory Group. At both sites the community was closely involved in the management and operation of the research program.

7. References

De Broekert, P. (1996) *An Assessment of Airborne Electromagnetics for Hydrogeological Interpretation in the Wheatbelt, Western Australia*, Division of Resource Management Technical Report No. 151, Agriculture WA, Perth.

Ferdowsian., R, George, R., Lewis, R., McFarlane, D., Short, R. and Speed, R. (1996) The extent of dryland salinity in Western Australia, in *Proc. 4th National Workshop on the Productive Use and Rehabilitation of Saline Lands*, Albany, March 1996, pp. 89-88.

George, R.J. (1999) *Evaluation of Airborne Geophysics for Catchment Management, Toolibin Report*, Agriculture Fisheries and Forestry Australia, Canberra.

George, R.J., Beasley, R., Gordon, I., Heislers, D., Speed, R.J., Brodie, R., McConnell, C. and Woodgate, P. (1999) *Evaluation of Airborne Geophysics for Catchment Management. National Report*, Agriculture Fisheries and Forestry Australia, Canberra.

George, R.J., Bennett, D.L., Wallace, K., and Cochrane, D. (1996) Hydrologic systems to manage salinity at Toolibin and Towerrinning Lakes, in *Proc. 4th National Workshop on the Productive Use and Rehabilitation of Saline Lands*, Albany, March 1996, pp 173-181.

George, R.J., McFarlane, D.J. and Nulsen, R.A. (1997) Salinity threatens the viability of agriculture and ecosystems in Western Australia, *Hydrogeology Journal* 5, 6-21.

Nulsen, R.A., Beeston, G. , Smith, R. and Street, G. (1996) Delivering a technically sound basis for catchment and farm planning, in *Proc. WALIS '96 Forum*, Perth, pp. 66-71.

Nulsen, R.A. (ed.) (1998) *Groundwater Trends in the Agricultural Area of Western Australia*, Division of Resource Management Technical Report 173, Agriculture WA, Perth.

Pracilio, G., Street, G.J., Nallan Chakravartula, Angeloni, J., Sattel, D., Owers, M. and Lane, R. (1998) *National Dryland Salinity Program, Airborne geophysics surveys to assist planning for salinity control; 3. Lake Toolibin Catchment*, Fugro Airborne Geoscience, Perth.

Raper, G. P., Guppy, L.M. and Argent, R.M. (1999) Innovative use of simulation models in farm and catchment in WA, this volume.

Sattel, D. (1998) Conductivity information in three dimensions, in *AEM'98, International Conference on Airborne Electromagnetics*, Sydney, Australia, February 1998, pp. 1-4.

AUTHORS

RICHARD J. GEORGE AND DONALD L. BENNETT

Catchment Hydrology, Agriculture Western Australia, PO Box 1231, Bunbury, WA 6231, Australia.

rgeorge@agric.wa.gov.au

CHAPTER 20

INNOVATIVE USE OF WATER BALANCE MODELS IN FARM AND
CATCHMENT PLANNING IN WESTERN AUSTRALIA

G.P. RAPER, L.M. GUPPY, R.M. ARGENTand R.J. GEORGE

1. Abstract

Soil salinisation in the agricultural regions of southern Australia is caused by the
replacement of the deep-rooted, perennial native vegetation with shallow-rooted annual
species. Its remediation requires significant changes to the water balance. The land
managers with the highest potential to influence the local water balance are farmers,
acting in catchment groups. In order to make changes to the water balance of a
sufficient magnitude to reverse or limit salinisation, farmers require an understanding of
the local effects of the treatment options available to them and the likely landscape
effects of applying those treatments over whole catchments. Agriculture Western
Australia is using two water balance models to assist farmer groups to understand these
effects. AgET is a simple, one-dimensional model that predicts the effects of soil and
vegetation type on the water balance. The groundwater model, MODFLOW, is used to
predict the landscape-scale effects of changes in vegetation type and other water
management practices. Using these models in an integrated way has many benefits: (1)
farmers' knowledge is included in the model calibration process; (2) farmers value the
opportunity to participate in the process and learn from it, and (3) water management
practices likely to succeed at the catchment scale are tested by the model simulations.

2. Introduction

The development of agriculture in Western Australia has resulted in the replacement of
deep-rooted, perennial native vegetation with shallow-rooted annual crops and pastures.
Annual agricultural species usually transpire less water than native vegetation, because
they do not grow during summer and their shallow rooting habit restricts access to soil
water. As less of the rainfall is being used by vegetation, recharge to groundwater in the
agricultural regions of Western Australia has increased.

Additional groundwater recharge has led to a trend of rising groundwater tables. As a
consequence, soluble salts accumulated in the sub-soils over geological time scales have
been mobilised by the rising water tables. This has increased the area of naturally saline
land and caused secondary salinisation of otherwise productive soil. By 1994, 1.8
million hectares, about 10% of privately owned cleared land, was salt affected and this
may expand to over 6 million hectares before a new hydrological equilibrium is reached
(George *et al.* 1997).

Between 1920 and 1980 farmers were encouraged to treat the symptoms, rather than
the causes of salinisation by planting salt tolerant plants on affected areas (Conacher *et
al.* 1983). Since that time, farmers and other land managers have been encouraged to

319

A.J. Conacher (ed.), Land Degradation, 319–331.

change landuse practices in order to reduce groundwater recharge and enhance discharge to control rising water tables. These changes include manipulation of traditional agricultural rotations, the inclusion of deep-rooted, perennial fodder species and agriforestry (George *et al.* 1997). The effects of changes to the paddock (field) and subcatchment scale water balance, however, are at present experimentally quantified for only a limited number of soil types by species combinations (Smith et al. 1998). The effects of tree planting on groundwater levels are known to depend on the scale of planting and the local hydrogeological conditions (George *et al.* 1999). They show that relatively small plantings of trees can ameliorate soil salinisation when planted on local aquifers or where they can intercept the full saturated thickness of the aquifer. Conversely, tree planting on intermediate or regional scale aquifers often leads to lower water tables only within 10 to 30 m of the belt or block of trees. Furthermore, no experimental observations of catchment scale changes to vegetation type, other than large-scale tree planting (Schofield *et al.* 1989), are currently available, nor are they likely to be in the near future. The urgent need to make significant changes to agricultural land management practices is therefore not matched by broadly applicable empirical information on the impact of land management change on groundwater levels.

Agriculture Western Australia (AgWest), in collaboration with the Blackwood Basin Group, is currently engaged in a process of providing catchment-scale advice on groundwater management options to catchment groups throughout the upper and middle reaches of the Blackwood Basin (Fig. 1). The Blackwood Basin covers 2.8 million hectares in total and encompasses 143 farmer-based catchment groups (Grein 1995). The process in non-prescriptive; that is, groundwater management strategies proposed by the catchment groups are assessed for their likely impact on future salinity risk rather than some ideal strategy proposed. It is intended that the process will also be iterative. The aim of this process to allow the catchment groups to develop groundwater management strategies which are compatible with their farming practices and which they are therefore most likely to implement.

The U.S. Geological Survey groundwater model, MODFLOW (McDonald and Harbaugh 1988), is being used to simulate groundwater response to changes in land management. MODFLOW, however, requires recharge as model input. It is common practice to assign a small proportion of annual rainfall to recharge (10 to 15%) and adjust the proportion during calibration. Alternatively, simple, one-dimensional models are sometimes used to estimate recharge under different land management for spatially distributed groundwater models, such as MODFLOW. In this case, AgET, which was designed both as a research and an education tool, is used.

This paper describes work in progress and is primarily concerned with the application of AgET. AgET is used to: (1) extract useful water balance information from farmers to assist in the modelling process; (2) provide farmers with an opportunity to learn about the processes which control the rates of groundwater recharge in their catchment, and (3) demonstrate the level of changes to land management practices likely to be required to halt or limit the spread of soil salinisation. The AgET model will be described and then a case study presented to show how AgET is used to provide recharge estimates for MODFLOW. Advantages and limitations of this modelling process will also be discussed.

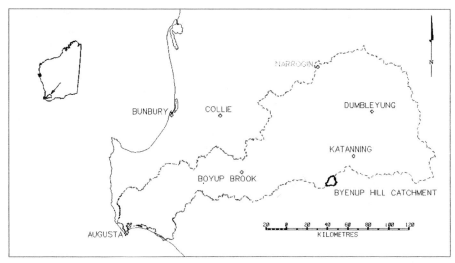

Figure 1. Location map showing the Blackwood River Basin and the Byenup Hill catchment

3. AgET Overview

WAttle was developed in 1996 as an educational and extension tool to allow quantitative comparisons of deep infiltration under different agricultural rotations and potential revegetation species (Argent and George 1997). Since 1997 the program has undergone a name change to AgET, due to a clash with the W.A. Department of Conservation and Land Management's database of *Acacia* species (wattles). It has also undergone some functional and cosmetic changes.

AgET is described in detail by Argent and George (1997) and recent model changes have been outlined by Raper *et al.* (1999). Therefore, only the main features of the program will be described here. The program has a Windows™ interface and comes with a Windows on-line help file. The innovative step in its development was in using default soil and crop data made easily accessible to the non-expert user with data selection screens. A second level of data editing is available for more accomplished users on separate screens. The model performs a 40 year simulation period in only a few seconds on a modern, low-end laptop computer (Pentium 75). The quick execution time allows AgET to be used interactively with farmers or groups at the kitchen table or in the shearing shed.

AgET is a simple cascading bucket model, with three layers. The three layers represent a duplex soil profile overlying a deeply weathered, clay-rich regolith, which is typical of the agricultural regions of Western Australia (Tennant *et al.* 1992). The model structure is similar to that used in the APSIM (McCown *et al.* 1996), PERFECT (Littleboy *et al.* 1989), and CERES models (Hodges *et al.* 1992).

3.1. PROGRAM SCREENS

AgET uses a map of the south-west of Western Australia divided into crop variety trial (CVT) regions (Fig. 2) to allow users to select their area of interest. The CVT regions are immediately recognisable by farmers and other potential users because the map appears in crop variety planting recommendations issued annually by AgWest. The

CVT regions are climate-based, and high quality, daily rainfall data are available from a number of sites within each region. Users select a region of interest then choose a rainfall station from a list. Daily rainfall data for the 40 years from 1 January 1954 to 31 December 1993 for the selected site are then loaded. Average daily values of pan evaporation from one station in each CVT region are also available.

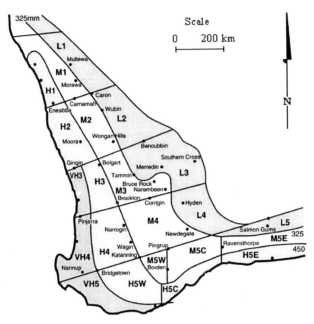

Figure 2. Crop Variety Trial (CVT) regions for Western Australia. The regions are derived from rainfall zones: VH, very high, >750 mm mean annual rainfall; H, high, 450-750 mm; M, medium, 325-450 mm; L, low, <325 mm and geographic zones: 1, north; 2, north central; 3, central; 4, south central; 5, south. Regions H5 and M5 are further subdivided into eastern (E), central (C) and western (W) zones. The Byenup Hill catchment falls in the H5W region.

The *Management Units* screen provides a pull-down menu list of soil groups for which default soil water storage and vertical flow parameters are stored. Up to eight management units can be selected and the parameters plus a schematic of the soil profile for each is shown on individual tabbed display screens (Fig. 3). The horizon thicknesses and initial water contents can be edited on the display screens. Further soil parameters, such as the saturation, drained upper (DUL) and lower (LOL) soil moisture contents, and effective daily hydraulic conductivities, are edited on a *Soil Detail* screen accessed from the *Management Units* screen. Soils data were obtained from AgWest (N. Schoknecht, pers. comm.), Moore (1998), Jensen et al. (1990), Maidment (1991) and Dingman (1993).

Following the *Management Units* screen is the *Crops* screen, which provides for the selection of two rotations for which water balances will be calculated and compared. Two pull-down lists of crops; one for the 'Current' rotation and one for the potential 'New' rotation, are shown. A rotation of up to eight years can be chosen for each.

Twenty-four crop or plant options are available for each year of each rotation, including bare soil, annual crops (summer and winter), annual and perennial pastures, and shrubs and trees. Crop factor and rooting depth parameters for each plant option can

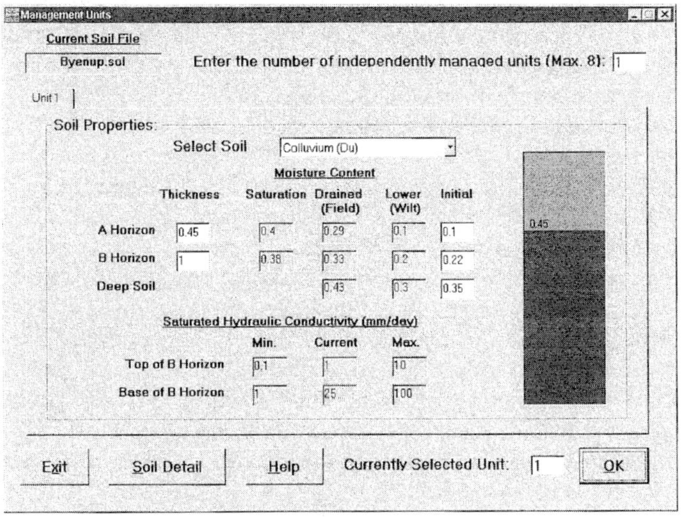

Figure 3. AgET main Management Unit screen

be individually edited in a separate screen accessed from the *Crops* screen. Crop and plant parameters were obtained from AgWest, Reid (1981), and Doorenbos and Pruitt (1984). Crop factors are marked with a colour coding according to the quality of the data used in their determination.

After selecting soil and crop parameters the user moves to the program *Run* screen where a simulation period is selected and the water balance for the 'Current' and 'New' rotations are performed. Monthly rainfall, runoff, evapotranspiration, and deep drainage are shown graphically on the screen as the simulations are performed. An option to graph the difference between 'New' and 'Current' rotations is included. The user can return to the *Management Units* or *Crops* screen or move to the *Report* screen from the *Run* screen.

The *Report* screen summaries the output data, including average and annual values of the water balance components. Average annual change in total profile water storage is also shown so that an assessment of the initial conditions can be made. This is particularly important for rotations including deep rooted species, because changes in storage may be large from year to year but the average value should approach zero. Twenty five, 50 and 75 percentile values for all water balance components are also reported. Daily, monthly and annual model output data can be exported to a text file for further analysis.

3.2. WATER BALANCE CALCULATION

AgET performs a simple water balance based on:

$$P = ET + R + D + \Delta S \tag{1}$$

where: P is daily precipitation (rainfall), ET is evapotranspiration, R is runoff, D is deep infiltration and ΔS is the change in soil storage.

Equation (1) is applied to the A horizon by firstly testing for rainfall in excess of the infiltration capacity; all excess rainfall is directed to runoff, and interception by foliage is ignored in all cases. The remainder of the daily rainfall is directed into the store. Maximum possible evapotranspiration is calculated as the product of daily pan evaporation, a pan factor (set to 0.8) and a daily crop factor. The pan factor is applied because the rate of evaporation from a Class A pan is only representative of the potential evaporation from a plant canopy (Brutsaert 1982). There is a linear reduction factor on evapotranspiration for soil moisture levels below 75% of saturation to account for the decreased availability of water to plants. Saturation excess runoff occurs whenever the horizon storage exceeds its maximum capacity. Deep drainage and evapotranspiration are also determined for the store, and then subtracted to give the new soil water storage. The calculations are based on the soil moisture available over the effective rooting depth of the crop in question or the layer thickness, whichever is the lesser. For example, if a crop has roots in the A horizon only, the balance is performed on the A horizon only, and any drainage from the A horizon goes to the deep infiltration component, irrespective of the presence or otherwise of a B Horizon.

Deep infiltration (D) from a soil store with a moisture content (ST) above the drained upper limit (DUL) is determined from:

$$D = Ksat*[(ST-DUL)/(SAT-DUL)]^2 \qquad (2)$$

where Ksat is the saturated hydraulic conductivity (mm.d^{-1}) and SAT is the saturation moisture content. This function sharply reduces D to values below Ksat once unsaturated conditions exist.

The simplifications made in the development of AgET restrict the physical conditions to which it can be legitimately applied and the level of confidence that should be placed on the accuracy of model predictions. Capillary rise or access to a water table is not considered, and evapotranspiration will be incorrectly estimated for sites where groundwater is within 2 m of the ground surface. Whether an under or over estimation of evapotranspiration results for sites with shallow water tables will depend on the salinity of the groundwater and the plant species in question. Similarly, plants are considered unaffected by waterlogging, and evapotranspiration will be overestimated for waterlogged soils. Furthermore, there is no run-on within the A and B horizons because water movement in AgET is vertical only, therefore the calculated water balance for soils in discharge zones or which are inundated is further compromised.

Root length is constant throughout the growing season and from year to year and the same crop factors are used from one year to the next. The effects of late or early breaks to the growing season or a lack of finishing rains, therefore, will not be simulated well. Using the same crop factors from year to year has minimal effect on the predicted water balance for years without finishing rains, because low water availability prevents unrealistically high predicted evapotranspiration. The consequences of these simplifications are more likely to be significant when there is an early break to the growing season. The associated errors are likely to be of the same order of magnitude as other inaccuracies resulting from the simple model structure. Furthermore, perennial vegetation is considered instantaneously mature; trees do not grow and cannot be used in rotations.

In practice, AgET is calibrated either against observed rates of groundwater rise under the chosen management unit and/or rotation combination, or against the user's expert judgment. Deep infiltration and runoff rates predicted for pre-clearing vegetation are commonly used where no site data are available. Long-term mean deep infiltration under pre-clearing vegetation is a convenient calibration figure, because catchment averaged groundwater recharge prior to clearing is known to be only a fraction of one percent of annual rainfall over most of the agricultural region of Western Australia (McFarlane *et al.* 1993)

4. Case Study Catchment

The Byenup Hill Catchment is situated in the Shire of Kojonup, about 270 km south-east of Perth, Western Australia (Fig. 1). It covers an area of approximately 7500 ha. The terrain is relatively undulating, and major drainage lines in the catchment have incised valleys. The stream flows to the Beaufort River and eventually into the Blackwood River. The orientations of many of the ephemeral, second and third order streams are influenced by geological structures including dykes and faults. The stream waters are moderately saline (1000 to 2000 mS.m⁻¹).

Byenup Hill experiences cool, wet winters and hot, dry summers. Average annual rainfall in the study area is 495 mm. Most of the rain falls in winter, though thunderstorms may provide localised rainfall during the summer. Figure 4 shows the annual rainfall and long-term trend for Cranham farm, which is an official Bureau of Meteorology rainfall station.

The area around Byenup Hill was first settled around 1890, at which time land clearing commenced and continued at varying rates until the 1980s. The remaining natural vegetation is mainly confined to rocky hilltops and upper slopes. Some remnants occur along major drainage lines. Current landuse at Byenup Hill is dominated by wool production, though the area under crop has increased in recent years.

Areas of soil salinity within the catchment have been mapped by the landholders and are almost exclusively situated in the valley floors. This suggests that the groundwater flow directions roughly follow the surface topography. Within the valleys, however, there are areas of more extensive salinity where mafic dykes and other geological features cross the natural surface drainage causing groundwater build up (Lewis 1991). There are isolated saline seeps on hillsides in the catchment, in particular on Byenup Hill itself. Hillside seeps are also the result of geological structures, mostly mafic dykes.

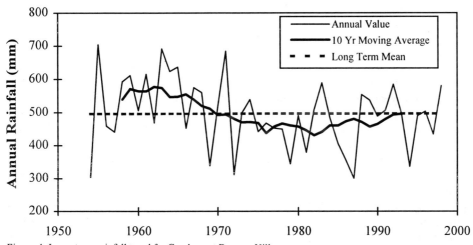

Figure 4. Long-term rainfall trend for Cranham at Byenup Hill

Twelve piezometers were drilled at Byenup Hill by AgWest in 1997. All holes were drilled to bedrock, which ranged from 8.1 to 22.7 m below ground surface. The Water and Rivers Commission drilled 12 cored boreholes in the immediate vicinity of the catchment during October 1996, and a further six in 1997. World Geoscience Corporation drilled four holes in 1995.

Regolith mapping at Byenup Hill was carried out from field reconnaissance, interpretation of bore logs and spoil, air-photo interpretation and remotely sensed radiometrics interpreted by CSIRO. Care was taken to select mapping units that were consistent with soils and land management units mapped by the landholders during an earlier farm planning exercise (Hardy 1993). Regolith units and their interpreted depths are shown in Table 1 and their spatial distributions are shown in Figure 5. Remnant vegetation at Byenup Hill was mapped from the landholders' farm plans and a black and white rectified air-photo mosaic.

4.1. AgET CALIBRATION AND APPLICATION

Several visits were made to the catchment during which the AgET program was demonstrated to the farmers. Once the physical meaning of the model parameters were explained in term they could relate to their experience, most could readily volunteer information that contributed to calibration of the model. Most time with the catchment group members was spent achieving acceptable calibrations for the colluvium unit (Du in Table 1), which the farmers commonly refer to as marri sandy duplex. The model was deemed to be adequately calibrated when deep infiltration under pre-clearing vegetation approached zero and deep infiltration under a one in three cereal-pasture rotation produced an average rate of groundwater rise consistent with locally observed values. No statistical calibration was possible. The calibrated model parameters are shown in Table 2. Note that AgET was not used to calculate recharge values for the valley floor units, because of the presence of a shallow water table under most of these areas. Consequently, no parameter values are shown in Table 2 for these units.

Recharge for each month during the period from June 1997 to December 1998 was then determined for each regolith unit using pre-clearing vegetation or the cereal-pasture rotation used for AgET calibration. No other rotations were used in MODFLOW calibration because the difference between rotations widely practised at Byenup Hill do not result in significantly different recharge estimates when compared to recharge under pre-clearing vegetation.

Figure 6 shows the spatially distributed recharge for June 1997 when the total rainfall at Byenup Hill was 86.4 mm. The spatial pattern of recharge is dominated by the distribution of the regolith units, as would be expected. Significant areas of very low recharge coincide with areas of remnant woodland vegetation, especially noticeable in the south-east of the catchment.

A satisfactory MODFLOW calibration has not yet been achieved. Predicted groundwater levels for a number of the piezometers in the catchment show similar responses to those observed. However, changes in groundwater levels elsewhere in the catchment change very rapidly during the initial time step. This is interpreted as a high degree of sensitivity to the initial groundwater head distribution, which is currently a poor estimate of true groundwater levels. This is in turn due to the relatively undulating terrain and high degree of complexity in the mapped regolith units. Several landholders also pump water from local fractured rock aquifers in the catchment and fully determining the spatial extent of these systems is beyond the resources of the current project.

Table 1. Regolith units mapped at Byenup Hill and representative depths to basement

Map Unit	Description	Depth (m)
	Valley Floor Units	
Cs	Channel Alluvium	7
Sa	Saline Channels	7
Fd	Floodplain Alluvium	9
	Granitoid Units	
Du	Colluvium	20
M	Micaceous Colluvium	18
Gr	Granitoid Eluvium	14
	Mafic Units	
Mf	Mafic Eluvium	19
Gb	Gabbro/Dolerite Outcrop	5
	Ferruginous Units	
Gv	Gravelly Eluvium	20
Fe	Nodular & Massive Ferricrete	20
Sl	Fe Slope Unit	21
Fcv	Fe Base Unit	8
	Low Permeability Units	
Lk	Lake or Swamp Alluvium	6
Ts	Tertiary Alluvial Sediments	8
Si	Silcrete	21
	High Permeability Units	
QV	Quartz Vein	5
Qtz	Quartz Eluvium	15.5
Ds	Deep Sand	7

5. Concluding Remarks

AgET provides recharge estimates of sufficient accuracy for input to a spatial groundwater flow model, such as MODFLOW. Rates of groundwater rise predicted by MODFLOW match some observed values well. There is evidence that, where an acceptable match is not currently achieved, factors other than recharge estimation are the cause.

Using AgET to estimate recharge for the catchment model has several advantages. First, explicitly including landholders in the calibration process makes use of valuable knowledge accumulated by farmers over several decades of observation. Second, the model is easily calibrated without collecting extensive soil hydraulic properties where none are available. Third, the calibration process is a valuable learning exercise for the landholder and most participated enthusiastically. Finally, there is the hope that the

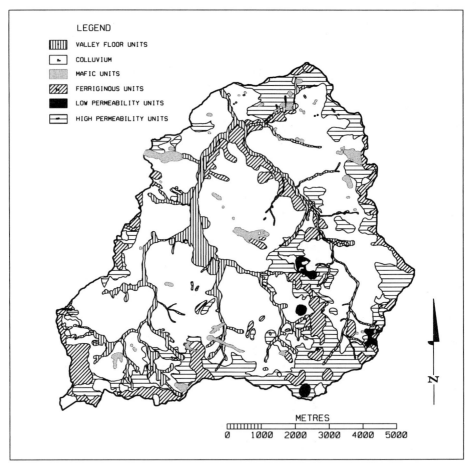

Figure 5: Spatial distribution of regolith units mapped at the Byenup Hill catchment, see Table 1 for regolith unit descriptions.

involvement of the farmers at this early stage will lend credibility to the catchment modelling exercise in the landholders eyes. Importantly, this should in turn, lead to early and widespread implementation of catchment management plans which will have a significant impact on the extent and severity of soil salinisation.

There are technical limitations in using AgET for this type of modelling exercise. Most importantly, AgET does not cope with sites that become waterlogged or where the groundwater is shallow. There are already significant areas of salt affected land in the catchment where water tables are shallow. Recharge rates predicted by AgET on these sites will invariably be over-estimates. Compared to the common practice of attributing a uniform proportion of annual rainfall to recharge, the use of even a simple model, such as AgET, is a significant improvement.

Table 2: Soil parameters required by AgET for regolith units mapped at Byenup Hill Catchment. See Table 1 for unit descriptions

Unit	A Horizon				B Horizon				Deep		Ksat	Ksat
	Depth (m)	Sat	DUL	LOL	Depth (m)	Sat	DUL	LOL	DUL	LOL	A/B	B/Deep
Granitoid Units												
Du	0.45	0.4	0.29	0.1	1.0	0.38	0.33	0.2	0.43	0.3	1.0	25
M	0.3	0.35	0.25	0.12	1.7	0.38	0.36	0.25	0.43	0.3	1.0	10
Gr	0.3	0.42	0.37	0.12	1.0	0.36	0.34	0.12	0.43	0.3	2.0	50
Mafic Units												
Mf	0.2	0.38	0.28	0.1	0.5	0.45	0.3	0.15	0.39	0.3	0.7	10
Gb	0.2	0.4	0.27	0.1	0.5	0.45	0.25	0.12	0.39	0.3	0.7	10
Ferruginous Units												
Gv	0.8	0.4	0.3	0.07	2.2	0.43	0.3	0.1	0.4	0.25	5.0	40
Fe	0.8	0.38	0.24	0.04	1.0	0.35	0.2	0.04	0.4	0.25	1.5	20
Sl	0.8	0.39	0.29	0.04	2.0	0.42	0.22	0.08	0.4	0.27	2.0	35
Fcv	0.8	0.39	0.29	0.06	1.2	0.44	0.23	0.08	0.4	0.27	1.5	30
Low Permeability Units												
Lk	0.5	0.35	0.25	0.21	1.0	0.4	0.38	0.28	0.43	0.3	0.1	5
Ts	0.3	0.34	0.28	0.1	0.7	0.38	0.34	0.21	0.43	0.3	1.0	10
Si	1.0	0.3	0.2	0.1	1.0	0.2	0.08	0.04	0.43	0.3	4.0	75
High Permeability Units												
QV	0.4	0.42	0.34	0.08	1.0	0.44	0.2	0.1	0.32	0.3	3.0	200
Qtz	0.8	0.42	0.32	0.1	1.8	0.37	0.34	0.1	0.43	0.33	2.0	35
Ds	1.0	0.38	0.22	0.08	2.0	0.4	0.32	0.1	0.4	0.28	4.0	40

6. Acknowledgments

Don McFarlane, AgWest, Albany, initiated the development of AgET. David Hall, David Tennant and Noel Schoknecht have provided expert advice on soil and crop parameters. Windows is a registered trademark of Microsoft Corporation. Most importantly, we thank the members of the Byenup Hill Catchment group for their involvement in this work.

7. References

Argent, R.M. and George, R.J. (1997) WAttle - A water balance calculator for dryland salinity management. MODSIM 97, International Congress on Modelling and Simulation, Hobart.

Brutsaert, W.H. (1982) *Evaporation into the Atmosphere: Theory, History, and Applications*, D. Reidel, Dordrecht.

Conacher, A.J., Combes, P.L., Smith, P.A. and McLellan, R.C. (1983) Evaluation of throughflow interceptors for controlling secondary soil and water salinity in dryland agricultural areas of southwestern Australia: I. Questionnaire surveys, *Applied Geography* 3, 29-44.

Dingman, S.L. (1993) *Physical Hydrology*, Prentice Hall, Englewood Cliffs.

Doorenbos, J. and Pruitt, W.O. (1984) Guidelines for Predicting Crop Water Requirements, FAO Irrigation and Drainage Paper 24, Rome.

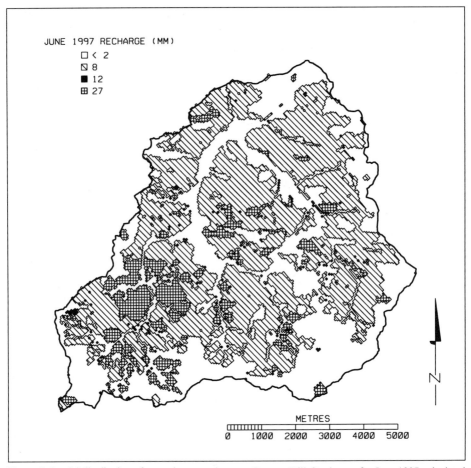

JUNE 1997 RECHARGE (MM)
□ < 2
◨ 8
■ 12
⊞ 27

METRES

0   1000   2000   3000   4000   5000

Figure 6. Spatial distribution of groundwater recharge at Byenup Hill Catchment for June 1997 calculated with AgET

George, R., McFarlane, D. and Nulsen, B. (1997) Salinity threatens the viability of agriculture and ecosystems in Western Australia, *Hydrogeoogy Journal* **5**, 6-21.

George, R.J., Nulsen, R.A., Ferdowsian, R. and Raper, G.P. (1999) Interactions between trees and groundwaters in recharge and discharge areas - a survey of Western Australian sites, *Agricultural Water Management* **39**, 91-113.

Grein, S. (1995) *Remnant Vegetation and Natural Resources of the Blackwood Catchment: an Atlas*, Agriculture Western Australia Misc. Publ. 9/95, Perth.

Hardy, J. (1993) *Byenup Hill Catchment Report*, Department of Agriculture Western Australia Misc. Pub. 22/93, Perth.

Hodges, T., Johnson, S.L. and Johnson, B. (1992) A modular structure for crop simulation models: implemented in the SIMPOTATO model, *Agronomy Journal* **84**, 911-915.

Jensen, M.E., Burman, R.D. and Allen, R.G. (eds.) (1990) *Evapotranspiration and Irrigation Water Requirements*, ASCE Manuals and Reports on Engineering Practice No. 70, ASCE, New York.

Lewis, M F (1991) Lineaments and salinity in Western Australia - carriers or barriers, *International Hydrology and Water Resources Symposium*, Perth, Western Australia· Institution of Engineers, Australia, pp. 202-209.

Littleboy, M., Silburn, D.M., Freebairn, D.M., Woodruff, D.R. and Hammer, G.L. (1989) *PERFECT: A Computer Simulation Model of Productivity Erosion Runoff Functions to Evaluate Conservation Techniques*, Department od Primary Industries, Queensland, Training Series QE93010, Brisbane.

Maidment, D.R. (1991) *Handbook of Hydrology*, McGraw-Hill, New York.

McCown, R.L., Hammer, G.L., Hargreaves, J.N.G., Holzworth, D. and Freebairn, D.M. (1996) APSIM: A novel software system for model development, model testing and simulation in agricultural systems research, *Agricultural Systems* **50**, 255-271.

McDonald, M.G. and Harbaugh, A.W. (1988) *A Modular Three-Dimensional Finite-Difference Ground-water Flow Model*, US Geological Survey, US Department of the Interior, Techniques of Water Resources Investigations Bk 6, Chap. A1, Washington DC.

McFarlane, D.J., George, R.J. and Farrington, P. (1993) Changes in the hydrologic cycle, In: R.J. Hobbs and D.A. Saunders (eds.), *Reintegrating Fragmented Landscapes: Towards Sustainable Production and Nature Conservation*, Springer-Verlag, New York, pp. 146-186.

Moore, G. (1998) *Soilguide: A Handbook for Understanding and Managing Agricultural Soils*, Agriculture Western Australia Bulletin 4343, Perth.

Raper, G.P., Argent, R.M. and George, R.J. (1999) AgET and Catcher - Complementary Water Balance Tools for Education and Farm and Catchment Planning, Paper presented at the MODSS'99 Conference, Brisbane, Qld, August 1999.

Reid, R. (1981) *A Manual of Australian Agriculture*, William Heinemann, Melbourne.

Schofield, N.J., Loh, I.C., Scott, P.R., Bartle, J.R., Ritson, P., Bell, R.W., Borg, H., Anson, B. and Moore, R. (1989) Vegetation strategies to reduce stream salinities of water resource catchments in south-west Western Australia, Water Authority of Western Australia Report WS 33, Perth.

Smith, A.D., George, R.J., Scott, P.R., Bennett, D.L., Rippon, R.J., Orr, G.J. and Tille, P.J. (1998) Results of investigations into the groundwater response and productivity of high water use agricultural systems 1990-1997: 6. Summary of all sites, Agriculture Western Australia Resources Management Technical Report 179, Perth.

Tennant, D., Scholz, G., Dixon, J. and Purdie, B. (1992) Physical and chemical characteristics of duplex soils and their distribution in the south-west of Western Australia, *Australian Journal of Experimental Agriculture* **32**, 827-843.

AUTHORS

G.P. RAPER [1], L.M. GUPPY [2], R.M. ARGENT [3] and R.J. GEORGE [4]

[1] Corresponding Author: Agriculture Western Australia, PO Box 1231, Bunbury, Western Australia, 6231, Australia. Email: Praper@agric.wa.gov.au

[2] Agriculture Western Australia, Katanning. Currently: Bureau Of Rural Sciences, PO Box E11 Kingston, Act 2604, Australia. Email: Lisa.guppy@brs.gov.au

[3] Centre For Environmental Applied Hydrology, The University Of Melbourne, Parkville, Victoria, 3052, Australia. Email: Argent@civag.unimelb.edu.au

[4] Agriculture Western Australia, PO Box 1231, Bunbury, Western Australia, 6231, Australia. Email: Rgeorge@agric.wa.gov.au

Part Five

Land Degradation and Policy

This final part comprises only three chapters, all from Australia.

The first two papers, by David Pannell, and Elizabeth Kington and Keith Smettem, are fairly specific in dealing with policy implications of and approaches to dealing with the major problem of secondary, dryland salinity in Western Australia. Nevertheless, their findings have potential application for other forms of land degradation and in other places. Pannell asks why, given the considerable amount of research which has been carried out over the past 30 years in particular, landholders have not adopted effective remedial measures. This question could well be asked of other landholders and managers confronting different forms of land degradation in many other regions. He searches for reasons for non-adoption from the empirical and theoretical literature dealing with adoption of innovations, decision making under uncertainty, the value of information, the economics of farm management, the theory of market failure, and transaction costs. Lack of awareness of the problem is not the major factor. Instead, he finds that the major impediments relate to the economic costs and benefits of current treatment options, the difficulties of trialling the options, long time scales, externalities, and social issues. Farmer reluctance to adopt the radical changes being recommended is therefore understandable. Unfortunately, policy measures and extension programs are doing little to change the underlying causes of non-adoption and are therefore largely ineffective.

This finding is reinforced by Kington and Smettem, who take a very different approach to the same problem. They examine a particular catchment in the south west of Western Australia and evaluate the effectiveness of three policy approaches to salinity management which have been attempted in the catchment sequentially since 1947: regulatory, co-operative and market driven. They also discuss methods for determining policy effectiveness. They consider that none of the policies has been particularly effective to date and that a more systematic policy approach (including social, environmental and economic aspects) is needed. They also ask whether policy strategies should now emphasise minimising the inevitable social and environmental impacts of salinity rather than focusing on the problem itself.

The final paper, by Jeanette and Arthur Conacher, extends the assessment of policies by reviewing Australia's numerous policy (and strategic) responses to the overall problem of land degradation. Similar approaches are being taken in many other countries as a result of increasing networking in our shrinking world: the Australian experience therefore has broad relevance. Nevertheless, Australia's federal system of government, with nine major jurisdictions (and a large number of local governments) makes this a complex task, particularly since the Australian Constitution gives the federal government few powers in relation to environmental and land use matters.

The authors summarise some of Australia's international obligations and the more important States' legislation, and discuss some of the areas of apparent, common agreement amongst the jurisdictions (notably the adoption of ecologically sustainable development principles). Specific, important policies and strategies have included the Decade of Landcare and the Natural Heritage Trust, and the development of integrated natural resource management, integrated catchment management, and regional planning. Constraints mitigating against the more effective development and implementation of policies designed to deal with land degradation are summarised.

Areas in which improvements are needed and (often) are being made are discussed. They include: development of nationally co-ordinated and improved data bases; an agreed set of national environment indicators and benchmarking processes; improved dissemination of knowledge to managers of natural resources and expansion of regulatory and economic measures; stronger environmental policy implementation, including mandatory approaches such as National Environment Protection Measures; encouragement of voluntary co-operation through education and incentives such as Environmental Management Systems, covenants and environmental awards; improved devolution of management to regional and local organisations; and strengthened public participation. Some models for improved natural resource and environmental management are also discussed. A considerably strengthened regional approach is recommended to achieve (some of) the above improvements and to accommodate the enormous problem of jurisdictional complexity.

CHAPTER 21

EXPLAINING NON-ADOPTION OF PRACTICES TO PREVENT DRYLAND
SALINITY IN WESTERN AUSTRALIA: IMPLICATIONS FOR POLICY

DAVID J. PANNELL

1. Abstract

In agricultural regions of Western Australia in the coming decades, dryland salinity will
result in the loss of millions of hectares of productive agricultural land, will severely affect
native vegetation and fauna, will continue to salinise almost all waterways and lakes, and
will cause great damage to roads, buildings and other infrastructure. Scientists believe that
to avert (or even to significantly reduce) this disaster, very large areas of current
agricultural land would need to be converted to perennial plant species, either trees or
perennial pastures. Although the farming community in Western Australia has become
much more aware of issues of natural resource conservation in the past two decades, its
response so far to the salinity problem has been on a scale which is orders of magnitude
smaller than recommended by scientists. This paper explores reasons for this, based on
empirical and theoretical literature concerned with adoption of innovations, decision
making under uncertainty, the value of information, the economics of farm management,
the theory of market failure, and transaction costs. Lack of awareness of salinity is probably
not a major factor explaining slow and low adoption of the recommended practices. Rather,
the major factors relate to the economic costs and benefits of current treatment options, the
difficulties of trialling the options, long time scales, externalities, and social issues. This
combination of factors means that the problem in many regions is extremely adverse to
rapid adoption, probably more so than for any other agricultural issue in Australia. In other
words, farmer reluctance to adopt the radical changes being recommended is completely
understandable and, indeed, reasonable from the farmers' perspectives. Current policy
measures and extension programs are doing little or nothing to change the underlying
causes of non-adoption. Measures which would begin to do so are urgently needed. The top
priority should be investment to develop new, profitable, perennial-plant-based agricultural
systems, but this area has, up to now, received relatively little funding.

2. Introduction

Dryland salinity is considered to be among the most serious environmental problems in
Australia (Anon. 1999; Murray-Darling Basin Ministerial Council 1999). Of the six
Australian states, Western Australia has by far the greatest area of land affected by dryland
salinity, with 1.8 m ha out of an estimated national total of 2.5 m ha in 1996 (Anon. 1996).
Figure 1 shows the region of crop-livestock farming in Western Australia where dryland
salinity is a growing problem.
  Within the current policy framework, prevention of a predicted further worsening in

335

A.J. Conacher (ed.), Land Degradation, 335–346.
© 2001 *Kluwer Academic Publishers. Printed in the Netherlands.*

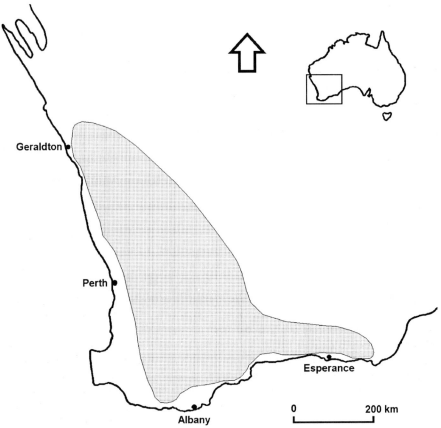

Figure 1. Crop and livestock production region in Western Australia where dryland salinity is a growing threat

salinity substantially depends on farmers voluntarily changing their farming practices away from a system based almost solely on annual plant species towards much greater use of perennial plant species. Even though rainfall for much of the region is low (ranging from 250 mm to 700 mm per annum), enough water evades capture and transpiration by annual crops and pastures to cause naturally saline ground waters to rise steadily in most of the region. A change to perennial species would reduce or eliminate this process of ground water rise.

Although this is well understood and widely discussed, the adoption by farmers of perennial species has been, in most districts, at a scale which is a small fraction of that recommended by hydrologists for prevention of salinity (Anon. 1996; Pannell *et al.* 1999). This is despite the devotion of substantial resources to attempts to promote farmer adoption, via extension, research, farmer group facilitation, and other similar means.

This paper reviews and discusses reasons for this low response by farmers. It draws on existing literature about adoption of innovations in agriculture, and evidence related specifically to dryland salinity in Western Australia. The aim is to identify reasons why the adoption problem is so much more acute for dryland salinity than for other farming issues.

3. Adoption of Agricultural Innovations

There is a wealth of empirical evidence on the factors which influence farmers' adoption of innovations in general (Feder and Umali 1993; Feder *et al.* 1985; Lindner 1987; Rogers 1995), and some evidence for land conservation practices in particular (Barr 1999; Cary and Wilkinson 1997; Pannell 1999a; Sinden and King 1990; Wilkinson and Cary 1992).

In general, it seems clear that most farmers are cautious in their adoption behaviour. They come to any innovation with scepticism, uncertainty, prejudices and preconceptions and with an existing farming system that may or may not be operating as they would wish, but is at least operating. Unless they are new to farming, they will have trialled other innovations in the past and concluded that at least some of them fell far short of the claims made for them. They will be particularly wary of a system which is radically different from that with which they are familiar and comfortable. They will almost certainly hold an attitude that the people advocating such a radical system do not understand the realities of farming, or at least of their farm.

In most situations, farmers will not commit to adoption of an innovation without successfully trialling it. If small-scale trials are not possible or not enlightening for some reason, the chances of widespread adoption are greatly diminished. Conducting a trial incurs costs of time, energy, finance and land which could be used productively for other purposes. Clearly, the larger the scale of the trial that is necessary, the larger is the cost of this information, and the less likely the farmer is to make the investment in trialling.

The outcome of a trial is, in most cases, an improved ability by the farmer to judge whether the innovation will advance his or her objectives. Lindner (1987), in a wide-ranging review of the adoption and diffusion literature, concluded that the objectives of individual farmers figure centrally in the adoption and diffusion process and that, "the final decision to adopt or reject is consistent with the producer's self interest" (p. 148). 'Self interest' in this context is considerably broader than merely 'profit'. It may, for example, include objectives related to risk, leisure and environmental protection. Nevertheless, profit is a particularly important element of 'self-interest'.

4. Low-Adoption of Salinity Management Practices

I discuss three categories of factors which have a pronounced negative impact on adoption of salinity management practices: factors relating to the potential to learn from trials of the practices; factors relating to the benefits and costs of the practices, and social factors. The discussion is based on trialling water-capturing technologies, such as perennial plants and drainage systems. Hydrologists believe that perennial plants (trees, shrubs and pastures) will need to replace a substantial proportion of existing annual crops and pastures if the final extent of saline land is to be reduced to any significant extent (Anon. 1996; George *et al.* 1999; Murray-Darling Basin Ministerial Council 1999). Drainage for management of surface waters will play an important complementary role but will be insufficient on its own.

4.1. DIFFICULTIES OF TRIALLING FOR SALINITY MANAGEMENT TECHNOLOGIES

For any farming innovation, a farmer's uncertainty about its performance is initially high. Off-farm information (for example, from scientists or extension agents) may help to reduce the uncertainty, but the key to reducing uncertainty is on-farm trialling, preferably on the farmer's own property (Abadi Ghadim 1999). For most innovations, such trialling is the normal course of action for farmers interested in evaluating a technology, and it leads to a relatively high quality decision about the technology. However, for dryland salinity, there are a number of reasons why trialling is more difficult and less helpful than for other management issues.

Given that the perennial plant species currently available to farmers are not profitable in their own right in most situations, the focus in this discussion is on trialling to determine the salinity-prevention benefits from perennials or from drainage. Some of the difficulties relate to both biological and engineering approaches to salinity management, some to perennial plants generally (both woody and herbaceous) and some specifically to woody perennials (trees and shrubs).

4.1.1. Observability is Low or Observations are Costly

For a trial to be worthwhile, the results of the trial must be observable. In the case of dryland salinity, observability can be a very substantial problem, especially if the practice being trialled is preventative rather than ameliorative. Making observations of the water table at all is difficult. It requires the installation and observation of piezometers, which are costly. Even when a piezometer reading is obtained, given the considerable complexity and heterogeneity of underground geological structures in agricultural regions of Western Australia, it can be difficult to know how representative the observation is.

There is also considerable difficulty in attributing any change in the water table to the practice which is being trialled. One difficulty is the absence of a suitable control against which the result can be compared. When trialling an innovation such as a new crop species, it is relatively easy to compare the crop's performance relative to other traditional enterprises in the same growing conditions, or when trialling an agronomic practice, results can easily be compared with and without the practice. However, for a perennial plant enterprise established in order to prevent rises in the groundwater table, such comparisons are all but impossible. The reasons for this include the following:

- In the case of trees, the time scale of growth is radically different to traditional annual crops and pastures, so comparisons of annual growth are of little relevance.
- The impacts related to salinity are not readily visible, as noted above.
- There is a time lag of uncertain length between water entering the soil and impacts on the groundwater level (R. Ferdowsian, Agriculture Western Australia, pers. comm. 1999).
- Salinity on different parts of the farm would potentially respond differently to the same type of perennial enterprise due to geological heterogeneity.
- There are a number of variable factors overlaid on the trial, and their impacts are confounded with any effect attributable to the trees. These additional factors include: (a) variability over time and space in climatic conditions; (b) variability over space in soils' physical structure, chemical composition, and topography; (c) variability over space and time in weeds, pests and diseases; and (d) variability over time in other

management practices, due, for example, to changing economic circumstances. Spatial variation is always an issue for interpretation of trials, but the combination of variations in space and time which affect observations of long-term trends in the groundwater table, makes them particularly difficult to interpret. At the very least, these issues increase the duration of the trial necessary to reach confident conclusions (even to decades). In the extreme, they may mean that a trial could never be conclusive. In either case, the prospective benefits of conducting a trial are reduced, potentially by so much that it is not worth conducting a trial.

It may be that the only available method for assessing the impact of an area of perennials on the water table would be to observe the deviation of the water table from a prior trend. The need to look for a deviation in an historical trend, rather than comparing two current treatments, would add to the delay before any conclusion could be reached confidently. This is because, in the absence of a control treatment, it is more difficult to determine whether any observed deviation is attributable to the new practice or to other factors, such as atypical rainfall. Uncertainty about the response lag length adds to the difficulty of interpreting any observed trend deviation.

4.1.2. Long Time Scales
Even if observability was high, and a control treatment was available for comparison, groundwater movements are slow, so it may be a very long time indeed before a farmer's uncertainty about the soundness of a water-capturing technology is sufficiently reduced to prompt widespread adoption. Generally, salinisation processes are slow relative to the time frames used for most management decision making. Obviously, the slower the salinisation process, the longer it will take to be convinced about differences in salinisation rates. Furthermore, slowness reduces the overall value of trialling and may lead to a judgment that the benefits of the trial do not outweigh the costs.

4.1.3. Externalities
In a survey of farmers in the upper Kent River catchment, Kington and Pannell (1999) found that 62% of farmers believed that their neighbours are contributing to their salinity problem. While the survey did not explore the proportion of the problem that was attributed to inter-farm flows, it appears that farmers are significantly over-rating the extent of the externality problem (R. Ferdowsian, Agriculture Western Australia, pers. comm. 1999). Indeed, Pannell *et al.* (1999) argue that farm-to-farm externalities have generally been over-emphasised as a cause of dryland salinity and of farmers' failure to prevent it. Reasons why externalities are not as critical as some have portrayed include both hydrological and socio-economic reasons:

- A substantial proportion of discharge occurs within the same farm as the recharge which causes it.
- Low transmissivity of soils means that treatments can be effective locally in some cases.
- Farm sizes continue to increase, so the occurrence of farm-farm externalities is steadily reducing.
- Time lags between establishment of perennials and impacts on water tables at some distance are very large, so discounting would dramatically reduce the external benefits relative to local benefits.

- Given the financial costs and benefits of current technologies, the optimal strategy in many situations is not to plant trees, even if externalities were fully internalised.

If farmers believe incorrectly that the rise in their water table is due to the management practices of their neighbours, their motivation to trial a water capturing technology is reduced. One reason for this is that the observed water table level in a region adjacent to a treated area may be considered to be attributable to a neighbour, rather than the treatment, so the relevance of the observed information to management decision making is reduced. Second, in an extreme case, there may be fears that the trial would become salinised due to a recalcitrant neighbour. A third reason for reduced motivation to trial is that the value of information from a trial is related to the profit advantage of the new practice. To the extent that the profit advantage is perceived to be reduced by externalities from a neighbour, the incentive to trial may be reduced.

4.1.4. Necessary Scale of Implementation
One element of the difficulty of trialling is the size of trial that is necessary. The larger this is, the less likely the farmer is to make the investment in trialling. In this regard, water-capturing technologies are a particular concern. Such technologies clearly require a minimum scale for their effects to be apparent and hydrological evidence indicates that the necessary scale for impacts to be apparent at any distance from the trial is very large relative to the usual size of trials (George *et al.* 1999). Indeed, it would appear that in at least some situations, little short of full adoption is necessary.

4.1.5. Quality of Implementation
For the information from a trial to have value for decision making, the trial needs to be indicative of the innovation's performance in the long run. If the technology used in the trial is implemented poorly, then the trial will clearly be less likely to meet this requirement. Poor implementation is more likely when the innovation is radically different from technologies with which the farmer is familiar, and this does appear to describe the situation when an annual crop/livestock farmer is trialling a tree-based enterprise, for example.

4.1.6. Resources Required for Trialling
I have noted already that where a large-scale trial is necessary, the cost of trialling is correspondingly larger, and trialling is therefore less attractive. A large trial consumes not only land but also labour and finance which could otherwise be used productively on the farm.

4.1.7. Risks of Trial Failure
It was noted earlier that farm-to-farm externalities may threaten the very survival of trees in a trial. Other threats include drought, diseases and pests, each of which would pose a threat of uncertain magnitude given the farmer's lack of familiarity with the tree enterprise. Trials of any innovation always face a risk of failure, but given the large scale over which a trial of a tree enterprise appears necessary to discern water table impacts, and the levels of other resources invested in such a trial, the potential losses from a trial failure are substantial. This provides further discouragement to a risk-averse farmer considering such a trial.

Finally, it should be noted that this dismal picture does not apply to all types of technologies relevant to salinity. In particular, items (a) to (d) would not apply to trials of plant species intended to make productive use of saline areas, such as salt bush. For these, the aspects of interest would be the direct plant productivity of the species and perhaps its value to stock, and these would be readily observable and measurable in a short time frame. The issues related to the difficulty of learning about impacts on groundwater would also be of less significance if the technologies were profitable in their own right, so that they might be adopted despite continuing uncertainty about their benefits for salinity prevention.

4.2. ECONOMIC COSTS AND BENEFITS OF CURRENT TREATMENT OPTIONS

The discussion thus far has focussed on perceptions, but the only way to create enduring positive perceptions about a practice is for the practice to be beneficial in fact. With large amounts of energy and resources devoted to persuasion, it may be possible to create temporarily an overly-optimistic perception of a system, but once farmers have personal experience with the system, they will certainly put more weight on this than on any amount of persuasion or exhortation. Thus, successful trials or successful adoption are necessary for favourable perceptions in the medium to long term.

4.2.1. Costs
One potential threat to the actual profitability of a new, complex system is that there are likely to be substantial costs in establishing and maintaining the new system. This is particularly true of systems involving trees. Even if labour and finance availability are not absolutely constrained, their high requirements are costs which must be at least offset by the benefits.

Second, trees, shrubs or perennial pastures occupy land which would otherwise have generated income from traditional crops or pastures, and this 'opportunity cost' must be added to the direct establishment costs and set against any benefits from the new practice. It is often not recognised that the opportunity costs of lost income involved in establishing trees are of a similar order to the substantial establishment costs, and for perennial pastures opportunity costs are by far the major cost.

Third, trees may compete directly with crops and pastures for light and nutrients.

Fourth, trees may physically obstruct other farm activities (for example, preventing easy movement of agricultural machinery, or constraining grazing by livestock in order to avoid damage to seedlings).

Fifth, the loss of flexibility involved in committing large areas of land to trees can be very costly. In Western Australia, a majority of agricultural profits is generated in a minority of very productive years. Farmers typically move to exploit these years by allocating more land to crops. If they are prevented from doing this because the land is occupied by trees, or because the available financial resources have been invested in tree establishment and so are unavailable for purchase of cropping inputs, the impact on farm viability could be fatal.

Whether or not they are fatal, these problems, and any others, must be set against the expected long- and short-term benefits of the new system for farmers to reach a realistic assessment of its value. For Western Australia, apart from high rainfall areas in southern regions (greater than 600 mm per year – approximately 10% of the shaded area in Figure 1), there are currently no tree enterprises that are clearly profitable to individual farmers even in the long run. There are prospects, particularly oil mallees (Cooper 1999) and *Pinus pinaster*, but substantial further development work is required to generate a range of profitable tree options suitable for all regions and soil types.

4.2.2. Long Time Scales and Discounting of Benefits

Tree-based systems are characterised by high up-front costs, and benefits which occur some time in the future. If farmers have to borrow money to pay the up-front costs, it is obvious that any direct comparison of the up-front costs with the eventual benefits will not be valid without allowing for the cost of interest. This is a simple version of the rationale which econo nis s use for discounting future benefits to make them comparable to current costs. Discounting is not about allowing for inflation; it is about allowing for the benefits that funds invested could have generated if invested instead in the best alternative use.

The impact of discounting on distant future benefits can be very large, making it difficult for individuals to justify investments with long-delayed paybacks. For example, suppose a farmer's discount rate is 10%. The farmer is considering a tree crop that is not harvested for 30 years (e.g. *Pinus pinaster*). Establishment and opportunity costs amount to the equivalent of A$2000 (US$1200) per hectare up front. To cover these costs, the value of harvested products at year 30 would have be at least A$35 000 (US$21 000) per hectare.

There is some controversy about the appropriate discount rate, with some arguing that high rates discriminate against future generations by discouraging investments with long pay-off times (Pearce and Turner 1990). This is a complex, philosophical argument about what is in the long-term public good. However, it relates only to public investments or public policies. For a private individual farmer, discounting is absolutely uncontroversial and will inevitably affect their investment decisions.

4.2.3. Externalities

It was noted earlier that the importance of externalities as a cause of dryland salinity has been greatly over-stated by some (e.g. by Hayes, 1997), but it is nevertheless true that externalities can mean that the benefit-cost trade-off considered by an individual farmer is different to the trade-off facing the community as a whole. Therefore the net benefits of adopting perennials as perceived by the farmer can be less than they would be from a community-wide perspective. This applies particularly to externalities from the farming sector to the natural resources or environmental resources valued highly by the broad community. For example, the Kent River is a potential source of potable water for residents of the south coast of Western Australia, but it faces a serious salinity problem arising from farms in its upper catchment. Most official reserves of native habitats in the agricultural region of Western Australia are under severe threat from rising saline water tables.

4.3. SOCIAL FACTORS

The National Landcare Program has been the primary policy instrument for reducing agricultural land degradation in Australia. At its core, it has been an attempt to harness social processes to promote adoption of changed practices. On the other hand, some other social processes do not act in favour of adoption of salinity prevention practices.

4.3.1. Concern About the Social Fabric

The scale of change implied by hydrologists' recommendations is so profound that it implies significant alterations to the social fabric of rural society. Farmers are understandably resistant to embracing changes which they see as threatening further declines in rural populations and concomitant declines in rural services. This is clearly evident in the south coast region of Western Australia, for which a profitable tree-based enterprise is available. Blue gums (*Eucalyptus globulus*) are now substantially more profitable than traditional agricultural enterprises on suitable soils with adequate rainfall (the only example of this type in the state). In some districts, many farmers have sold their farms to tree production companies. Many of the remaining farmers in these districts are fearful and resentful of the changes that are occurring, and are resistant to the planting of trees for this reason, despite their evident profitability.

4.3.2. *High Transaction Costs*

In some cases, externality problems could be solved by direct negotiation amongst farmers. However, there are costs of various types in any negotiation for which the stakes are high:

- costs of time taken in the negotiations;
- psychological costs to the participants;
- social costs to the participants, their families and possibly others;
- possibly costs from legal action or mediation, and
- costs of monitoring and enforcing any agreements reached.

Collectively these are transaction costs, where the transaction in this case is an agreement.

In some catchments, the process of negotiation is particularly difficult because there are multiple sources of the problem, or there is uncertainty about who is the source. In such a case, the transaction costs outlined above are likely to be substantially greater. High transaction costs mean that any externality problem is unlikely to be solved without outside government intervention.

4.3.3. *Fairness*

Because of the existence of substantial off-farm impacts of salinity, there are serious questions about who benefits from salinity prevention, and who should pay for it. Most of the off-farm benefits do not accrue to easily identifiable individuals or groups, but to the community as a whole. Therefore, it is very difficult for individual farmers to resolve concerns they have about the fairness of the allocation of benefits and costs. It is likely that such concerns will remain unresolved without specific attention from government, with the result that farmers choose not to adopt.

This has not been a complete list of the factors affecting adoption of innovations generally, but a review of factors that make adoption unusually difficult for salinity prevention practices. Other factors are outlined in the more general adoption literature cited earlier.

5. Policy Implications

The National Landcare Program (NLP) started with the premise that land degradation in agriculture could be solved by awareness-raising and education programs for farmers (Curtis and De Lacy 1997; Vanclay 1997). This paradigm has been the dominant force in Australia shaping policies for prevention of natural resource degradation in agriculture. The NLP approach has been very successful in raising awareness of resource conservation issues among farmers, and in some cases this awareness has led to changed practices.

However, for dryland salinity in Western Australia, the changes have been much too small to prevent the steady increase in area of saline land.

There are at least four major flaws embodied in the thinking which has led to the current structure of salinity policy in Australia.

5.1. FLAW 1

The key factor inhibiting change is farm-to-farm externalities, because these mean that farmers need to act in a coordinated way in order to manage salinity. In reality, even if it were possible to completely internalise all existing farm-to-farm salinity externalities, the impact on incentives faced by farmers would be sufficient only to prompt minimal changes in management (Pannell *et al.* 1999).

5.2. FLAW 2

Assuming that the externality problem can be overcome or dismissed, viable solutions exist which farmers could beneficially adopt if they chose to. The persistence of this view must be due either to negligence among policy makers in failing to conduct proper economic evaluations of the available practices, or willful self-deceit. It is remarkable the extent to which one still hears the view expressed that there must surely by now be sufficient information available, and that the only requirement is to communicate the information to farmers. In reality, the problem is not lack of information, but lack of options. There is sufficient information about the existing options to know that in most cases they are not sufficiently beneficial to individual farmers even in the long run to offset their direct and indirect costs.

5.3. FLAW 3

Even if economically viable solutions do not exist, farmers will voluntarily make the sacrifices involved in adopting changed management if we can raise environmental awareness among them and inculcate a stewardship ethic. Such a view fails to take account of the level of sacrifice that is implicitly being expected of farmers – it is very substantial indeed. Barr (1999) emphasises the inadequacies of relying on voluntarism and a stewardship ethic. He comments that, "there is a significant body of research that demonstrates that links between environmental beliefs and environmental behaviour are tenuous" (p. 134). He notes that the NLP involves only a minority of farmers (albeit a "substantial" minority), and that, "it is probably unrealistic to expect any voluntary policy to achieve any greater degree of penetration of the farming community than has been achieved by Landcare" (p. 135).

Perhaps even more importantly, the proposition that we should encourage farmers to adopt practices which are not in their own best interests raises serious ethical and moral questions. Even if it somehow were made practically effective in promoting adoption, the fairness of such a policy would require serious consideration.

5.4. FLAW 4

Group-based extension methods, based on joint decision making to create catchment plans and harnessing of peer pressure, are sufficient. While group-based approaches undoubtedly do have important advantages, they do not in themselves address the fundamental difficulties involved in learning about the on-farm performance of salinity management practices, as described earlier. They also do little if anything to alter the financial incentives for adoption faced by farmers. To the extent that benefits for salinity prevention are an important and necessary feature of a particular practice, its promotion to the farming community will require innovative new methods to facilitate farmer observation and learning from trials. This seems to be an area ripe for further research.

I will comment briefly on the emphasis given to the development of catchment plans, as many Landcare officers devote a substantial proportion of their time to this task. In my observation, and from discussion with Landcare officers, the elements that go into most catchment plans have not been properly evaluated in terms of their likely economic benefits. In this situation, it is not surprising that farmers' commitment to implementation of the plan is difficult to maintain once they are faced with the reality of the time and expense involved. This reality is faced outside the public glare and peer pressure of the group. It is naïve to expect that a mere plan, even if agreed to verbally by members of the farm group, will be implemented unless farmers believe that it is in their interests to do so. In relation to salinity, the difficulties for farmers in coming to believe that any plan could be in their interests have been presented at length earlier.

It is clear that by far the most important need from salinity policy is to alter the financial incentives for adoption of perennial production systems. Persuasion, education and extension will remain inadequate while the available options are so financially unattractive. A small proportion of public resources devoted to the salinity problem is now allocated to development of new, profitable perennial enterprises. The necessary work includes screening perennial species for productivity, processing ability and salable products; research on markets and required marketing infrastructure; and establishment of systems to finance the establishment of new industries. Given the critical importance of these activities, they have been, and continue to be, grossly under-funded. Such biological and industrial development work is not certain to succeed, but without it we seem certain to fail in our battle against salinity.

Apart from its increased attractiveness to farmers, a perennial species that is profitable in its own right has an additional important advantage: it can avoid some of the major difficulties of trialling which arise when the objective is focused on water table management. If the focus is on profitable production of perennials, impacts on the poorly observable groundwater table can be accepted as a beneficial side benefit whose magnitude does not need to be determined accurately before adoption can be justified.

6. Acknowledgments

The author is grateful to Sally Marsh, Ruhi Ferdowsian, Don McFarlane, Elizabeth Kington, Dan Carter, Steven Schilizzi, Nicole Glenn, Simone Blennerhassett, Bob Lindner, Ted Lefroy, John Pickard and an anonymous reviewer for helpful discussion, advice or comments. The funding support of the Grains Research and Development Corporation and the Rural Industries Research and Development Corporation is gratefully acknowledged.

7. References

Abadi Ghadim, A.K. (1999) *Risk, Uncertainty and Learning in Farmer Adoption of a Crop Innovation*, Unpublished PhD thesis, The University of Western Australia.
Anonymous (1996) *Salinity: A Situation Statement for Western Australia*, Government of Western Australia,

Perth.

Anonymous (1999) *Dryland Salinity and its Impact on Rural Industries and the Landscape*, Prime Minister's Science, Engineering and Innovation Council, Occasional Paper Number 1, Department of Industry, Science and Resources, Canberra.

Barr, N. (1999) Social aspects of rural natural resource management, in *Outlook 99*, Proceedings of the National Agricultural and resources Outlook Conference, Canberra, 17-18 March, Vol. 1, Commodity Markets and Resource Management, ABARE, Canberra, pp. 133-140.

Cary, J.W. and Wilkinson, R.L. (1997) Perceived profitability and farmers' conservation behaviour, *Journal of Agricultural Economics* **48**, 13-21.

Cooper, D. (1999) *An Economic Analysis of Oil Mallee Industries in the Wheatbelt of Western Australia*, Unpublished thesis, Master of Science in Natural Resource Management, Faculty of Agriculture, The University of Western Australia.

Curtis, A. and De Lacy, T. (1997) Examining the assumptions underlying Landcare, In: S. Lockie and F. Vanclay (eds.) *Critical Landcare*, Key Papers Series 5, Centre for Rural Social Research, Charles Sturt University, Wagga Wagga, pp. 185-199.

Feder, G., Just, R. and Zilberman, D. (1985) Adoption of agricultural innovations in developing countries: a survey, *Economic Development and Cultural Change* **33**, 255-298.

Feder, G. and Umali, D. (1993) The adoption of agricultural innovations: a review, *Technological Forecasting and Social Change* **43**, 215-239.

George, R.J., Nulsen, R.A., Ferdowsian, R. and Raper, G.P. (1999) Interactions between trees and groundwaters in recharge and discharge areas – a survey of Western Australian sites, *Agricultural Water Management* **39**, 91-113.

Hayes, G. (1997) *An Assessment of the National Dryland Salinity R,D&E Program*, LWRRDC Occasional Paper No. 16/97, Land and Water Resources Research and Development Corporation, Canberra.

Kington E. and Pannell, D.J. (1999) Dryland salinity in the upper Kent River catchment of Western Australia: Farmer perceptions and practices. Sustainability and Economics in Agriculture Working Paper 99/09, Agricultural and Resource Economics, University of Western Australia, http://www.general.uwa.edu.au/u/dpannell/dpap9909f.htm

Lindner, R.K. (1987) Adoption and diffusion of technology: an overview, In: B.R. Champ, E. Highley and J.V. Remenyi (eds.), *Technological Change in Postharvest Handling and Transportation of Grains in the Humid Tropics*, ACIAR Proceedings No. 19, pp. 144-151.

Murray-Darling Basin Ministerial Council (1999) *The Salinity Audit of the Murray-Darling Basin, a 100 Year Perspective*, 1999, Murray-Darling Basin Commission, Canberra.

Pannell, D.J. (1999a) Social and economic challenges in the development of complex farming systems, *Agroforestry Systems* **45**, 393-409.

Pannell, D.J., McFarlane, D.J. and Ferdowsian, R. (1999) Rethinking the externality issue for dryland salinity in Western Australia, Sustainability and Economics in Agriculture Working Paper 99/11, Agricultural and Resource Economics, University of Western Australia, http://www.general.uwa.edu.au/u/dpannell/dpap9911f.htm

Pearce, D. W. and Turner, R.K. (1990) *Economics of Natural Resources and the Environment*, Harvester Wheatsheaf, New York.

Rogers, E.M. (1995) *Diffusion of Innovations*, Free Press, New York.

Sinden, J.A. and King, D.A. (1990) Adoption of soil conservation measures in Manilla Shire, New South Wales, *Review of Marketing and Agricultural Economics* **58**, 179-192.

Vanclay, F. (1997) The social basis of environmental management in agriculture: a background for understanding Landcare, In: S. Lockie and F. Vanclay (eds.) *Critical Landcare*, Key Papers Series 5, Centre for Rural Social Research, Charles Sturt University, Wagga Wagga, pp. 9-27.

Wilkinson, R.L. and Cary, J.W. (1992) *Monitoring Landcare in Central Victoria*. School of Agriculture & Forestry, The University of Melbourne.

AUTHOR

DAVID J. PANNELL

Agricultural and Resource Economics, University of Western Australia, Nedlands WA 6907, Australia. David.Pannell@uwa.edu.au

CHAPTER 22

EVALUATION OF POLICY APPROACHES TO DRYLAND SALINITY MANAGEMENT IN THE KENT RIVER CATCHMENT

ELIZABETH A. KINGTON and KEITH R.J. SMETTEM

1. Abstract

To date, the Australian government has attempted to use various legislative and policy initiatives to manage the spread of dryland salinity and protect the natural environment. Despite these initiatives, the area of land affected by dryland salinity continues to increase and may now be difficult to control using existing management capabilities.

This paper evaluates three quite different approaches to dryland salinity management which have been attempted in the Kent River catchment, located in the southwest of Western Australia. The approaches are; regulatory, co-operative and market driven. The implementation and impact of each management approach are assessed within their respective historical, social and environmental contexts. The assessment reveals that existing regulatory and co-operative approaches have implementation problems and have not been effective in controlling dryland salinity at the catchment scale.

It is concluded that although current policy evaluation methods are somewhat underdeveloped, it is possible to combine a number of evaluation approaches in order to gain insight into the advantages and disadvantages of economic and behavioural incentives to manage salinity problems.

2. Introduction

The upper reaches of the Kent River catchment (Fig. 1) are significantly affected by dryland salinity and waterlogging problems, in common with many other catchments in the south west of Western Australia (WA).

Widespread land clearing since the 1960s has resulted in rising groundwater tables and mobilisation of salt stored in the regolith. The visible symptoms vary from obvious salt scalds of bare, unproductive land, to seasonally waterlogged areas where the effects of salinity can be more subtle, resulting in changes to pasture species composition, plant senescence, and lost productivity. Land clearing has caused saline and silted creek-lines, degraded river quality, destroyed the riparian habitat and led to the demise and local extinction of native ecosystems and species (Hobbs 1998).

The full extent of this degradation and its trends over time are poorly known. Outflow of water and salt from the entire Kent River catchment has been measured at the Styx Junction gauging station (No. S604053) and from the upper catchment at Rocky Gully gauging station (No. S604001) since 1956 (Fig. 1 shows gauging station locations).

347

A.J. Conacher (ed.), Land Degradation, 347–361.

Parts of this region have been described as brackish swampland since the first European explorations in the early 19th Century. Two of the first explorers of the southwest, Dr Wilson in 1829 and Captain Septimus Roe in 1835, both noted that the Kent River was fast flowing, deep and fresh and the surrounding country was mainly flat open grassland. However, watering holes were very often saline, making drinking water difficult to find (Wilson 1968). These accounts represent the oldest (and some of only a few) existing records of the Kent River catchment environment in its pristine condition, and identify natural expressions of salinity before land was settled and cleared. The relevance of this within a policy evaluation is that no matter how effective, salinity management will never remove all salt expressions.

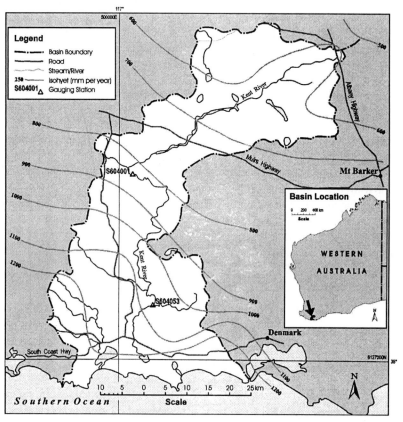

Figure 1. Location of the Kent River catchment in Australia. Diagram supplied courtesy of Water and Rivers Commission (2000).

This region of Western Australia was one of the first settled (Glover 1979), and the Kent river water remained potable until 1965 (Steering Committee 1989). By this time agriculture was yielding large harvests, propelled by generous bank loans, good rainfall, agricultural mechanisation and improved plant varieties (Crowley and De Garis 1969), encouraging large-scale land clearance.

The Kent River drainage basin covers 73 553 hectares and is characterised by a Mediterranean climate of temperate winters and hot summers. Average annual runoff is 80 905 ML and nearly 40% percent of the total flow comes from the upper catchment, where the land use is predominantly pastured for production of wool from merino sheep. Since 1961, 40% of the native vegetation in the lower catchment has been protected from clearing within a State Forest. The upper catchment was zoned for agricultural land and largely cleared, and there are now about 120 land-title holders, mostly farmers. The Kent River today is too saline to drink, salinity threatens up to 20% of farmland, and waterlogging problems currently affect 75% of the land area (Ferdowsian and Ryder 1997).

The salinity problem can be traced to the upper catchment by performing hydrograph and salt load separation on data from the river gauging stations. By deducting the upper catchment flow and salt load from the total catchment flow and salt load and then dividing the salt load by the flow to obtain the concentration, it can be shown that since the 1980s (at least), water running off or discharging from the lower catchment has been more or less potable (Fig. 2). This shows that policies aimed at reducing river salinity levels should focus on reducing the salinity input to the river from the predominantly cleared upper catchment.

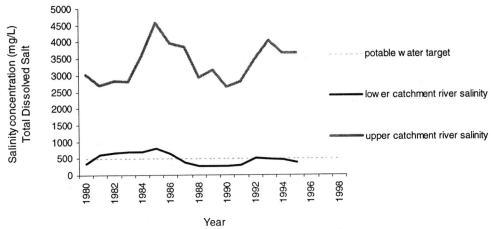

Figure 2. River salinity concentrations for the upper and lower catchments of the Kent river from 1979 to 1995. Data supplied courtesy of Water and Rivers Commission (2000).

This paper evaluates the development and implementation of dryland salinity management policies in the upper Kent River catchment. The evaluation forms part of the PhD study of the first author and seeks to ascertain if the policies have aided dryland salinity management and, if not, whether it is because the policies are flawed or because the environmental problem is intractable due to other reasons. It is useful to commence by determining the technical tractability of the problem.

3. Tractability of the Salinity Problem

Human-induced dryland salinity is the product of a physically altered landscape ecology that has unbalanced the hydrological cycle on a large scale. Land clearing allows additional water to reach the groundwater table, which then mobilises the salt in the soil, increasing the salinity levels of rivers. The effect on the land may not be expressed for many decades after native vegetation has been cleared and for regional aquifers, a new hydrological equilibrium may not be reached for up to 200 years (Government of Western Australia 1996), although in some cases, salinity expressions can occur far more rapidly (Conacher *et al.* 1983).

Dixon. *et al.* (1998) modelled the hydrology of the upper Kent River catchment. They concluded that in order to reduce discharge and salt loads from the upper catchment, 37% of cleared areas required additional revegetation with trees on the lower slopes. This gives a total of 67% of the catchment under trees. The focus on lower slope plantings is to remove water moving laterally by subsurface flow from upslope. If this proportion of lower slope planting is not achieved then it is suggested that a much larger percentage of the upper catchment might need to be revegetated to achieve similar reductions in salt loads entering the river. Within policy, there is still is no consensus on the technical processes and priorities for management.

The modelling work highlights a need for catchment farmers to manage land strategically in order to reduce saline discharge to the river, but management of river salinity problems is not necessarily the same as management required to address the spread of soil salinity problems. Implementation of the proposed replanting strategy to address soil salinity problems would radically alter existing agricultural practices (Hatton and Nulsen 1999), and farmers are unlikely to voluntarily implement the necessary land use changes when both economic and behavioural incentives are lacking (Barr and Cary 1984; Pannell 1999).

Together, all this suggests that dryland salinity management is faced with some inherently intractable technical problems, which differ from irrigation salinity management problems where water flow potentially can be more easily identified and controlled, as has been identified in the Murray-Darling Basin in Eastern Australia (Walker et al. 1999).

4. How can Dryland Salinity Policies be Evaluated?

Policy analysts have defined many difficulties with evaluating environmental policy, declaring the need for a new inter-disciplinary approach (Nelson 1974; White 1979; Wallace *et al.* 1995). In general, it is suggested that the nature of environmental management is rarely accommodated adequately within policy approaches. The lack of information about the status of the environment makes evaluation especially difficult (Dovers 1995). The traditional policy or program evaluation methodology has been criticised as top-down (Elmore 1979) and single disciplinary in perspective (Syme and Sadler 1994), often resulting in only partial description of the whole policy picture. Disparate evaluation methodologies add only *ad hoc* knowledge to the discipline of policy evaluation research (White 1979). Consequently, as a research discipline, environmental policy evaluation is particularly underdeveloped (Bellamy *et al.* 1999) and no adequate framework or methodologies exist for examining the whole environmental policy process (Wallace *et al.* 1995).

Traditionally, policy implementation has been described as a top-down or linear extension process consisting of three simple stages; the policy, its implementation and policy consequences. Policy evaluations of this process have generally sought to assess its appropriateness, efficiency and effectiveness (Department of Finance 1994). Bellamy (1999) suggests that Australian Government perception of evaluation as the systematic assessment of programs has been aimed at assisting managers to provide a basis for fund allocation and accountability. This paper attempts to evaluate the policy, its implementation constraints and its effectiveness at managing dryland salinity problems in the Kent River catchment. The evaluation of dynamic policy implementation processes being so inherently difficult, analysis using the 'building blocks' described by Mitchell (1991) for Integrated Catchment Management (ICM) will be useful to assess the adequacy of policies to fulfil key policy requirements.

Within environmental policy analysis, biophysical and economic indicators and evaluation against benchmarks are popular because they have the least transaction cost. They are most easy to define and use and can be simple to quantify (Nelson 1974). International development organisations push economic policy evaluation tools to the forefront of environmental evaluation research (for example, Smith 1997). Evaluation methods such as Benefit Cost Analysis (BCA), Environmental Impact Assessment (EIA) and, to some extent, Social Impact Assessment (SIA), which are based on an attempt to sum everyone's utility, are simple, efficient and cost less to apply compared with other, more cumbersome evaluation techniques. Benefit Cost Analysis remains unrivalled in its popularity as a policy evaluation tool (Walker 1994).

As a consequence of bio-physical complexity, system interconnectedness and community ethics, the evaluation of environmental policy requires something more multidimensional than these simple, utilitarian and linear assessment tools. Evaluation methods should be developed more according to the problem because too often "the methodological dog is used to wag the problematic tail" (Fox 1987:138). Applied policies in the social and political environment share also the same historical context, an important consideration for a complete understanding of policy (Fox 1987). As a consequence of the growing severity, urgency and inherent complexity of environmental problems, use of detailed policy 'mapping' to identify the most appropriate analytical methodology is often not possible. Instead, techniques to assess policy implementation are increasingly used as an evaluation tools to pinpoint environmental policy failure.

5. Dryland Salinity Policy

Most policies designed to protect the environment tend to be formulated in reaction to environmental crises and lack well-defined objectives and sufficiently long time-frames for effective change. Dovers (1995) lists 20 components required for an effective environmental policy program and points out that in Australia, implemented policies failed to satisfy most of these components.

Policy auditors have declared that environment policy performance and financial accountability is too difficult to evaluate, because the evolutionary nature of natural resource management policy objectives has not been easy to measure using traditional performance auditing methods (Australian National Audit Office 1997).

5.1 THE THREE POLICY APPROACHES IN THE KENT RIVER CATCHMENT

Since the 1970s, three policy approaches have been implemented to address growing salinity problems. They are identified here by the driving policy mechanism behind each as, regulatory, co-operative and market-driven. Other policies of significance have also influenced salinity management in this and other areas. They are the WA Soil and Land Conservation (SALC) Act (1945) and in particular amendments in 1982 when Land Conservation District Committees were set up, and The National Decade of Landcare Program (NDLP) which extended through the 1990s (and its predecessor the National Soil Conservation Program (NSCP)). While each of the policies being considered here has been developed in isolation and has its own internal objectives, which can be assessed, they also overlap significantly (Fig. 3).

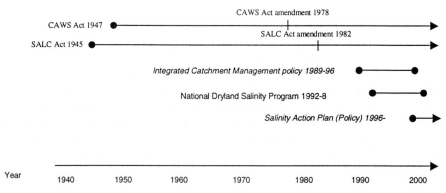

Figure 3. The periods of State policies to address dryland salinity since 1975, affecting the Kent River catchment.

5.1.1. The Regulatory Approach: Country Areas Water Supply Act 1947
In 1947, the Kent River water was designated as a future potable supply and the catchment came under the jurisdiction of the Country Areas Water Supply Act. Under the Act, the Governor may define catchment boundaries, extend or alter water catchment boundaries, unite two or more catchment areas and abolish catchments. The Act also gives the Minister the powers to divert, intercept and store all stream water and to take water found on or under the land. Any person who causes or permits the indigenous undergrowth, bush, or trees on land to be removed, destroyed or extensively damaged will receive a fine not exceeding A$1000 (US$570), and would be made to restore and land with tree cover (see the *Country Areas and Water Supply Act* 1947 (WA) Sections 9,11,11A,12A, 12B(I&II) & 17).

As an extension to potable water appropriation for the State in 1978, the Government selected the Kent and three other river catchments with the objective of bringing them back to potable levels. They were included in an amendment to the 1947 *Country Areas Water*

Supply Act. As part of the amendment, a policy was implemented to protect these rivers from rapidly increasing salinity through a catchment clearing ban. Until the 1960s most of these river salinity levels were below the World Health Organisation drinking water standard of 500 mg L^{-1}. Other southwest rivers which extended further inland were already brackish and considered beyond recovery. The State's southwest 'intermediate' rivers were still not appropriated, but only a few were located in areas of consistent flow, had a low population and could be dammed.

To encourage restoration of the catchment land, the government initially purchased some farms in an already dammed catchment and re-planted the farmland with trees (Batini 1980). Such an expense, however, would have been high and undesirable for the whole or newly selected catchments. By this time 64% of the Kent River catchment had been cleared for agriculture, which was more than the other chosen catchments. When a land-clearing ban was extended to the three new catchments 'overnight' it caused uproar amongst the farming community. The newspapers of 1979 declared that farmers would be defiant and clear lands anyway despite the warnings of legal action (*The National Farmer*, 1979).

Compensation costs to farmer *not* to clear more land in the chosen catchments had by 1996 cost the taxpayer a total of A\$34 million (US\$19.4 million), with a further A\$10 million (US\$5.7 million) in reforestation and land purchase costs (Viv Read, WRC Salinity Management Program Coordinator, pers. comm., 1999).

The Public Works Department (PWD) monitored and evaluated this clearing control policy through stream quality testing and aerial photography. This role was passed on to the Water Authority of Western Australia (WAWA) when it replaced the PWD in 1985, and then to the Water and Rivers Commission (WRC) in 1996. Unfortunately, the aerial monitoring of tree cover to assess illegal land clearing over the years did not pick up the evident destruction of the forest understorey, which was to prove to be an unfortunate short-fall of both the policy and the remote monitoring method. In 1989, the Steering Committee for Research on Land Use and Water Supply (1989) published a report stating that the Kent (and other) river salinity levels had increased dramatically. Meanwhile, the percentage of the upper catchment land that had been cleared had remained the same.

The effect of this clearing control policy on salinity levels was significantly diminished because only one fifth of the uncleared land had been protected. The understorey was over-grazed or completely removed from 79% of uncleared farmland and most uncleared areas were left unfenced, allowing livestock access (Strawbridge 1997). The removal of this lower and middle storey vegetation eliminated up to half of the recharge control potential of the intact native forest (Greenwood *et al.* 1985a). Information about the condition of this uncleared land was not assessed and published until 1992 (True *et al.* 1992). Moreover, compliance with the clearing policy was barely enforced with one notable case of a Kent River farmer fined for illegal clearing. It was well known by the farming community that other illegal clearing was taking place. Farmers were not expected to spend any of the compensation money to fence-off uncleared land or limit stock access, and many used the lump sum payments to retire debt.

In summary, despite working as a policy to prevent land-clearing, policy enforcement and compliance were not well administered and the policy was rushed in without acknowledging or addressing implementation problems in time. This policy was an attempt by the State water authority to impose a clearing ban on privately managed agricultural land. Farmers considered the policy to be excessive government interference, and although it may have slowed the rate of salinity increases it did not prevent further rises in river salinity levels. The key criticism was that farmers were never required to fence to protect uncleared land from stock grazing.

5.1.2. The Co-operative Policy Approach: Integrated Catchment Management 1989
In WA, the need for integrated natural resource management first became apparent when eutrophication problems emerged in populated estuary and harbour areas in the southwest during the 1970s. It was realised that to address such problems would require control of land use over large areas to better manage catchment run-off. Farm fertiliser and industrial effluent draining into rivers were identified as major sources of estuary pollution (Bradby 1997).

The WA Government Integrated Catchment Management (ICM) policy document was introduced in 1989 as the first attempt at providing an overarching 'umbrella' to natural resource management (Government of Western Australia 1989). It sought to integrate land, water and vegetation management and to balance resource use and conservation needs. It did not take long for this policy to lose favour with politicians, and implementing agencies were reluctant to accommodate it within existing management structures. Within seven years the ICM policy had become marginalised as an environmental management policy in WA. Committed and dedicated agency personnel, community catchment groups and the small government Office of Catchment Management were powerless, in the face of larger interests, to maintain the momentum of institutional re-structuring and community development needed to deliver an effective ICM approach. The Office of Catchment Management was officially disbanded in 1996 after being shunted around through various government agencies.

ICM was described as an educational, co-operative, philosophical and managerial policy process which would embrace and resolve environmental decision-making and management, and ensure resolution of implementation problems. It was hoped that it would lead to reduced land degradation, resolve conflicting government policies, lessen agency statutory responsibilities and increase public involvement in decision-making (Syme *et al.* 1994).

In 1990, an evaluation of ICM policy implementation suggested five 'building blocks' which should be accommodated for successful implementation (Mitchell 1991). Essentially, ICM should incorporate a systems approach which is integrated, stakeholder driven, has balanced ecological and social goals and has well defined partnerships between stakeholders.

As the chosen WA focus catchment for the National Dryland Salinity Program (NDSP) in 1993, a major Kent River catchment project would use the ICM policy approach to focus a Land and Water Resources Research and Development Corporation (LWRRDC) five-year research program. It was to be developed as an integrated research, development and extension project which was customer-driven, involved stakeholders, was collaborative, multi-disciplinary, avoided duplication, and was transferable on a State basis (Boyd 1993).

Within the Kent ICM program many individual, small-scale on-ground and large-scale research projects were conducted as part of focus catchment and National Program objectives. Catchment committees defined goals and implemented dryland salinity management strategies. A program budget of A\$300 000 (US\$182 800) for dryland salinity management was channelled into numerous government research projects and publications, while three community groups obtained federal Natural Heritage Trust (NHT) funding for landcare projects as a result of their involvement in the National Dryland Salinity Program (NDSP).

As an educational and promotional policy approach, ICM policy is inherently difficult to evaluate, especially so soon after implementation. The evaluator must somehow take into account, or measure the effect of, the many networks of people which have been formed and ideas which have been spread. Fox (1987) suggests the impact of such processes cannot be known for decades.

The benefit-cost analysis (BCA) conducted as part of the ICM plan determined that sale of the potable water would be necessary to make catchment rehabilitation cost effective (an option which other WA river catchments do not enjoy). Without this option, the BCA results indicated that it would only be economic to plant the best land to commercial tree crops, leaving a large proportion of the upper catchment in a degraded state which would cost more to repair than the land was worth.

By evaluating ICM against the approach proposed by Mitchell (1991), we suggest that the program was not community-driven. Members of State agencies dominated the committee decision-making process. The focus was really on research for, rather than community-driven, catchment management. There were no social research or management process developed for the implementation of integrated management options. The lack of a common vision or language between government and community stakeholders was not exposed, therefore conflicting issues went unresolved within an ICM plan which could not then be implemented.

Analysis of the policy framework revealed firstly, a deficiency in understanding holistic management (which ICM should have been addressing). This, combined with the lack of experience on how to implement it, and the scarce resources which were provided, essentially left a few dedicated individuals the task of implementing a policy which was largely institutionally unsupported. Secondly, without an implementation process or legal framework in place to enforce policy, centralised catchment authorities lacked the tools to implement catchment plans. Finally, we also suggest that the natural resource agency managers found the praxis of ICM philosophy to be potentially antagonistic to their individual program agendas, and they therefore provided minimal co-operation. Government policy fails when it is not institutionally supported (Mazmanian and Sebatier 1983).

Towards the end of the NDSP in the Kent River catchment, the Water and Rivers Commission (WRC) took over as lead agency from Agriculture Western Australia (AgWA). The WRC initially had no management structure or procedure in place to instigate salinity management strategies along ICM lines, despite the LWRRDC and the NDSP having supposedly endorsed an ICM process for the upper catchment. The critical mass needed from community participation for catchment salinity management had not been obtained and major conflicting interests remained unresolved.

In 1997, during the fourth year of the five-year program, a farmer survey was undertaken in the Kent River catchment as part of this study. Farmers were asked to quantify salinity management abatement activities within the catchment and to assess farm community participation and adoption levels. Survey questionnaires were sent out to all 120 upper catchment farmers in May 1996 and by October, 53% had returned completed forms - representing 69% of the upper catchment land area.

The survey results showed that although the adoption of landcare practices by Kent River farmers increased from 1995 to 1997 (Fig. 4), there was still insufficient implementation to abate or recover saline land, either for themselves or as part of an integrated catchment management plan (Kington and Pannell, unpublished). Most importantly, against the time frame of hydrological equilibrium and spatial spread of cause and effect, the co-operative policy was clearly not providing enough incentive (economic or behavioural) for farmers to halt the spread of land degradation problems.

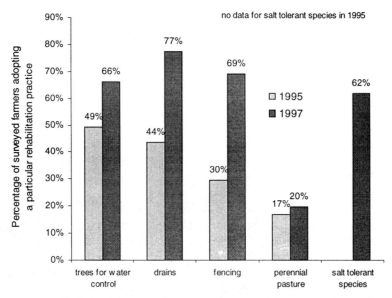

Figure 4. Adoption of dryland salinity management options for 1995 and 1997 by Kent River catchment farmers.

5.1.3. Commercial Opportunism: the WA Salt Action Plan, 1996

A new policy strategy designed specifically to address dryland salinity problems was introduced before the end of the National Dryland Salinity Program in the Kent River catchment. Initiated as a result of community concerns about the lack of Government salinity policy, the WA Salinity Action Plan (SAP) was orchestrated by the executives and scientists of four natural resource government agencies; Agriculture Western Australia (AgWA), Conservation and Land Management (CALM), Water and Rivers Commission (WRC), and to a lesser extent, the Department of Environmental Protection (DEP). Within this new policy strategy the Kent River was designated a public drinking water recovery area and could now obtain significant funding to achieve this objective over the ensuing 30 years. Within the SAP, the primary policy mechanism targeting the Kent River catchment was to encourage farm forestry and prepare and implement catchment plans in cooperation with the community. The Asian paper industry preference for bluegum (*Eucalyptus gobulus*) tree woodchips, which in WA grow to maturity within only 10 years, created a profitable new market opportunity for government, businesses and farmers within this high rainfall region of WA. The promotion of commercially-driven farm forestry within the Kent River catchment was already underway, supported by CALM's woodchip industry (and WRC river water recovery) objectives. However, catchment farmers did not emulate this enthusiasm.

By surveying government records, farmer and other information sources, we have found that by 1998, up to 12% of farmers had sold or leased the whole farm property to plantation companies (Fig. 5). There was commercial interest in a further 8% and it was predicted that, in total, up to 30% of farms could sell-out to the tree plantation companies in the future (Burdass *et al.* 1998). Although this would result in revegetation of over 37% of the upper Kent catchment, the revegetation would be in *ad hoc* locations and not strategically located on lower slope positions. In consequence, it seems likely that further replanting would be required in order to meet the salt reduction target of 550 mg L^{-1}.

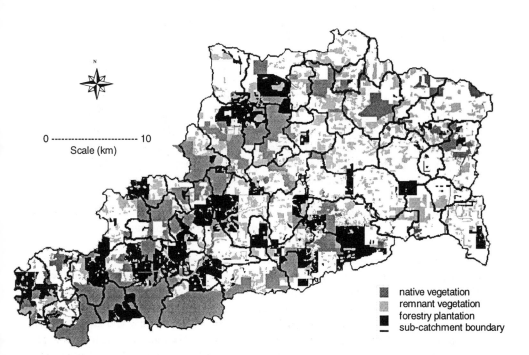

Figure 5. Bluegum plantation development in the upper Kent River catchment above gauging station S564001. Diagram supplied courtesy of Water and Rivers Commission (2000).

The adjacent Denmark River catchment is another of the water recovery catchments where bluegum tree plantations (planted since 1990) now cover 80% of the catchment (Fig. 1). The Denmark River salinity level is now decreasing and, assuming that it continues dropping, would make this policy approach effective as a catchment treatment for controlling river (and land) salinity problems. However, as a higher rainfall catchment this option has been entirely commercially driven because of relatively high potential plantation yields over the entire catchment.

A BCA of profit per hectare from bluegums compared with wool explains why some (especially older) farmers have sold their whole farms to the new bluegum plantation companies. The social costs and benefits to the local community and economy of a forestry plantation environment are not measured against the apparent salinity management success. There is growing concern that corporate, absentee land ownership, employment changes and vacated farms will inevitably change the social and economic infrastructure (York Gum Services 1995), just as it has affected communities in New Zealand plantation areas (Smith 1981). When assessed using the methodology of Mitchell (1991), this market-driven policy lacks an integrated approach, and is not balanced with social or ecological goals. The apparent onus of responsibility has been with the government and not the farmers to develop on-farm policy approaches to prevent land degradation, whereas the developing Landcare ethics under the National Decade of Landcare encouraged on-farm environmental responsibility by the farmers themselves.

Policies may not always be chosen because they are the most appropriate. Political acceptability plays a major role and at present, the trend is towards market mechanisms. Such mechanisms are focused more on seeking ways to develop products from the land rather than addressing fundamental land management problems.

6. Conclusions

Salinity problems in the Kent River catchment have been addressed sequentially since 1947 by three very different policies. We have attempted to analyse and compare these using a variety of evaluation approaches and methodologies and to move away from the traditionally narrow approaches to policy evaluation. Mitchell's (1991) guiding ICM principles have also thrown some light on dryland salinity policy design and operation. Refinement to methods of determining policy effectiveness allowing broader insights, improved management systems, as well as creating a 'policy learning' environment, will ultimately lead to more informed policies.

Dryland salinity policies implemented in the Kent River catchment have been based on very different compliance mechanisms, and it was found that often policies themselves are inadequate because they have not been well planned and implemented, not necessarily because their instruments are inherently weak.

It was identified that there was little responsibility by community and government to endorse a purely regulatory approach. The policy prevented further clearing but in the Kent river catchment, clearing was already so far advanced that neither regulation nor money spent by the WRC (until 1996) compensating farmers not to clear land in the Kent River catchment, were effective in reducing river salinity levels to acceptable drinking levels.

The ICM approach has been ineffective because the government agencies, farmers and markets do not necessarily work for the collective good against their individual interests. ICM could be improved if behavioural change leading to increased environmental responsibility by the community was embraced by all stakeholders.

Market opportunities (namely plantation forestry) may be the most effective in terms of adoption. However, commercial trees are often grown on the best soils where salt risk is least. Their use can also lead to long-term local social and environmental problems with what is essentially a new monoculture. The social needs of existing farm communities are not necessarily accommodated within a purely commercially-driven or regulatory policy which provides financial incentives to ensure widespread adoption, but lacks appropriate incentives to change land-degrading behaviour. In contrast, promotional and educational policies which have no financial incentives to offer do not overcome the problem of policy enforcement and therefore compliance does not reach critical mass.

What has been noted about all three approaches to salinity management is that they have not been sufficiently systematic in terms of following the criteria of Dovers (1995) - they do not even follow a simple policy checklist, such as; policy formation, implementation and evaluation. Despite this, selected catchments have experienced some success in slowing down salinity expression, but it is proposed that success will only come when the government is committed to a systematic policy approach (including social, environmental and economic). Without a planned approach incorporating strategic replanting of high recharge areas, it seems likely that salinity control will only be achieved by major replanting of most of the present agricultural land.

Finally, given the long time scales required to contain the spread of dryland salinity, it could be asked whether policy strategies should now focus on minimising the inevitable social and environmental impacts, and therefore extending current policy beyond its almost exclusive focus on dryland salinity management.

7. Acknowledgments

Funding for this research was provided by the Land and Water Resources Research and Development Corporation, Canberra, Australia. Ian Parker, Dr Geoff Syme, Honorary Fellow Dr Henry P. Schapper, Dr John Pickard, Mr Viv Read of the WA Water and Rivers Commission, and Associate Professor David Pannell are thanked for their comments and support.

8. References

Australian National Audit Office (1997) Commonwealth Natural Resources Management and Environmental Programs: Australia's Land, Water and Vegetation Resources Performance Audit. http://www.anao.gov.au/summaries_97/sum_36.html

Barr, N. F. and Cary, J. W. (1984) *Farmer Perceptions of Soil Salting: Appraisal of an Insidious Hazard*, School of Agriculture and Forestry, The University of Melbourne.

Batini, F. E. (1980) Regional Management - A case study the Wellington Dam Catchment in Western Australia, in N.M. Oates, P.J. Greig, D.G. Hill, P.A. Langley and A.J. Reid (eds.), *Focus on Farm Trees. A National Conference on the Decline of Trees in the Rural Landscape*, Capitol Press, Box Hill, Victoria, pp. 77-88.

Bellamy, J. A., Walker, D. H. and McDonald, G. T. (1999) Tracking progress in natural resource management: A systems approach to evaluation. *unpublished*.

Boyd, D. (1993) *Rehabilitation of agricultural catchments affected by land and stream degradation: an integrated research approach*, Hydrology and Land Use Research, The Water Authority of Western Australia.

Bradby, K. (1997) *Peel-Harvy: The Decline and Rescue of an Ecosystem*, Greening the Catchment Taskforce (Inc.), Mandurah, Western Australia.

Burdass, J., Grieve, R. and Robinson, B. (1998) *An Integrated Catchment Management Plan for the Upper Kent River Catchment. Albany, WA*, The Kent Steering Committee, The National Dryland Salinity Program.

Crowley, F. K. and De Garis, B. K. (1969) *A Short History of Western Australia*, Macmillan, South Melbourne.

Department of Finance (1994) *Doing Evaluations: A Practical Guide*, Department of Finance, Canberra.

Dovers, (1995) Information, sustainability and policy, *Australian Journal of Environmental Managemnet* 2:

142-156.

Elmore, R. (1979) Backwards mapping: implementation research and policy decisions, Policy Science Quarterly **94**, 601-616.

Ferdowsian, R. and Ryder, A. (1997a) Salinity and hydrology of Wamballup Swamp catchment, *The National Dryland Salinity Program Conference, 13-15th October. Mount Barker, WA*, LWRRDC and Kent River catchment Technical Team, Western Australia, pp. 31-33.

Fox, C. J. (1987) Biases in public policy implementation evaluation, *Policy Studies Review* **7**, 128-141.

Glover, R. (1979) *Plantagenet: A History of the Shire of Plantagenet, Western Australia*, The University of Western Australia Press, Nedlands, W.A.

Government of Western Australia (1989) *Working Together: Integrated Management of Western Australia's Lands and Waters*, Perth.

Government of Western Australia (1996) *Salinity: a Situation Statement for Western Australia*, Perth.

Greenwood, E.A.N., Klein, L., Beresford, J.D., Watson, G.D. and Wright, K.D. (1985a) Evaporation from the understorey in the Jarrah (*Eucalyptus marginata*) forest, southwestern Australia, Journal of Hydrology **80**, 337-349.

Hatton, T. J. and Nulsen, R.A. (1999) Towards achieving functional ecosystem mimicry with respect to water cycling in Southern Australian agriculture, *Agroforestry* **45**, 202-214.

Hobbs, R. J. (1998). Impacts of land use on biodiversity in southwestern Australia, *Ecological Studies* **136**, 81-106.

Kington, E. A. and Pannell, D.J. (1999) Dryland salinity in the upper Kent River catchment of Western Australia: Farmer perceptions and practices, SEA Working Paper 99/10, http://www.general.uwa.edu.au/u/dpannell/dpap9909.htm.

Mazmanian, D.A. and Sabatier, P.A. (1983) *Implementation and Public Policy*, Scott, Foreman, Glenview, Illinois.

Mitchell, B. (1991) *Integrated Catchment Management in Western Australia: Progress and Opportunities*, WA. Centre for Water Research, University of Western Australia, Nedlands, W.A.

Nelson, R. R. (1974) Intellectualising about the moon-ghetto metaphor: a study of the current malaise of rational analysis of social problems, *Policy Sciences* **5**, 375-414.

Pannell, D. J. (1999) Explaining non-adoption of practices to prevent dryland salinity in Western Australia: implications for policy, COMLAND, September 21-25 1999.

Smith, B. N. P. (1981) Forestry and rural social change: a comment, *New Zealand Journal of Forestry* **26**, 103-107.

Smith, S., Vos, H.B. and Organisation for Economic Co-operation and Development (1997) *Evaluating economic instruments for environmental policy*, Organisation for Economic Co-operation and Development, Paris.

Steering Committee for Research on Land Use and Water Supply (1989) *Stream Salinity and its Reclamation in Southwest Western Australia*, Water Resource Directorate, Water Authority of Western Australia, Report No. WS 52, Perth.

Strawbridge, M. (1997) The extent and condition of remnant vegetation in the upper Kent catchment. *The National Dryland Salinity Program Conference, 13-15th October. Mount Barker, WA*, LWRRDC and Kent River catchment Technical Team, Western Australia, pp. 78-79.

Syme, G. J., Butterworth, J.E. and Nancarrow, B.E. (1994). *National Whole Catchment Management: A Review and Analysis of Process*, The Land and Water Resources Research and Development Corporation, Canberra.

Syme, G. J. and Sadler, B.S. (1994) Evaluation of public involvement in water resources planning. A researcher-practitioners dialogue, Evaluation Review **18**, 523-542.

The National Farmer (1979) Row over 'secret' farm clearing ban, Western Australia, *The National Farmer* 7th March 1979, p. 25.

True, D. E., Kikiros, G. and Froend, R. (1992) *Preliminary assessment of the effects of grazing on the remnant vegetation in the Kent River water reserve*, Water Authority of Western Australia, Leederville, W.A.

Walker, G., Gilfedder, M. and Williams, J. (1999) Effectiveness of current farming systems in the control of dryland salinity, CSIRO Land and Water, Canberra. http://www.mdbc.gov.au/News/index.html.

Walker, K. (1994) *The Political Economy of Environmental Policy: an Australian Introduction*, University of New South Wales Press, Kensington, NSW.

Wallace, M.G., Cortner, H.J. and Burke, S. (1995) Review of policy evaluation in natural resources, *Society and Natural Resources* **8**, 35-47.

White, I. L. (1979) An interdisciplinary approach to applied policy analysis, *Technological Forecasting and Social Change* **15**, 95-106.

Wilson, T.B. (1968) *Narrative of a voyage around the world*, Dawsons of Pall Mall, London.
York Gum Services (1995) Farm Forestry in the Peel, South West and Great Southern Regions. Optimising the Potential by Addressing Planning Issues and Community Concerns, Discussion paper prepared for the South West Development Commission's Timber and Pulp Working group.

AUTHORS

ELIZABETH A. KINGTON[1] and KEITH R.J. SMETTEM[2]

PhD Research student[1] and Associate Professor[2]
Department of Soil Science and Plant Nutrition,
The University of Western Australia
Nedlands, WA 6907, Australia

All correspondence to Elizabeth A. Kington at the above address or by email to ekington@agric.uwa.edu.au

CHAPTER 23

POLICY RESPONSES TO LAND DEGRADATION IN AUSTRALIA

JEANETTE CONACHER and ARTHUR CONACHER

1. Abstract

Australia is experiencing severe and growing problems of land degradation, affecting agricultural lands, aquatic systems and indigenous flora and fauna. Policies and strategies have been developed at the national level to deal with agriculture, land use, water management, rangelands, biodiversity and environmental impact assessment. However, there is no all-encompassing national legislation directed to land degradation. State laws, policies and strategies affect the management of soil erosion, vegetation clearance, soil and water pollution, flora and fauna conservation, protection of prime agricultural land, control of feral animals and weeds, and pastoral lands. International agreements under which Australia has obligations relate to biodiversity, desertification, greenhouse and environmental pollution. All levels of government (federal, State, Territory and local) have endorsed an ecologically sustainable development (ESD) strategy, recognising inter-generational equity, adopting precautionary and user-pays principles, and involving the community in natural resource decision-making processes. Specific, important policies and strategies have included the Decade of Landcare and the Natural Heritage Trust, and the development of integrated natural resource management, integrated catchment management, and regional planning.

Constraints affecting the implementation of policies and strategies designed to mitigate land degradation include: the size and complexity of the country; its small population and skill base; the surfeit of overlapping, conflicting and costly governments, Acts, regulations and agencies; and a lack of resources and skills, particularly in local governments. Recommendations to deal with the constraints include strengthening and grouping local governments at regional levels, and replacing current State and Territory agencies dealing with the environment, land-use, natural resources and planning with new (or equivalent federal) agencies devolved to strengthened, regionally-based structures.

2. Introduction

Despite a considerable period of occupation by Aborigines, land degradation did not become a problem until European settlers chopped down the first tree and ploughed the first furrow (Powell 1976). Whilst some aspects of land degradation have been treated with varying degrees of success, in general the problem has continued to develop (State of the Environment Advisory Council 1996). A wide range of biophysical, political, legal, institutional, social and cultural constraints has impeded governments' abilities to deal

363

A.J. Conacher (ed.), Land Degradation, 363–385.

effectively with land degradation: some are more amenable to reform or treatment than others. A major constraint is Australia's federal (and mostly bicameral) system of government: a single House of Parliament cannot decide on a course of action and implement it. Yet failure to respond effectively will lead to even more serious environmental problems in the future.

This paper draws on work completed recently (Conacher and Conacher 2000) to identify and analyse some of the *main* policies and strategies which have been developed to tackle the problems of land degradation in Australia. It considers how the policies are implemented and asks whether their implementation has been successful. The paper then discusses the main constraints to implementation and suggests means of overcoming them.

3. The Problem

For the purposes of the discussion, and to distinguish the problem from *natural hazards*, land degradation is defined as:

> alterations to all aspects of the biophysical environment by human actions to the detriment of vegetation, soils, landforms, water, ecosystems and human well-being (modified from Conacher and Conacher 1995:3).

Land degradation is not new in Australia - Aboriginal occupation and use of fire over a period of more than 60 000 years almost certainly caused modifications of natural systems, although these impacts are probably impossible to distinguish from those caused by climatic fluctuations over the same period (Kirkpatrick 1994; Zimmer 1999). However, any such modifications pale into insignificance when compared with the range, scale and intensity of the adverse environmental effects of European land use practices, especially over the past 80 years.

Almost all agricultural land and a proportion of rangeland require some form of treatment for erosion, salinisation, waterlogging, soil structural decline, acidity, pollution, loss of biodiversity and other problems (Table 1). Some problems are common to all farmed areas; others differ in their degree of severity or locations. The annual economic costs of land degradation - in works, loss of production and offsite impacts - are considerable (Table 1), yet the figures exclude the costs of damage to ecosystems, aesthetics, natural heritage or other environmental values. While some of the estimates may be unreliable, it is clear that there is a major problem. Efforts to deal with it are often well beyond the resources of individual farmers or even government agencies.

The causes of land degradation are as many and complex as the problems themselves. Not least has been clearing of the original, deep-rooted vegetation, resulting in hydrological disturbances which include rising water tables, waterlogging, flooding, and soil and water salinisation. Other causes are linked to: European-type agricultural practices mismatched to a largely harsh physical environment; chemical, biological and physical pollution of natural systems; the increased frequency of fire, and the introduction of exotic pests and diseases. Government policies, economic forces and peoples' attitudes towards the environment are also important influences which affect the ways in which people use the land. But in a sense it is futile to try and apportion such costs - or blame - to the

Table 1. Extent and costs of land degradation in Australia. Sources: House of Representatives Standing Committee (1989); LWRRDC (1995); State of the Environment Advisory Council (1996); ABS (1996, 1997); Industry Commission (1998); *Australian Landcare* (Sept 1999, pp. 26-29).

DESCRIPTION	AREA AFFECTED/ CONDITION	COSTS	COMMENTS
Land degradation Australia	over half all agric land affected (farmer est. 1993 16.4 m ha land affected - 3.9 erosion - 1.12 salinity - 11.3 other)	over $2 b/yr requ in works $1.2 b in lost production	estimates made in early 1980s - could be greater now. Off-site costs not included
Land cover change	over 70% native veg cleared or modified since 1788 - 40% forests cleared - 90% temperate woodlands/mallee - 75% rainforests est. > 3000 km²/yr still being cleared		clearing contributes about 15% of CO_2 emissions
Rangelands	2% severely damaged 15% in need of destocking		guidelines for rangeland management released mid 1999
Plants and animals	Percentages extinct, endangered or vulnerable: 5% higher plants; 23% mammals; 9% birds; 7% reptiles 16% amphibians; 9% freshwater fish		
Pests and weeds	over 200 nuisance weed spp insects 18 introduced feral mammal spp - rabbits	$3.3 b/yr lost prodn, quality and control (excl. other costs) $3.1 b/yr lost prodn $90 m - $600 m/yr lost prodn & land values (agric and pastoral areas)	 probably declined since release of calicivirus

Table 1 Continued.

Salinity	> 2.5 m ha affected (4.5% of intensive farmland). Potential 15 m ha in 50 yrs 80 regional towns affected 230 km of roads affected in WA	$270 m/yr cost in envtl degradation, degraded water supplies, lost prodn, infrastructure damage; $100 m/yr cost of damage to urban infrastructure; highway repair costs $1 m/yr	1999 Stage 2 of National Dryland Salinity Program provided $15 m over 5 years for R & D, env. manage
Irrigation salinity	156 000 ha affected	>$10 m/yr lost production	650 000 ha irrigated land have water table within 2 m of surface
Waterlogging	250 000 ha high water tables in NSW;	yield losses <30% common;	
	MD Basin 500 000 ha waterlogging & salinity	$65 m/yr lost prodn; downstream costs $37 m	MD Basin water table rises 100-500 mm/yr
Soil structural decline	large parts W Vic, W NSW, S Qld , SW WA affected	$200 m/yr lost production	
Wind and water erosion	approx 4 m ha/yr (variable)	$30 m/yr offsite effects in Qld	
Acidity	29 m ha	$134-300 m/yr lost production	2 m ha potential disturbance of natural acid sulfate soils
Eutrophication	1991 >1000 km Darling R (NSW) affected with algal bloom	probably >$10 m/yr in Austr for management	in NSW in 1991-92, $9.4 m loss from tourism and recreation due to a severe algal bloom
Prime agric land	thousands of ha/yr lost to suburban and other developments		In Vic, about 0.5 m ha prime land lost in 1973-75; 167 000 ha 1977-81; >50 000 ha 1986-91
Pesticides	>15 m ha/yr treated (14.9 m ha herbicides; 3.1 m ha insecticides; 0.8 m ha fungicides) 160 m stock treated with vet chems	expenditure on pesticides >$1 b/yr (mostly agric) insect pests in stock cost >$600 m/yr	costs vary widely according to location and enterprise

Table 1 Continued.

Farm expenditure on envt manage & protection measures	$178 m 1993-94	$72 m pest/weed control; $69 m works (erosion, salt, waterlogging); $12 m revegetation; $16 m fencing; $6 m farm plans; $3 m self education

problem. What we do know is that an urgent and concerted public and private effort is needed to deal with it.

National and State government structures and roles in natural resource management are now examined.

4. Government and Community Roles

Australia has a three-tiered, federal system of government comprising a central Commonwealth (or federal) government, six States and two Territories, and over 700 local governments. There are also a number of offshore responsibilities (such as part of Antarctica and several large islands) but these are not considered here. Non-statutory arrangements are numerous. Detailed structural and statutory environmental management arrangements for Australia can be read in Bates (1995) and Gardner (1999). State, Commonwealth and environment websites also provide current information on legislation and institutions.

4.1. COMMONWEALTH

In general, the Commonwealth has exercised its environmental responsibilities indirectly through its financial, trade, export and foreign affairs powers rather than through the Australian Constitution, which makes no reference to environmental obligations or land-use matters. Nevertheless, there are several Commonwealth agencies which develop policies and standards either separately or in conjunction with the States.

The prime Commonwealth Department with environmental responsibilities is the *Department of Environment and Heritage* (DEH), with a number of bodies under its control. One of these is *Environment Australia,* responsible for overseeing the key Commonwealth environment laws and policies covering the seas, biodiversity, environment protection and national and international heritage. Environment Australia links up with the *Department of Agriculture, Fisheries and Forestry* (AFFA) in the administration of two important environment and natural resource management programs - the Natural Heritage Trust (NHT), and projects directed to the Murray-Darling Basin (discussed below). AFFA also has environmental responsibilities related to quarantine regulations and the use, registration and monitoring of agricultural chemicals, although administration mainly takes place at State levels. Regional planning matters are generally the responsibility of the Department of Transport and Regional Services, which incorporates the National Office for Local Government.

A number of Ministerial Councils with Federal and State representation are responsible for the development and co-ordination of national environmental and natural resource policies and programs. Some relevant Councils are: the Australia and New Zealand Environment and Conservation Council (ANZECC); the Agriculture and Resource Management Council of Australia and New Zealand (ARMCANZ), and the National Environment Protection Council (NEPC). The Councils, in turn, are advised by a wide range of specialist committees or groups with their members drawn from agency, industry, public or research sectors. Examples include the Murray-Darling Basin Commission, the Australian Landcare Council, Co-operative Research Centres (CRCs), the Commonwealth Scientific and Industrial Research Organisation (CSIRO) and the Land and Water Resources Research and Development Corporation (LWRRDC). The Council of Australian Governments (COAG) is an important body. Comprising the Prime Minister, State Premiers, Territory Chief Ministers and local government representatives, it has played a key role in overseeing policy and structural reform, and improved co-operation and co-ordination amongst governments. The *Inter-Governmental Agreement on the Environment* (IGAE) (COAG 1992: discussed below) and a national water reform policy (COAG 1995) have been some major environmental initiatives by COAG.

4.2. STATES AND TERRITORIES

States and Territories are the main land and environmental managers in Australia. Their responsibilities are accomplished through various natural resource management and environment protection agencies, laws, regulations and policies (refer 'policies', below) which apply to private and public lands. Relevant States' legislation covers air, soil, minerals, water, flora and fauna protection, agriculture, forestry and fisheries, as well as land use controls under planning, development and lands laws. Control of pollution, and environmental impacts of major developments, take place via specific environment protection statutes and authorities. By way of illustration, an excellent summary of some of the foregoing can be read in the Western Australian Environment Defender's Office book, *The Law of Landcare* (Clement and Bennett 1998). Agencies are generally expected to make reference to State (and national, where relevant) planning and environment protection laws in carrying out their responsibilities. In some situations (such as public nuisance) common law may apply.

4.3. LOCAL GOVERNMENTS

The prime land management responsibilities of local governments (known as Shires or Counties in rural areas, and administered by Councils) focus on service delivery and infrastructure, and the execution of planning schemes. Local plans must conform to State legislation and policies, plans and strategies which have statutory backing. In general, roles in rural planning are limited in relation to environmental and natural resource management, but some significant changes are underway. For example, a growing number of local governments are voluntarily formulating their own regional or local environmental and land management policies and plans (example: Evans 1999). Others are responding to Commonwealth and State planning policies and regional strategies (SROC 1996) which refer to matters such as: ecologically sustainable development; Local Government Agenda 21; State of the Environment reporting and benchmarking; erosion or pollution control; and protection of biodiversity, good quality agricultural lands and groundwater or surface water catchments (Greening Australia 1995; Wright 1995; Dore and Woodhill 1999).

Another strategy has been to reduce the number of local governments. For example, Victoria reduced theirs by more than half during the early 1990s and also attempted to rationalise the new boundaries within biogeographic regions. However, proposals for mandatory amalgamations of local authorities proved electorally unpopular in Tasmania, and contributed to the party in power at the time losing the subsequent election (John Pretty, Manager, Planning Services, Department of Primary Industries, Water and Environment, pers. comm. 1999).

5. Policies and Strategies

In general, it is sometimes difficult to distinguish between policies and strategies since the two are inter-linked. Inability to recognise the differences - or confusion between the two - may underlie some past policy failures in the management of land degradation and natural resources in Australia.

In broad terms, *policy* refers to interventions by governments (including agencies and local governments) which may be economic, political, social or environmental in nature. The objective is to achieve change or reform (theoretically for public benefit). Policies can be framed, for example, to address certain constraints (such as market failures or environmental problems), to achieve greater efficiencies (social, economic or structural), or to promote new technologies. Statements of policy might be general in content, offering only a 'guiding' framework for decision makers, planners and managers; or they can be quite specific in intent - as exemplified by COAG's water reforms (discussed later).

Strategies provide the detailed means whereby policies are implemented (and are sometimes incorporated within the policy framework). Specific goals and targets are specified, action plans detailed and review processes incorporated.

Policies may be articulated in other ways, sometimes as part of strategic documents. Alternative approaches include the use of economic instruments (taxes, subsidies, pricing), structural reform (of agencies, industry), legislation or regulation, or incentives, awards and research grants.

With reference to land degradation and natural resource management policies in Australia, some explanations for past policy failures centre around: an inability to translate them into practice (perhaps they are too general or too vague); too narrow (or sectoral) a focus which ignores other important natural resource or environmental needs; or inexperience in framing policies which appropriately address inter- and intra-sectoral issues demanding integrated, responsive and strategic approaches to policy formulation and implementation.

Numerous, often overlapping, policies and strategies relating to land degradation have been developed over the years, and only some of the more important ones are discussed in this paper.

A number of specific and general policies and strategies relating to land degradation and natural resource management have emanated from Federal Ministerial Councils and other bodies from the early 1980s. They cover areas such as sustainable agriculture, water and pricing, wetlands, rangelands, algal blooms, pests and weeds, agricultural chemicals and natural heritage (Table 2). Specific States' strategies and policies which affect the management of natural resources refer to similar matters, such as soil conservation, vegetation clearance, water quality (including salinity), environmental pollution, and protection of prime agricultural land. Many of these have been specific in focus. Some more general policies and strategies have covered conceptual and strategic areas such as Agenda 21, biodiversity, ecologically sustainable development, environmental impact assessment, integrated catchment management and Landcare (Table 2). Most recently, national policies have indicated the importance of broader, integrated natural resource management and regional planning approaches to deal with land degradation, natural resource and other land use problems (discussed below).

As signatory to over 60 international environmental conventions, a number of environmental obligations are discharged by the Commonwealth government and generally administered at the State level: some are now incorporated in the new Commonwealth environment protection legislation, the *Environment Protection and Biodiversity Conservation Act 1999*. Such agreements guide the development of many of Australia's environmental laws and policies. Insofar as land management and degradation are concerned, agreements relating to habitat protection, desertification, greenhouse and environmental pollution are of particular note (Table 3). While of considerable importance, there has been some unease over the degree to which some of the international obligations are discharged. Difficulties relate to: the slow rates of adoption and development of action plans; problems in dealing with general or abstract terms; cursory reference given to responsibilities; 'protected' wetlands placed under developmental pressure, and a paring back of greenhouse targets, to quote some examples (see Comino 1997 for an example relating to Australia's implementation of the Ramsar agreement).

On the positive side, however, in 1992 all levels of government endorsed a national *Ecologically Sustainable Development* Strategy (ESD) (Commonwealth of Australia, 1992). This was one of the Commonwealth's major commitments arising from the Rio world conference on environment (World Bank 1997), and an extension of objectives identified in a *National Conservation Strategy* in 1984 (Department of Home Affairs and Environment 1984). The prime objectives were:

• to ensure the protection and sustainable management of natural resources and ecosystems to the benefit of the economy, social welfare and the environment, and
• to recognise inter-generational equity, and adopt precautionary and user pays principles.

Governments were to harmonise their activities in legislation and administration, involve communities in decision-making processes, and incorporate ESD into their environmental and land management policies. Some years on, these objectives are largely being met around Australia, although there is room for improvement. In particular, it seems that difficulties are being encountered in converting broad principles into actions (Dovers 1997; Calver *et al*. 1999; Productivity Commission 1999a).

Table 2. Some national, state and territory environmental protection policies and strategies and other measures

POLICIES, AGREEMENTS etc.	REFERENCE	EXAMPLES OF STATE/LOCAL RESPONSES
Intergovernmental Agreement on the Environment	COAG 1992	
National Conservation Strategy	Department of Home Affairs and Environment 1984	Government of Victoria 1987; DCE 1987
National Soil Conservation Strategy	ASCC 1988	
Agenda 21	UNCED 1992	Local Agenda 21 www.peg.apc.org/~counciln et/clnet/members/LA21.htm
National Strategy for Ecologically Sustainable Development	Commonwealth of Australia 1992c	
Sustainable agriculture	ESD Working Group 1991	DARA 1990; RIRDC 1997
National Greenhouse Response Strategy	Commonwealth of Australia 1992a DEST 1995	Government of Victoria 1990 DEP 1994
National Water Quality Management Strategy - fresh and marine waters	ANZECC/ARMCANZ 1999	EPA SA 1997
National Water Quality Management Strategy - groundwater	ARMCANZ/SCARM 1996	DWR 1991
Australian Drinking Water Guidelines	NHMRC/ARMCANZ 1995 (under revision)	
Water reform, water policies	COAG 1994, 1995; ARMCANZ 1994a	DENR 1995
Salinity	National Dryland Salinity Program	MDBC 1993a; State Salinity Council 1998
National Rangelands Strategy	ANZECC/ARMCANZ 1996	Government of WA 1997b
National Drought Policy	Taskforce 1991	
National Weeds Strategy	ARMCANZ/ANZECC 1997	Government of Tasmania 1994
National Strategy for Land Information Management	ALIC 1990	
National Waste Minimisation & Recycling Strategy	CEPA 1992	Government of NSW 1994; Dept Commerce & Trade 1993

Table 2 Continued.

Beef feedlots	SCA 1992	
Decade of Landcare, Integrated Catchment Management	Hawke 1989; Commonwealth of Australia 1991	Working Group 1991 Government of Victoria 1992 NSWSCS (n.d.) ACT 1991

Table 3. Some international environment, conservation and heritage treaties and conventions to which Australia is a signatory. Listed in chronological sequence according to the date of the instrument (not included). From State of the Environment Advisory Council 1996, Appendix 1. The first date shown is the year the treaty was signed by Australia; the second the year in which the treaty entered into force generally.

Convention on Wetlands of International Importance especially as Waterfowl Habitat (the Ramsar Convention) (1974; 1975)

Convention for the Protection of the World Cultural and Natural Heritage (under the auspices of UNESCO) (1974; 1975)

Convention on International Trade in Endangered Species of Wild Fauna and Flora (CITES) (1973; 1975)

Japan-Australia Agreement for the Protection of Migratory Birds and Birds in Danger of Extinction and their Environment (JAMBA) (1981; 1981)

China-Australia Agreement for the Protection of Migratory Birds and their Environment (CAMBA) (1986; 1988)

Convention for the Protection of the Natural Resources and Environment of the South Pacific Region (SPREP) (1987; 1990)

Montreal Protocol on Substances that Deplete the Ozone Layer (1988; 1989)

United Nations Framework Convention on Climate Change (1992; 1994)

Convention on Biological Diversity (1992; 1993)

United Nations Convention to Combat Desertification in those Countries experiencing Serious Drought and/or Desertification, particularly in Africa (1994; not yet in force)

Following the development of the national ESD strategy, political recognition of the more practical needs for environmental and institutional reform was addressed through the Council for Australian Governments' *Intergovernmental Agreement on Environment* (IGAE) in 1992 (COAG 1992). Clarification of government roles and responsibilities, harmonisation of laws and policies, and better environmental protection were among the Agreement's chief objectives. Such reforms have been pursued through the consolidation of Commonwealth and State environment protection laws, and the development of consistent, national environmental data sets and standards, the latter to be incorporated in State laws and policies. Most States have undertaken extensive structural rearrangement of their resource agencies to achieve greater efficiencies - though some have argued at a considerable cost (Christoff 1998).

However, the IGAE appears to focus more on environment protection and development than on bettering natural resource management *per se* and addressing broader land degradation issues. The latter have been the subject of recent government attention - by the Industry Commission (1998), a House of Representatives Standing Committee on Environment and Heritage inquiry into catchment management (in progress in July 1999), and through the formulation of a national integrated resource management policy (discussed below). Thus, there now appears to be a stronger political commitment to improving natural resource management laws, strategies and structures than a few years ago, when similar proposals were deferred or largely ignored (for example, Western Australia's 1997 Taskforce review of natural resource management (Taskforce 1997), commented on by Robinson and Humphries 1997).

Some specific agricultural and land use policies and programs are now discussed, in relation to Landcare, integrated natural resource management, integrated catchment management and regional planning and management.

5.1. DECADE OF LANDCARE AND THE NATURAL HERITAGE TRUST

Probably the most influential move in Australian politics to affect the comprehensive management of land degradation was contained in the Prime Minister's *Statement on the Environment* in 1989 which announced the *Decade of Landcare* (Hawke 1989). This initiative brought together two unlikely partners in the Australian Conservation Foundation (ACF) and the National Farmers' Federation (NFF), but it was the culmination of several discrete national and State programs and community projects from the previous decade (Northrop 1999). The ACF and NFF combination was a fortuitous one in their common concern over land degradation and the desire for a broader but integrated national policy, with strong financial and research support from government.

Drawing on the success of the *National Soil Conservation Program* (NSCP) set up by the Commonwealth government in 1984 (Bradsen 1988), and on ESD principles, the joint proposal was accepted by government and implemented as a national Landcare program. Funds totalling $320 million over 10 years were to support community-based landcare groups, education projects and several national water and vegetation initiatives, with the objectives of planting one billion trees over the decade and arresting land degradation and improving water quality around the nation. Inappropriate or distorting rural policies such as water subsidies, incentives for clearing natural vegetation and drought relief were to be reviewed, as was agricultural chemicals legislation (these reviews have been completed). One of Australia's largest catchments and agricultural producers, the Murray-Darling Basin, received support for its Natural Resource Management Strategy, and a new Land and Water Resources and Research Development Corporation (LWRRDC) was set up to examine natural resource management issues and promote integrated management. Decade of Landcare program plans for the Commonwealth and States were released progressively between 1991 and 1992.

In 1992 and again in 1996, additional Commonwealth support was given to the Landcare program (Keating 1992, 1996). This emphasised on-ground achievements for projects, responding in part to criticisms that too much funding was being absorbed into administrative and agency costs rather than being directed to on-ground works. Water and pest management projects were added to the program.

The part sale in 1997 of the public communications utility, Telstra, injected a further $1.5 billion into a wider, national environmental management program. It saw the Landcare and associated programs taken in under the umbrella of the *Natural Heritage Trust* (NHT). Further sales of Telstra in 1999 added another $250 million to the NHT, reducing some programs such as Rivercare and extending others - National Landcare, Farm Forestry and the National Land and Water Resources Audit (Table 4). Focus shifted towards funding of larger projects within a catchment or regional context, to farm forestry programs (linked to the abatement of greenhouse gas emissions, salinity remediation, and the needs of the pulp and paper industry) and to greater accountability in project performance (see below). In 1998 the Commonwealth government embraced urban environmental and land use issues in a *Living Cities* program which drew in management of waterways, coasts, and chemical pollution (Liberal Party 1998).

Table 4. The Coalition Government's Environment Programs through the NHT: $1.15 billion over five years. Source: the Coalition's *Saving our Natural Heritage* policy document.

BUDGET ($ million)	PROGRAM
318	Natural Vegetation Initiative to tackle land and water degradation
32	National Land and Water Resources Audit to provide a national appraisal of the extent of land and water degradation
163	Murray/Darling 2001 Project in partnership with relevant States
85	National Rivercare Initiative
100	Coasts and Clean Seas Initiative
80	Development of a National Reserve System to preserve Australia's biodiversity
8	National Wetlands program
16	Additional assistance to the Endangered Species program
12	Additional assistance to manage Australia's World Heritage Areas
16	Implementation of the National Strategy for Feral Animal Control
19	National Weeds Strategy formulation and implementation
16	Various measures to address city air pollution
5	Community-based waste management awareness
164	National Landcare Program

The establishment of a *National Land and Water Resources Audit* (NLWRA) in 1998, under the NHT, was a significant development towards understanding and managing land degradation in Australia (refer website www.nlwra.gov.au). The objective of the Audit is to provide a comprehensive and consistent appraisal of Australia's land resource base through scientific, economic, biophysical environmental and social assessments. These will provide the basis for future government decisions on policy directions, funding priorities and baseline data for monitoring change.

Western Australia provides an interesting example of meeting the challenges of handling urban/rural natural resource management and environmental issues. In 1999, the Urban Hills Land Conservation District (LCD) located east of Perth became the first partly urban LCD to be set up in the State. Incorporating three local government areas, it faces a range of difficult land management issues - wetland management, water and air pollution, loss of agricultural land, wildfire and prescribed burning, pest and weed invasions and pressures on remnant natural vegetation in a fragmented and overlapping administrative structure - exacerbated by urban/rural dichotomies. With agency, industry, local government and community representatives, the LCD committee expects to frame a management strategy in what is effectively a regional arrangement. The process has been advantaged by the prior formulation of two local government environmental management plans (Shire of Kalamunda 1995; Shire of Mundaring 1996) and a draft regional management plan for the Eastern Metropolitan Region nearing completion (Evans 1999).

5.1.1. Reviewing Landcare

One of the most striking aspects of this government program has been the extraordinary enthusiasm with which it was adopted by rural and urban communities. There has also been strong international interest (*Australian Landcare*, September 1999, pp. 14-15). The notion of 'caring for the land' appears to have tapped deep into the psyche of many Australians, and by 1999 there were over 4500 Landcare groups around the country (*Australian Landcare*, Dec. 1999:21). During the mid 1990s it was estimated that more than 30 000 farm properties belonged to these or similar groups (ABS 1996:139), and that about 150 000 people were involved in various community-based landcare activities such as tree planting, riverine restoration and environmental monitoring (Alexandra et al. 1996). Since 1994, over 70 corporate sponsors have donated more than $14 million to at least 530 projects through Australian Landcare Ltd. (*Australian Landcare* Dec. 1999:22). There has been a noticeable shift, too, in agencies' attitudes in terms of working with one another and interacting with communities: this would have been unthinkable 20 years ago, as would have been the willingness of farmers to allow groups of people to walk over their land (Alexander 1995; Lockie and Vanclay 1997).

There are, less fortunately, some weaknesses. In terms of program performance and accountability, some audits and reviews have revealed deficiencies in program management and delivery (Government of Western Australia 1995; Curtis et al. 1995; ANAO 1997). In addition, particular individuals or groups can exert powerful influences on decision making, skewing some programs or locking out other legitimate players. Another concern is that despite the huge Landcare effort, more land around Australia is still being cleared of natural vegetation than being replanted (Hawarth 1997). The rate of clearing has been estimated by Campbell (1999:53) at three times the rate of replanting. There is the matter, too, of the considerable focus on tree planting (especially farm forestry), drainage lines and salinity, deflecting attention from other less visible but important land degradation issues. *Integrated* management approaches to the problems, supported by good research and balanced allocation of resources, are imperative if sound outcomes are to be achieved.

Delivery of government policies in land management relies on public education, effective communication and expert technical support to modify land managers' attitudes and practices. This is being done notably through Landcare (or equivalents), integrated catchment management (ICM) groups (see below) and extension services. Even so, national surveys have found that only about one third of all farmers belong to a Landcare group, and that there is wide regional and industry variation (Mues *et al.* 1994, 1998). And although farmers often possess farm plans, these are not always implemented.

The question then arises as to why there are relatively low proportions of farmer participation in Landcare, given the huge resources and publicity allocated to it. Some reasons canvassed in Lockie and Vanclay's (1997) *Critical Landcare* relate to questions of attitude, socio-economic considerations, lack of resources and skills, and resistance to change, amongst others: see also Pannell, and Kington and Smettem, this volume.

But the question also needs to be asked, by what criteria should Landcare 'success' be measured? Should they be the number of trees planted, number of groups or individuals in groups, funds expended, kilometres of earthworks or fences constructed? On-ground evidence of Landcare works in the WA wheatbelt is disappointingly insignificant relative to the extensive areas of stressed natural vegetation and salinised land - which is not to say that valuable work is not being done. Is it a question of time needed for impacts of works to become evident, or a problem of scale, or should more qualitative or complex performance measures be used, such as landscape appearance, health of plants, animals and soils, or ecosystem resilience? In this context, the development of appropriate sets of environmental indicators (including socio-cultural variables) are a valuable contribution to monitoring changes to the natural resource base (ANZECC 1998; see also Conacher 1998). This process, with the NLWRA and other national initiatives such as the development of integrated environmental data bases, will facilitate the improved assessment of land condition, performance and trends, and enable effective management decisions to be made.

5.2. INTEGRATED NATURAL RESOURCE MANAGEMENT

Natural resource management laws are usually administered sectorally by purpose-specific agencies (agriculture, mining, fisheries, forestry, water, national parks and wildlife). However, many of these have merged or are undergoing varying degrees of amalgamation into natural resource, and/or environment, super-agencies. Some autonomy is still present among the natural resource agencies, but there is a growing emphasis on co-operative arrangements through whole-of-government and partnership approaches to land and natural resources management. This includes resource-sharing and a refocussing on common management objectives which incorporate ecologically sustainable principles, integrated planning and regional administration. Community consultation and participation are now an accepted part of decision-making processes and policy development.

Up to the 1970s, land use and natural resource policies around Australia generally focussed on economic development and production rather than environmental or natural resource protection or conservation. This development orientation was reflected, for example, in vast land releases, closer settlement schemes, pastoral lease arrangements and irrigation projects, all with considerable, albeit unintended, adverse environmental impacts

(Heathcote 1965; Davidson 1969; Powell 1976). Since the 1970s, however, international developments, growing public environmental awareness, major environmental disputes and improved scientific knowledge have seen the evolution of a wide range of environmental protection and management policies, laws and national initiatives in Australia.

Such changing emphases are well exemplified in Agriculture Western Australia's shift of focus from 'production' to 'sustainable' agriculture. Following a restructuring in 1992, almost half the Department's annual budget is now expended on resource protection and sustainable rural development. *The Sustainable Rural Development Program* is achieved through regional partnership groups and strategic regional planning. This takes place in seven 'sustainable rural development' regions. With agency, community and industry inputs, regional (or catchment) strategies address natural resource management issues within broader social, economic and environmental frameworks. Strategies are to be consultative, realistic and committed to ongoing evaluation (AgWA 1996).

Additionally, environmental law reforms have sought the consolidation, repeal or revision of laws and the introduction of specific environmental objectives which were previously lacking. Traditionally, the thrust of natural resource laws tended to be prescriptive, although it was uncommon for penalties to be applied. In this regard, bad land managers often saw themselves as beyond the reach of the law. Upgrades to environment protection legislation and specific policy instruments now ensure that serious pollution offences and major developments come under greater scrutiny, although there is a preference for self regulation and 'duty of care'. The same is true of soil conservation legislation.

There is no comprehensive Commonwealth natural resource management legislation equivalent to New Zealand's *Resource Management Act 1991* (Bührs and Bartlett 1993). The prospects for umbrella legislation and agencies are hindered by Australia's Constitution and federal system of government with its multiplicity of laws and responsibilities. It has been argued by some States and agencies that such a reform would be costly, hugely disruptive and difficult to achieve; and that in fact considerable change has already taken place (Industry Commission 1998). For those who have already suffered the strictures of economic rationalisation, corporatisation and down-sizing of government, such proposals would understandably be greeted with considerable dismay and suspicion. Having said that, there is merit in the idea of unifying natural resource management laws and policies to achieve more integrated and efficient approaches to managing land and resolving land degradation problems. This does not mean a 'cobbling together' of existing laws: reform needs to go deeper than that.

A recent government inquiry into ecologically sustainable land management recommended the introduction of a comprehensive, single, unifying statute in each jurisdiction (including the Commonwealth) to replace existing *ad hoc* land and natural resource management and environmental protection legislation (Industry Commission 1998). The statutes would contain statements of principles for the management of resources and would centre around a 'duty of care' obligation on land users and managers and a reliance on self-regulation. Each State would have an independent agency to administer the statute within a constructive rather than prohibitive framework. In fact, several States have already made moves in this direction.

Queensland has framed a natural resource management policy and new legislation which replaces six natural resource laws. Tasmania's statutory Resource Management and

Planning System (RMPS) integrates national and State natural resource management and planning policies and eleven core pieces of legislation under one framework. Western Australia has incorporated 16 pieces of agricultural legislation under its new *Agriculture Management Act*.

A further important development has been the release of a national, natural resource management policy discussion paper, formulated in close consultation with the States and Territories (AFFA 1999). Its objective is to provide policy directions for natural resource management in rural Australia over the next 10-15 years and to guide decision making by landholders, local regional groups, industry and governments. It is to be responsive to new management environments, recognising land degradation problems and the need for a follow-on from the Decade of Landcare. While the main focus is on the rural sector, wider environmental and sectoral issues are addressed. Key policy thrusts centre on achieving sustainable production, healthy ecosystems and viable rural communities through better management, acceptance of shared responsibilities, regional approaches to community empowerment, and capacity building.

5.3. INTEGRATED CATCHMENT MANAGEMENT (ICM)

Most States possess some form of integrated catchment management (ICM) strategy or policy (Second National Workshop on ICM 1997) or, in the case of New South Wales, an Act. The fundamental objectives of ICM are to manage the land resources of drainage basins ('catchments') in an integrated and sustainable manner so as to minimise land degradation, protect water quality and biodiversity, and recognise social, economic and cultural values. However, it needs to be noted that, in a country of generally low relief, and in areas with few surface water features, water catchments are not always appropriate units for land management. Bioregions (DEST 1996) or agroecological regions (ABS 1996:10) offer alternative spatial frameworks. There is also the difficulty of merging regions delimited on a biophysical basis with those devised for administrative, political or statistical purposes. Many electoral and local government boundaries, for example, coincide with water courses and not catchment boundaries.

In general, the States have set up ICM management committees, boards or co-ordinating groups with broad community representation to develop regional strategies and local action plans. Government funding is available to the groups through various national and State natural resource management or environmental programs, and is usually supplemented by 'in kind' support from agencies and land owners. Some boards have the power to raise local levies. A heavy reliance is placed on voluntary contributions of time and resources from community groups. This often acts as a constraint on the successful functioning of such groups and therefore on the achievement of ICM goals; on the other hand, community participation in land planning and management is a highly desirable social and democratic objective and can work well with good leadership and resourcing.

5.4. REGIONAL PLANNING AND MANAGEMENT

Some States are now moving towards integrated, regional approaches to land and natural resource management, sometimes incorporated in existing regional planning frameworks, or evolving from catchment and other land management structures. A particular difficulty (as referred to above in the context of ICM) has been how to determine the most appropriate regional management boundaries. Should they be administrative (as most are), biophysical, centred around a dominant local problem (as are some of Western Australia's Land Conservation Districts), or set up for funding purposes such as regional catchment groupings under the NHT.

There has been a history of regional planning in Australia, but usually with strong economic development or sectoral (agency) foci (examples: Country 1987; Department of Commerce and Trade 1999; Dore and Woodhill 1999). The plethora of such regions and overlapping functions frustrates efforts to achieve real integration. Several attempts at developing comprehensive regional or rural environmental strategies with Commonwealth support during the 1980s and 1990s appear to have fallen by the wayside due to insufficient resources or follow-up (see further, Conacher and Conacher 2000). Examples include the South West Environmental Strategy in Western Australia (Bradby and Pearce 1997) and the Southern Region of Councils Strategy in South Australia (SROC 1996).

Nevertheless, there have been some promising developments which appear to enjoy greater levels of political or statutory support than formerly. Sometimes located within State planning strategic frameworks or economic development regions, these plans acknowledge the importance of non-metropolitan areas and the need to manage natural resources in an integrated and sustainable manner in order to mitigate or avoid land degradation problems, and recognise socio-economic imperatives. Some examples are found in Victoria's rural regional strategies (example: DNRE 1997), and WA's Peel Region Scheme (Western Australian Planning Commission 1999). Their key elements include incorporation of ecologically sustainable development, employment creation and provision of leadership, management of land degradation, environmental protection, protection of agricultural land, biodiversity conservation and natural resource management. Several other States are pursuing similar initiatives.

6. Constraints Affecting the Implementation of Policies and Strategies

Many of the constraints which detract from or prevent the effective implementation of policies and strategies designed to mitigate land degradation in Australia have been referred to in the preceding sections and are now summarised. This leads in turn to the final section of the paper, which considers some means of overcoming those difficulties.

There are three overriding constraints. The first is the nature of the country itself - the sheer size of the continent (larger than the USA, excluding Alaska), its extensive range of mostly harsh biophysical environments, its small population and skill base. The second is the multiplicity of governments (at all levels), Acts, regulations and agencies: there are more than 600 environmental Acts in Australia (OECD 1998; Industry Commission 1998:93). With a population approaching 20 million and its nine State, Territory and Federal jurisdictions - each with full parliaments and Ministerial responsibilities for environment, land use, planning, education, health, transport, minerals, energy and so on, and the duplicatory array of agencies, public servants and local governments dealing with these matters - Australia is seriously over-governed. The associated costs and inefficiencies, not to mention the considerable difficulty of devising and implementing nation-wide policies and standards in the above portfolios and obtaining suitably skilled and experienced personnel to provide effective leadership in all the duplicated areas, pose enormous challenges. The third constraint is the spatial and economic scale and complexity of the land degradation problems themselves, and the inability of individuals or governments to deal effectively with the problem as a whole.

Other constraints - many of which have been considered in Conacher and Conacher (1995:Ch. 6; 2000) and Dore and Woodhill (1999:Part F) - include: indirect Constitutional allocation of responsibility for land use and environmental matters to the States (now made more explicit in the new Commonwealth *Environment Protection and Biodiversity Conservation Act 1999*); the difficulty of synchronising agreements amongst States and Territories when they are governed by political parties of differing political persuasions; inappropriate or conflicting inter- and intra-government policies or standards; overlapping, conflicting and out-dated laws; lip service paid to both democratic procedures and community participation in decision making; people's sometimes inappropriate perceptions of and attitudes towards the environment; inadequate data and scientific assessment (now being addressed); the large proportions of funds still being absorbed into administration and research rather than being directed to appropriate on-ground actions, and economic rationalist and competition policies resulting in priorities being given to the economic rather than environmental (and social) needs of regional or rural areas. With regard to the last point, many of the adverse impacts on regional Australia of the National Competition Policy have been strongly denied in a recent report (Productivity Commission 1999b). The fact remains that many of these areas are seriously afflicted by depopulation, withdrawal of services, declining farm incomes and a deteriorating resource base. Environmental considerations come low on the list of survival strategies.

With reference to local governments, major constraints are lack of resources (funds and personnel) and inadequate skills to deal with areas not previously their responsibility. Additionally, they are often subjected to the discretionary, overriding powers of State Ministers. Moves to develop regional groupings to overcome some of these difficulties have often encountered strong resistance for a range of political and other reasons.

As is the case with their State and Federal political counterparts, local government councillors generally have skills and experience often unrelated to the tasks of planning, land-use allocation, environmental or catchment management and conflict resolution. On the other hand, they have the benefit of being close to the communities they serve. Similar deficiencies amongst politicians at State and federal government levels are compensated for by a competent and skilled permanent public service providing expert, continuing and unbiased advice to governments through their Ministers. Unfortunately, with economic rationalism the public service has been politicised, down-sized and contracted-out with continuity of knowledge and skills severely impaired. Nevertheless and in (still) marked contrast, except in the largest local authorities, local government councillors generally have the advice of only a single CEO, an engineer, a planner, a health and recreation officer and, if they are fortunate, perhaps an over-worked and under-resourced environment officer, Landcare or catchment co-ordinator, often shared amongst several local governments. Parochialism is paramount. The drawbacks of such a limited base for dealing with highly complex environmental, Landcare, catchment, land-use and community conflict issues are obvious. Much is therefore left to reliance on often inexpert, but keen, individual land owners or voluntary catchment and 'friends' groups for on-the-ground management, with sometimes environmentally counter-productive activities being undertaken. Issues pertaining to public and private rights and responsibilities loom large and are not easily addressed. Public education and developing a sense of ownership of places and processes are some avenues for achieving change, while strong and skilled leadership is fundamental.

7. What is Needed?

According to the national inquiry into ecologically sustainable land management in Australia (Industry Commission 1998) and the recent OECD environmental performance review of Australia (OECD 1998), there are several areas in need of attention. They include the need for development of nationally co-ordinated and improved data bases. This is now being partly addressed by the *National Land and Water Resources Audit* and the establishment of national environmental data bases. With all States (and some local governments) now producing State of the Environment (SoE) reports, better SoE reporting is being assisted in part by the development of an agreed set of national environment indicators and benchmarking processes. Increasing adoption of Agenda 21 and other policies at local government levels facilitates this process. Improved dissemination of knowledge to managers of natural resources and expansion of regulatory and economic measures is being achieved through better networking and, under the auspices of the NCP, through areas such as water pricing and allocation. Stronger environmental policy implementation is under way, covering mandatory approaches such as National Environment Protection Measures (NEPMs), and encouragement of voluntary co-operation through education and incentives such as Environmental Management Systems (EMS), covenants and environmental awards. Improved devolution of management to regional and local organisations is developing slowly, but a greater focus and momentum are needed. And strengthened public participation in frank and open forums demands more than the lip service it sometimes receives; despite the many encouraging developments, cultures of secrecy and suspicion persist in some quarters.

Ideally, the enormous constraints posed by the way in which Australia is governed could be (partly) resolved by abolishing the States and Territories - which are historical hangovers from the colonial era, before Federation in 1900 - and by considerably strengthening the powers and responsibilities of regional groupings of the democratically-elected local governments. Pragmatically, State parliaments are unlikely to vote themselves out of power, and restructuring would be hugely expensive, disruptive, and no doubt unpopular in many quarters. However, it may still be possible to:

- merge local governments into regional groups with enhanced powers and resources;
- strengthen federally-based government environment, land-use, resource and planning agencies;
- create devolved, regionally-based structures for those federal agencies, allied directly with the regionally-grouped local governments, and
- reorganise the resources and personnel of the State agencies to the new regionally-devolved, federal agencies.

Another model for improved natural resource management has been proposed by Gardner (1999). In general it appears to follow the proposition of the Industry Commission (1998) referred to earlier. Gardner's model gives strong support to regionally-based structures under the guidance of one agency. This body merges all the present natural resource management agencies in each State into one structure responsible for natural resource management and supported by one statutory regime. Similar to New Zealand, Regional Councils would have considerable powers and be responsible for regional planning, delivering funds, and monitoring and auditing functions. However, the regulatory and policy functions of agencies would be kept separate. Local governments would still perform their traditional functions but with increased powers and with regional council representation.

A similar approach has been advocated in a more recent, national, natural resource management discussion paper (AFFA 1999). The paper also strongly supports a devolution of natural resource management powers to regional (or catchment management) structures. These structures are seen as most effective when generated by the community with appropriate levels of government support. Such frameworks are considered to be responsive to specific local needs as well as being appropriate to meeting regional-scale priorities through co-ordinated and negotiated decision-making processes.

It is now clear that natural resource management policies and attempts to deal with land degradation problems in Australia have come a considerable distance in recent years. Moves towards rationalising policies and structures have occurred, but the passage to a more strongly regionalised administration is not expected to be an easy one.

8. Acknowledgments

We thank Ian Douglas and Michael Stocking for their constructive comments.

9. References

ABS (1996) *Australian agriculture and the environment*, Catalogue No. 4606.0, Australian Bureau of Statistics, Canberra.

ABS (1997) *Environmental protection expenditure*, Catalogue No. 4603.0, Australian Bureau of Statistics, Canberra.

AFFA (1999) *Managing natural resources in rural Australia for a sustainable future*, A discussion paper for developing a national policy, National Natural Resource Management Taskforce, Agriculture, Forestry and Fisheries Australia, Canberra.

AgWA (1996) *Primary focus*, Newsletter of Agriculture Western Australia, July 1996, November 1996, Perth.

Alexander, H. (1995) *A framework for change. The state of the community Landcare movement in Australia*, National Landcare Facilitator's Annual Report, National Landcare Program, Canberra.

Alexandra, J., Hassenden, S. and White, T, (1996) *Listening to the land: a directory of community environmental monitoring groups in Australia*, Australian Conservation Foundation, Melbourne.

ANZECC (1998) *Core environmental indicators for reporting on the state of the environment: discussion paper for public comment*, ANZECC State of the Environment Reporting Task Force, Department of Environment, Canberra.

ANZECC/ARMCANZ (1996) *Draft national strategy for rangeland management*, Australia and New Zealand Environment and Conservation Council, and Agriculture and Resource Management Council of Australia (ARMCANZ) Working Group, Department of Environment, Sport and Territories, Canberra

ANZECC/ARMCANZ (1999) *National water quality management strategy. Australian and New Zealand guidelines for fresh and marine water quality*, prepared under the auspices of Australia and New Zealand Environment and Conservation Council and Agriculture and Resource Management Council of Australia and New Zealand, public comment draft July 1999.

Bates, G.M. (1995) *Environmental law in Australia*, 4th ed., Butterworths, Sydney.

...⁴ʰᵥ K. and Pearce, D. (1997a) *South west environmental strategy*, South West (WA) Local Government

‸·ᐧlation in Australia: report for the National Soil Conservation
ᵣᵤgᵣamme, racuity of Law, University of Adelaide.

Bührs, T. And Bartlett, R.V. (1993) *Environmental policy in New Zealand: the politics of clean and green*, Oxford University Press, Auckland.

Calver, M.C., Bradley, J.S. and Wright, I.W. (1999) Towards scientific contributions in applying the precautionary principle: an example from southwestern Australia, *Pacific Conservation Biology* **5**, 63-72.

Campbell, A. (1999) Bushcare: turning things around, *Australian Landcare*, December, pp. 52-57.

Christoff, P. (1998) Degreening government in the Garden State: environment policy under the Kennett government, *Environmental and Planning Law Journal* **15**, 10-32.

Clement, J.P. and Bennett, M. (1998) *The law of Landcare in Western Australia*, Environmental Defender's Office WA (Inc.), Perth.

COAG (1992) *Inter-governmental agreement on the environment*, Council of Australian Governments, Australian Government Publishing Service, Canberra

COAG (1995) *Water reform agreement*, Council of Australian Governments, Canberra.

Comino, M.P. (1997) The Ramsar convention in Australia: improving the implementation framework, *Environmental and Planning Law Journal* **14**, 89-101.

Commonwealth of Australia (1992) *National strategy for ecologically sustainable development*, Australian Government Publishing Service, Canberra.

Conacher, A.J. (1998) Environmental quality indicators: where to from here? *Australian Geographer* **29**, 175-89.

Conacher, A.J. and Conacher, J.L. (1995) *Rural land degradation in Australia*, Oxford University Press, Melbourne.

Conacher, A.J. and Conacher, J.L. (2000) *Environmental planning and management in Australia*, Oxford University Press, Melbourne.

Country, J.D. (ed.) (1987) *Australian regional developments*, Australian Government Publishing Service, Canberra.

Davidson, B.R. (1969) *Australia wet or dry? The physical and economic limits to the expansion of irrigation*, Melbourne University Press, Melbourne.

Department of Commerce and Trade (1999) *A regional development policy for WA: setting the direction for regional Western Australia, a policy framework discussion paper*, Synectics Creative Collaboration and Department of Commerce and Trade, Government of Western Australia, Perth.

Department of Home Affairs and Environment (1984) *A national conservation strategy for Australia*, Australian Government Printing Service, Canberra.

DEST (1996) *The national strategy for the conservation of Australia's biodiversity*, Commonwealth Department of Environment, Sport and Territories, Canberra.

DNRE (1997) *Victorian regional catchment strategies: Port Philip*, Department of Natural Resources and Environment, Melbourne.

Dore, J. and Woodhill, J. (1999) *Sustainable regional development: final report*, Greening Australia, Canberra.

Dovers, S. (1997) Institutionalising ecologically sustainable development. What happened, what did not, why and what could, Academy of Social Sciences Symposium, Canberra.

Doyle, T. and Kellow, A. (1995) *Environmental politics and policy making in Australia*, Macmillan, Melbourne.

EDO (1997) *A guide to environmental law in Western Australia*, Environmental Defenders Office, Perth.

Evans, S. (1999) *Regional environmental strategy*, Draft, Eastern Metropolitan Regional Council, Perth.

Gardner, A. (1999) The administrative framework of land and water management in Australia, *Environmental and Planning Law Journal* **16**, 212-57.

George, R.J., Nulsen, R.A., Ferdowsian, R. and Raper, G.P. (1999) Interactions between trees and groundwaters in recharge and discharge areas - a survey of Western Australian sites, *Agricultural Water Management* **39**, 91-113.

Greening Australia (1995) *Local Greening Plans, a guide for vegetation and biodiversity management*, Greening Australia Ltd., Canberra.

Hawke, R.J.L. (1989) *Our country, our future: PM's statement on the environment July 1989*, 2nd ed., Australian Government Publishing Service, Canberra.

Heathcote, R.L. (1965) *Back of Bourke: a study of land appraisal and settlement in semi-arid Australia*, Melbourne University Press, Melbourne.

House of Representatives Standing Committee (1989) *The effectiveness of land degradation policies and programmes*, House of Representatives Standing Committee on Environment, Recreation and Arts,

Australian Government Publishing Service, Canberra.

Industry Commission (1998) *A full repairing lease: inquiry into ecologically sustainable land management*, final report, Commonwealth of Australia, Canberra.

Johnson, A., McDonald, G., Shrubsole, D. and Walker, D. (1998) Natural resource use and management in the Australian sugar industry: current practice and opportunities for improved policy, planning and management, *Australian Journal of Environmental Management* **5**, 97-108.

Keating, P. (1992) *Australia's environment: a national asset. Prime Minister's statement on the environment*, Australian Government Publishing Service, Canberra.

Keating, P. (1996) *On conserving natural Australia*, Statement by the Prime Minister, 24 January 1996, Australian Government Publishing Service, Canberra.

Kirkpatrick, J. (1994) *A continent transformed, human impact on the natural vegetation of Australia*, Oxford University Press, Melbourne.

Lockie, S. and Vanclay, F. (eds.) (1997) *Critical Landcare*, Centre for Rural Social Research, Charles Sturt University, Wagga Wagga, NSW.

LWRRDC (1995) *Data sheets on natural resource issues*, Occasional paper No. 06/95, Land and Water Resources Research and Development Corporation, Canberra.

Mues, C., Chapman, L. and van Hilst, R. (1998) *Promoting improved land management practices on Australian farms: a survey of Landcare and land management related programs*, Australian Bureau of Agricultural and Resource Economics, Canberra.

Mues, C., Roper, H. and Ockerby, J. (1994) *Survey of Landcare and land management practices 1992-1993*, ABARE Research Report 94.6, Australian Bureau of Agricultural and Resource Economics, Canberra.

Northrop, L. (1999:) Looking back on the origins of the National Landcare Program, *Australian Landcare*, December, pp. 17-18.

OECD (1998) *Environmental performance reviews. Australia*, Organisation for Economic Co-operation and Development, Paris.

Powell, J.M. (1976) *Environmental management in Australia, 1788-1914. Guardians, improvers and profit: an introductory survey*, Oxford University Press, Melbourne.

Productivity Commission (1999a) *Implementation of ESD by Commonwealth departments and agencies*, draft, Productivity Commission, Canberra.

Productivity Commission (1999b) *Impact of competition policy reforms on rural and regional Australia. Draft report May 1999*, Productivity Commission, Canberra.

Robinson, S. and Humphries, R. (1997) Towards Best Practice: observations on western legal and institutional arrangements for ICM, 1987, 1997, in *Advancing integrated resource management: processes and policies*, 2nd National Workshop on ICM, Australian National University, 29 Sept to 1 Oct 1997, Canberra. Not paginated.

Second National Workshop on ICM (1997) *Advancing integrated resource management: processes and policies*, Australian National University, Canberra.

Senate Standing Committee on Science, Technology and Environment (1984) *Land use policy in Australia*, Australian Government Publishing Service, Canberra.

Shire of Kalamunda (1995) *District conservation strategy*, Kalamunda, Western Australia.

Shire of Mundaring (1996) *Environmental management strategy*, Mundaring, Western Australia.

SROC (1996) *Strategy for a sustainable Southern Region*, Prepared by Angela Hale, Environmental Planning Project officer for the Australian Local Government Association and the Southern Region of Councils, South Australia. 2 vols.

State of the Environment Advisory Council (1996) *State of the environment Australia 1996*, CSIRO Publishing, Collingwood.

Taskforce (1997) *Taskforce for the review of natural resource management and viability of agriculture in Western Australia, Draft, May 1997*, Minister for Primary Industries, Government of WA, Perth.

Western Australian Planning Commission (1999) *Peel region scheme*, Perth. Draft for public comment. 10 documents.

World Bank (1997) *Five years after Rio: innovations in environmental policy*, Environmentally Sustainable Development Studies and Monographs Series No. 18, The International Bank for Reconstruction and Development/The World Bank, Washington D.C.

Wright, I. (1995) Implementation of sustainable development by Australian local governments, *Environmental and Planning Law Journal* **12**, 54-61.

Zimmer, C. (1999) New date for the dawn of dream time, *Science* **284**, 1243-1246.

JEANETTE CONACHER[1] and ARTHUR CONACHER[2]

[1] Corresponding author. 4 Mitchell Road, Darlington, WA 6070, Australia
     ajconach@geog.uwa.edu.au
[2] Department of Geography, University of Western Australia, Nedlands, WA 6907,
     Australia

AGRICULTURE, LAND DEGRADATION AND DESERTIFICATION: CONCLUDING COMMENTS

MICHAEL STOCKING

This COMLAND Conference examined the current status of research and knowledge on several interconnected themes of land degradation, agriculture and desertification. The ultimate objective was to derive useful practices for land degradation control and appropriate policy responses. Topics upon which important contributions have been made include:

- soil erosion and land degradation processes, including the use of indicators, process modeling and geographical information systems;
- secondary salinisation of soil and water as exemplified so visibly in Western Australia;
- the ecological consequences of land degradation, especially in key areas such as 'soil health', pasture improvement and the heavy use of agro-chemicals;
- the social and economic consequences of land degradation, and how this impacts on policy issues such as land reform, catchment planning and economic incentives;
- technical remedial measures, employing airborne geophysical survey methods as well as field methods for land recovery;
- structural and agency solutions to land degradation, through the evaluation of policy approaches that have been found to work and the monitoring of specific practices for their adoptability; and
- social, economic and political solutions in SE Asia generally and Australia in particular.

The range of topics is commendably broad. Most, but not all, contributions came from geographers. The topics brought out some real areas of advantage for the involvement of geographers in practical issues of land degradation control. However, it needs to be recognised that many other disciplines are active in land degradation research. There are several competing groups for attention and funding, especially amongst soil scientists, ecologists, agronomists, engineers and hydrologists. The Second International Land Degradation Conference had been held in Khon Kaen, Thailand only eight months before COMLAND. It involved many non-geographers, and this reviewer was the only person at the COMLAND Conference to have been involved in both. So what are the areas of advantage for the involvement of geographers and for the work programme of an IGU Task Force?

First, geographers are well equipped in developing and compiling databases on biophysical and socio-economic aspects. Spatial techniques are obviously the *forte* of geographers. However, land degradation is also the study of process and change, and databases to monitor such changes are vital (and, it has to be said, usually missing from the work of other disciplines). For example, I find many of the land degradation mapping projects of international and some national agencies (it would be undiplomatic to name names) sterile, inaccurate and misleading, not least for their inability to capture process and change. As I write these concluding notes, yet another request, this time from a leading international soil scientist in the US, has come for me to help identify 'hotspots' of land degradation in Africa and estimate

387

A.J. Conacher (ed.), Land Degradation, 387–390.
© 2001 *Kluwer Academic Publishers. Printed in the Netherlands.*

erosion rates. The information base is scattered, uncoordinated and founded on dubious technical assumptions. Second, geographers have been adept at applying theories, experiments and models to land degradation. Empirical models such as the Universal Soil Loss Equation derived mainly from agricultural engineers are well past their 'Sell-by Date'. The more physical and rational models which attempt to capture process have far greater potential to be applied without the heavy hand of years of experimentation to derive accurate factor values for the models. There are dangers, however. Some physical models for erosion prediction seem to take on an objective of their own, and become an end in themselves rather than a practical tool. Third, geographers as evidenced in this COMLAND Conference have the necessary breadth of view to address the agro-ecology of land degradation. I was impressed at the Conference, for example, at how a spatial understanding of landscape – an especial area of expertise for geographers – was so vital in developing techniques of pasture improvement in a particularly difficult part of Spain. Finally, geographers, provided they do not come from departments obsessively engaged in separating human from physical geography, are good at integrating social and technical issues. Several papers, such as those tackling land reform and the many that touched on policy issues, demonstrate admirably how social factors **have to** be embedded in an understanding of biophysical relations and technical practicalities. Similarly, studies of chemical, physical and biological processes are of little relevance to improvement in human livelihoods without a broad contextualisation in social, political and economic realities.

At the same time, however, geographers must be aware of new agendas. Sometimes, I feel that geographers do often live nice integrated lives on their pleasant but small desert islands. Coming from a background of development studies, the practical implementation of aid policies and the influence of sociologists and economists, I am fortunate that I see new bandwagons in their construction. New frameworks for a 'Capital Assets Approach to Livelihoods' and for 'Biodiversity in Areas of Land Use', deriving respectively from micro-economics and ecology, were presented in my Keynote Paper. The Capital Assets Framework, where sustainable rural livelihoods are seen in terms of natural, social, physical, human and financial capital, has been hugely influential since 1998 in guiding development aid policy for the UK Government. It is now to be seen in policy reviews for the big multi-laterals such as the World Bank, UNDP and UNEP. To ignore such bandwagons, especially when they provide an analytical framework which brings social and technical issues together, is to court rejection of our work by the 'movers and shakers' of this world. Additionally, to include such developments as 'agrodiversity' shows how pragmatically we care about people as well as the environment.

So, how has COMLAND, a stable essentially of geomorphologists and physical geographers, demonstrated its interest in people? In many ways – the following are topics in the verbal presentations that I found particularly interesting to people-centred development:

- land reform and its effect on conservation in South Africa;
- changes in river hydrology, erosion rates and sedimentation in semi-arid Australia that affect settlements and infrastructure;
- Forest User Groups in Nepal in which rights are accorded to manage forest resources for the benefit of local people;
- agricultural terraces in Peru, which are both spectacular and the only way to farm sustainably a very difficult environment;
- pointing the finger of blame for environmental degradation in Kenya, which shows yet again how land degradation concerns can be hijacked by unscrupulous people for their own political agendas;
- how to mislead people with data outputs from runoff and erosion plots, and
- pasture improvement in SW Spain and how the role of the 200-year storm event at the wrong

time can affect people's reactions.

Lest, however, it should be feared that these physical geograph
their traditional roots, environmental concerns also featured i

- deep weathering of granite rocks and dolerite dykes;
- the amazing plant biodiversity of southern Western Australia (only .
- the role of logging and fires in the ecology and landscape history;
- *Mitika* populations in arid central Australia, and
- climatic change and non-linear response mechanisms of vegetation and erosio.

Regrettably not all these topics have translated to written papers in this volume. But they do show the vitality in the repertoire of geographers working on land degradation issues. However, they also reveal some challenges, which I feel COMLAND needs to take on board in order to make its contribution much better directed to mainstream issues. First, at a global level, geographers must make a more consistent input, especially theoretical and multi-disciplinary, to some of the debates such as global climate change, biodiversity, carbon sequestration and food security. These are all on the agenda of the major multi-lateral funders, such as the Global Environment Facility (GEF). I see economists, ecologists, bureaucrats and development planners at the GEF, but very rarely a geographer with academic credentials. Geographers could and should be able to make more balanced and realistic assessments of degradation and associated issues. Second, there has been a huge international effort in developing key indicators of degradation. Frankly, I am appalled at some of the results. Far from being true indicators, the lists that have been developed look like *à la carte* menus of every possible parameter of soils, water and vegetation. Geographers could have done so much better. Third, issues of scale in land degradation have rendered many experimental results of dubious relevance and accuracy – geographers have a much better handle on scale than some of my natural scientist colleagues. Finally, geographers are in a good position to facilitate local application of natural science, through *Landcare*, community action, economic analysis, appropriate valuation approaches, subsidy regimes, and so on.

The potential role of geographers and this IGU Task Force will need resources, as well as new strategic approaches. The better linkage with other like-minded groups, such as those who work in my field in development aid, is a prerequisite. Work within or alongside implementation agencies, such as the bilateral aid organisations (AusAid; DFID, for example), would make our work far more practical and pragmatic. In agricultural extension and research, participation and community partnerships are fully in vogue, and are essential components of any funded work. However, we do need to keep in mind the areas of advantage of geographers in GIS, mapping, remote sensing, and field assessment. Geographers are also better than any other discipline in understanding process, dynamics and change. Agriculture and land degradation is all about change, as evidenced in recent studies on how populations react to increasing numbers by intensifying land use and developing more conservative and sustainable practices. Until only five years ago, land carrying capacity was a fixed and immutable quantity. Sustainable intensification was a contradiction in terms. We now know better and have a base of empirical evidence, such as the Machakos (Kenya) experience. Geographers have a far more flexible response to 'new wisdom', and we should seek to capitalise on it.

AUTHOR

MICHAEL STOCKING

School of Development Studies
University of East Anglia
Norwich NR4 7TJ
United Kingdom

E-mail: m.stocking@uea.ac.uk